MEDICAL TRAVEL BRAND MANAGEMENT

Success Strategies for Hospitality Bridging Healthcare (H2H)

Advances in Hospitality and Tourism Series

MEDICAL TRAVEL BRAND MANAGEMENT

Success Strategies for Hospitality Bridging Healthcare (H2H)

Edited by

Frederick J. DeMicco, PhD, RD

Ali A. Poorani, PhD

APPLE ACADEMIC PRESS

First edition published 2023

Apple Academic Press Inc.
1265 Goldenrod Circle, NE,
Palm Bay, FL 32905 USA

760 Laurentian Drive, Unit 19,
Burlington, ON L7N 0A4, CANADA

CRC Press
6000 Broken Sound Parkway NW,
Suite 300, Boca Raton, FL 33487-2742 USA

2 Park Square, Milton Park,
Abingdon, Oxon, OX14 4RN UK

Library and Archives Canada Cataloguing in Publication

Title: Medical travel brand management : success strategies for hospitality bridging healthcare (H2H) / edited by Frederick J. DeMicco, PhD, RD, Ali A. Poorani, PhD.

Names: DeMicco, Frederick, editor. | Poorani, Ali A., editor.

Series: Advances in hospitality and tourism book series.

Description: First edition. | Series statement: Advances in hospitality and tourism book series | Includes bibliographical references and index.

Identifiers: Canadiana (print) 20220162840 | Canadiana (ebook) 20220162883 | ISBN 9781774637272 (hardcover) | ISBN 9781774637326 (softcover) | ISBN 9781003277392 (PDF)

Subjects: LCSH: Medical care—Marketing. | LCSH: Medical tourism—Marketing. | LCSH: Branding (Marketing)

Classification: LCC RA410.56 .M43 2022 | DDC 362.1068/8—dc23

Library of Congress Cataloging-in-Publication Data

Names: DeMicco, Frederick, editor. | Poorani, Ali A., editor.

Title: Medical travel brand management : success strategies for hospitality bridging healthcare (H2H) / edited by Frederick J. DeMicco, Ali A. Poorani.

Other titles: Advances in Hospitality and Tourism Series

Description: First edition. | Palm Bay, FL : AAP, Apple Academic Press, 2022. | Series: Advances in hospitality and tourism series | Includes bibliographical references and index. | Summary: "This new volume, which complements the editors' earlier volume Medical Travel Brand Management: Success Strategies for Hospitality Bridging Healthcare (H2H), explores the multitude of medical travel services and discusses the integration of traveling medical guests with destination providers, hospitality/healthcare professionals, and travel service providers. The editors also address the impact the COVID-19 pandemic has made on the travel industry, which has motivated them to bring together major players, renowned authors, practitioners, and researchers to create this book to help prepare the medical tourism market to not only recover from the devastating effects of the pandemic but also to provide tools and cases that will help to structure successful destinations for medical travel. From chapters on branding to assessing accreditation and post-care quality metrics, Drs. DeMicco, Poorani, and their fellow contributors take the reader through the critical phases of the medical travel journey: pre-visit, travel, on-site care, discharge and follow-up care. The authors address critical issues facing medical, health, and wellness travel from both macro and micro perspectives. Presentations of best practices and strategies demonstrate how some destinations have built, renewed, or engaged various stakeholders to construct or enhance their medical tourism destination. Medical Travel: Hospitality Bridging Healthcare (H2H) © also showcases best practices and innovative ways of designing and operating a profitable and entrepreneurial practice. Quality issues, aesthetics, and legal issues related to inbound and outbound medical tourism are also presented. The book explores the evolving nature of hospital design and the complex relationship between people and medicine as manifested in the relationship of hospital aesthetics to patient satisfaction. Importantly, the book also includes a chapter addressing medical health travel during the pandemic which describes how the pandemic has revolutionized telehealth and the medical travel industry, which can leverage the advances made in digital health and telemedicine. This volume will be an important resource for the four main players at the center of medical travel: medical travelers themselves, government agencies, intermediaries, and health and wellness providers. The selected best practices, research, cases, innovative strategies, SWOT analysis, and toolkits address the aims of all stakeholders"-- Provided by publisher.

Identifiers: LCCN 2022004850 (print) | LCCN 2022004851 (ebook) | ISBN 9781774637272 (hardback) | ISBN 9781774637326 (paperback) | ISBN 9781003277392 (ebook)

Subjects: LCSH: Medical tourism. | Medical care. | Hospitality industry.

Classification: LCC RA793.5 .M432 2022 (print) | LCC RA793.5 (ebook) | DDC 362.1--dc23/eng/20220411

LC record available at https://lccn.loc.gov/2022004850

LC ebook record available at https://lccn.loc.gov/2022004851

ISBN: 978-1-77463-727-2 (hbk)
ISBN: 978-1-77463-732-6 (pbk)
ISBN: 978-1-00327-739-2 (ebk)

ABOUT THE ADVANCES IN HOSPITALITY AND TOURISM BOOK SERIES

Editor-in-Chief:
Mahmood A. Khan, PhD
Professor, Department of Hospitality and Tourism Management,
Pamplin College of Business,
Virginia Polytechnic Institute and State University,
Falls Church, Virginia, USA
Email: mahmood@vt.edu

This series reports on research developments and advances in the rapidly growing area of hospitality and tourism. Each volume in this series presents state-of-the-art information on a specialized topic of current interest. These one-of-a-kind publications are valuable resources for academia as well as for professionals in the industrial sector.

BOOKS IN THE SERIES:

Food Safety: Researching the Hazard in Hazardous Foods
Editors: Barbara Almanza, PhD, RD, and Richard Ghiselli, PhD

Strategic Winery Tourism and Management: Building Competitive Winery Tourism and Winery Management Strategy
Editor: Kyuho Lee, PhD

Sustainability, Social Responsibility and Innovations in the Hospitality Industry
Editor: H. G. Parsa, PhD
Consulting Editor: Vivaja "Vi" Narapareddy, PhD
Associate Editors: SooCheong (Shawn) Jang, PhD,
Marival Segarra-Oña, PhD, and Rachel J. C. Chen, PhD, CHE

Managing Sustainability in the Hospitality and Tourism Industry: Paradigms and Directions for the Future
Editor: Vinnie Jauhari, PhD

Management Science in Hospitality and Tourism: Theory, Practice, and Applications
Editors: Muzaffer Uysal, PhD, Zvi Schwartz, PhD, and
Ercan Sirakaya-Turk, PhD

Tourism in Central Asia: Issues and Challenges
Editors: Kemal Kantarci, PhD, Muzaffer Uysal, PhD, and Vincent Magnini, PhD

Poverty Alleviation through Tourism Development: A Comprehensive and Integrated Approach
Robertico Croes, PhD, and Manuel Rivera, PhD

Chinese Outbound Tourism 2.0
Editor: Xiang (Robert) Li, PhD

Hospitality Marketing and Consumer Behavior: Creating Memorable Experiences
Editor: Vinnie Jauhari, PhD

Women and Travel: Historical and Contemporary Perspectives
Editors: Catheryn Khoo-Lattimore, PhD, and Erica Wilson, PhD

Wilderness of Wildlife Tourism
Editor: Johra Kayeser Fatima, PhD

Medical Tourism and Wellness: Hospitality Bridging Healthcare (H_2H)©
Editor: Frederick J. DeMicco, PhD, RD

Sustainable Viticulture: The Vines and Wines of Burgundy
Claude Chapuis

The Indian Hospitality Industry: Dynamics and Future Trends
Editors: Sandeep Munjal and Sudhanshu Bhushan

Tourism Development and Destination Branding through Content Marketing Strategies and Social Media
Editor: Anukrati Sharma, PhD

Evolving Paradigms in Tourism and Hospitality in Developing Countries: A Case Study of India
Editors: Bindi Varghese, PhD

The Hospitality and Tourism Industry in China: New Growth, Trends, and Developments
Editors: Jinlin Zhao, PhD

Labor in Tourism and Hospitality Industry: Skills, Ethics, Issues, and Rights
Abdallah M. Elshaer, PhD, and Asmaa M. Marzouk, PhD

Sustainable Tourism Development: Futuristic Approaches
Editor: Anukrati Sharma, PhD

Tourism in Turkey: A Comprehensive Overview and Analysis for Sustainable Alternative Tourism
Editor: Ahmet Salih İKİZ

Capacity Building Through Heritage Tourism: An International Perspective
Editor: Surabhi Srivastava, PhD

Medical Travel Brand Management: Success Strategies for Hospitality Bridging Healthcare (H2H)
Editors: Frederick J. DeMicco, PhD, RD, and Ali A. Poorani, PhD

Strategic Management for the Hospitality and Tourism Industry: Developing a Competitive Advantage
Editor: Vincent Sabourin, PhD

Tourist Behavior: Past, Present, and Future
Editors: Narendra Kumar, PhD, Bruno Barbosa Sousa, PhD, and Swati Sharma, PhD

The Hospitality and Tourism Industry in ASEAN and East Asian Destinations: New Growth, Trends, and Developments
Editors: Jinlin Zhao, PhD, Lianping Ren, D.HTM, and Xiangping Li, PhD

Event Tourism in Asian Countries: Challenges and Prospects
Editor: Shruti Arora, PhD

Small Island and Small Destination Tourism: Overcoming the Smallness Barrier for Economic Growth and Tourism Competitiveness
Robertico Croes, PhD

ABOUT THE SERIES EDITOR

Mahmood A. Khan, PhD, is a Professor in the Department of Hospitality and Tourism Management, Pamplin College of Business at Virginia Tech's National Capital Region campus. He has served in teaching, research, and administrative positions for the past 35 years, working at major U.S. universities. Dr. Khan is the author of several books and has traveled extensively for teaching and consulting on management issues and franchising. He has been invited by national and international corporations to serve as a speaker, keynote speaker, and seminar presenter on different topics related to franchising and services management. Dr. Khan has received the Steven Fletcher Award for his outstanding contribution to hospitality education and research. He is also a recipient of the John Wiley & Sons Award for lifetime contribution to outstanding research and scholarship; the Donald K. Tressler Award for scholarship; and the Cesar Ritz Award for scholarly contribution. He also received the Outstanding Doctoral Faculty Award from Pamplin College of Business. He has served on the Board of Governors of the Educational Foundation of the International Franchise Association, on the Board of Directors of the Virginia Hospitality and Tourism Association, as a Trustee of the International College of Hospitality Management, and as a Trustee on the Foundation of the Hospitality Sales and Marketing Association's International Association. He is also a member of the several professional associations.

ABOUT THE EDITORS

Frederick J. DeMicco, PhD, RD
Executive Director and Professor,
School of Hotel and Restaurant Management,
Northern Arizona University, USA,
E-mail: Frederick.DeMicco@nau.edu

Frederick J. DeMicco, PhD, RDN, is the Executive Director and Professor in the School of Hotel and Restaurant Management at Northern Arizona University, USA. Formerly, he was Professor and the Aramark Endowed Department Chair in the Lerner College of Business & Economics in Hospitality Business Management at the University of Delaware. He was Associate Director of the School of Hotel, Restaurant and Recreation Management (HRRM) at Penn State University, where he was Professor-in-Charge of the Hotel & Restaurant Management undergraduate program and also was the Professor-in-Charge of the Master of Science and PhD graduate programs for five years. He presently is a Conti Distinguished Professor at the Pennsylvania State University, USA. Dr. DeMicco is the President of the International Academics and the Dean of Business Executive and Professional Programs at the Learning Village in Tuscany (SIAF) Campus in Volterra, Tuscany, the home of Innovation and Imagineering. Dr. DeMicco was also a Darden Eminent Scholar Chair Visiting Professor to the University of Central Florida–Rosen College, the USF Sarasota- Manatee, and Colorado State University; and currently is an Adjunct Professor in the School of Public Policy & Administration at the University of Delaware, USA. Dr. DeMicco has worked in healthcare at the Massachusetts General Hospital, Boston, and at Walt Disney World in hotel management where he obtained a Ductorate Degree from Disney University, Orlando, Florida. Dr. DeMicco was the President of International CHRIE–NENA (Northeast North American Federation). He was presented the International CHRIE/Cornell Howard B. Meek Award of Excellence for 2017.

Ali A. Poorani, PhD

Associate Professor of Hospitality Business Management,
Alfred Lerner College of Business & Economics,
Delaware University, Newark, USA

Ali A. Poorani, PhD, is an Associate Professor with expertise in leadership and entrepreneurship. He is the Director of Hospitality Associates for Research & Training (HART), which offers a variety of research and training programs for the hospitality and healthcare industries. He developed the University of Delaware's Patient Experience Academy, in conjunction with Christiana Care Health System. He has published several scholarly articles, including Hospitality Bridging Healthcare. Dr. Poorani has worked in the hospitality industry for more than 14 years. He has conducted many workshops, focus groups, and training programs nationally and internationally, bringing hospitality to healthcare with a vision to make hospitals truly hospitable. His long-term engagement with the International Council on Hotel, Restaurant and Institutional Education (ICHRIE) includes: past board member of ICHRIE Finance Committee; chair, ICHRIE Finance SIG; board member for the Association of Hospitality Financial Management Educators (AHFME); member of ICHRIIE Strategic Planning Committee; ICHRIE Finance Committee; and Hospitality Industry Professionals-SIG. He also serves on the advisory board for Medical Tourism and Global Healthcare Research Summit. Dr. Poorani is a recipient of various community service awards, such as the 2005 Champion of Club Education Award, 2009 Alfred Lerner College of Business Outstanding Service/ Outreach Award, the 2012 University of Delaware Excellence in Advising and Mentoring Award, and the 2014 Leon and Margaret Slocomb Professional Excellence Award. Dr. Poorani received his PhD in leadership and human behavior from United States International University (USIU) in San Diego and his MBA from the University of Central Oklahoma.

CONTENTS

CONTRIBUTORS

Amanda Adkins
Alfred Lerner College of Business and Economics, Department of Hospitality Business Management, University of Delaware, 14 W., Main Street, Newark, DE–19716, USA

Janet Bruno
Arizona State University, Phoenix, Arizona, USA

M. Cetron
Forecasting International, 22 Commerce Road, Newtown, CT 06470, USA

Katie Contrera
Johns Hopkins University, Krieger School of Arts and Sciences, Advanced Academic Programs, Washington DC – 20036, USA, E-mail: katie.contrera@gmail.com

O. Davies
Forecasting International, 22 Commerce Road, Newtown, CT 06470, USA

Frederick J. DeMicco
School of Hotel and Restaurnat Management, the W.A. Franke College of Management, Northern Arizona University Flagstaff, Arizona, USA, E-mail: FDeMicco@udel.edu

Adel Eldin
Managing Director, ProntoCare Cardiologist Tampa, Florida, USA

Yuan Fang
Alfred Lerner College of Business and Economics, Department of Hospitality Business Management, University of Delaware, 14 W., Main Street, Newark, DE–19716, USA

Rita Fearon
School of Hospitality and Tourism Management, Oklahoma State University, USA

John La Forgia
Chief Marketing Officer, the Mayo Clinic (Retired), Rochester, Minnesota, United States

Yushu Fu
Alfred Lerner College of Business and Economics, Department of Hospitality Business Management, University of Delaware, 14 W., Main Street, Newark, DE–19716, USA

Ilan Geva
President at Ilan Geva and Friends, and Adjunct Professor at the Graham School, University of Chicago and DePaul University, Chicago, Illinois, USA; 2252 Washington Dr., Northbrook, IL–60062, USA, E-mail: ilan@ilanandfriends.com

Jackie Guzman
Michigan State University, USA

Terri Hollen
Alfred Lerner College of Business and Economics, Department of Hospitality Business Management, University of Delaware, 14 W., Main Street, Newark, DE–19716, USA

Nan Jaing
Alfred Lerner College of Business and Economics, Department of Hospitality Business Management, University of Delaware, 14 W., Main Street, Newark, DE–19716, USA

Martin Jeffrey
Professor, The Swiss School of Tourism and Hospitality (SSTH), Switzerland

Soo Kang
Colorado State University, Fort Collins, Colorado, USA

Carolyn Lammersfeld
Vice President of Integrative Oncology at Cancer Treatment Centers of America Global, Inc., USA

Mengyu Li
University of Delaware, USA

Sophia Linn
Assistant Director, Geospatial Centroid at Colorado State University, Colorado, USA

Claudia Mika
Temos International Healthcare Accreditation, Cologne, Germany

Ananth Mohan
Senior Vice President, Innovation, and Network Operations at Cancer Treatment Centers of America Global, Inc., USA

Kelly Murray
Manager, Enterprise Innovation at Cancer Treatment Centers of America Global, Inc., USA

Lali Odosashvili
The Lerner College of Business and Economics, University of Delaware, Newark, DE–19711, USA, E-mail: lalio@udel.edu

Radesh Palakurthi
Professor and Dean, I.H.G. Chair of Excellence, The Kemmons Wilson School of Hospitality and Resort Management University of Memphis, 3700 Central Ave, Suite 140, Memphis, Tennessee–38152, USA, Tel.: (901)-678–3430 Fax: (901)-678–0034, E-mail: rplkrthi@memphis.edu

Abraham Pizam
Professor and Linda Chapin Eminent Scholar Chair in Tourism Management,
Rosen College of Hospitality Management, University of Central Florida, Orlando, Florida, USA

Ali A. Poorani
Alfred Lerner College of Business and Economics, Department of Hospitality and Sport Business Management, University of Delaware, 14 W., Main Street, Newark, DE–19716, USA,
E-mail: poorani@udel.edu

Kathryn Raborn
Alfred Lerner College of Business and Economics, Department of Hospitality Business Management, University of Delaware, 14 W., Main Street, Newark, DE–19716, USA

Tammy Richmond
CEO/President, Go 2 Care, Inc., USA, Registered/Licensed, Fellow of the American Occupational Therapy Association, E-mail: tammy@go2care.com, www.go2care.com

Mardelle M. Shepley
Cornell University, NY 14853, USA

Peter P. Tschirky
Former Managing Director Grand Resort Bad Ragaz, Switzerland

Stephen M. Weiner
Founding Chair, Heath Law Practice, Mintz, Levin, Cohn, Ferris, Glovsky and Popeo, P.C, Boston, Massachusetts

Yan Cao
Alfred Lerner College of Business and Economics, Department of Hospitality and Sport Business Management, University of Delaware, 14W, Main Street, Newark, DE–19716, USA,
E-mail: poorani@udel.edu

Peter C. Yesawich
Former Chief Growth Officer of Cancer Treatment Centers of America Global, Inc., USA

Akiko Yoshimi
Colorado State University, Colorado, USA

Elizabeth Ziemba
Temos International Healthcare Accreditation, Cologne, Germany

ABBREVIATIONS

ACA	affordable care act
AI	artificial intelligence
ARB	angiotensin receptor blockers
ASCO	American Society of Clinical Oncology
BMI	body mass index
CA	content-analysis
CABG	coronary artery bypass graft
CAGR	compound annual growth rate
CCC	cannabis consumer coalition
CCS	complementary care services
CDS	clinical decision support
CHM	Chinese herbal medicine
CMS	Center for Medicare and Medical Services
CTCA	Cancer Treatment Centers of America
CTO	Colorado Tourism Office
DMO	destination management organization
DPC	direct primary care
EBD	evidence-based design
ED	emergency department
EMPM	early mobility and patient monitoring packages
FDI	foreign direct investment
FMT	Florida Medical Tourism
G.M.	general managers
GAO	Government Accountability Office
H2H	hospitality bridging healthcare
HCAHPS	Hospital Consumer Assessment of Healthcare Providers and Systems
HIMSS	Healthcare Information and Management Systems Society
HP	home-based physician
HR	human resources
ICT	information and communication technologies
IEEA/ISQua	International Society for Quality in Health Care External Evaluation Association
JAMA	Journal of the American Medical Association

JCI	Joint Commission International
MASCC	Multinational Association of Supportive Care in Cancer
METI	Ministry of Economy, Trade, and Industry
MHTC	Malaysia Healthcare Travel Council
ML	machine learning
MMJ	medical marijuana
MT	medical tourism
MTFs	medical tourism facilitators
MTIs	medical tourism intermediaries
MTSEZs	medical tourism special economic zone
NCCN	National Comprehensive Cancer Network
NPS	net promotor score
OAS CAHPS	Outpatient and Ambulatory Surgery Consumer Assessment of Healthcare Providers and Systems Survey
OECD	Organization for Economic Cooperation and Development
PCP	primary care providers
PMA	patient-mix adjustment
POS	point of service code
PP	practicing physician
PPE	personal protective equipment
PPP MT	product dimension, performance dimension, promotion dimension, and medical tourism dimension
PRSO	patient self-reported outcomes
QOPI	quality oncology practice initiative
RIC	Rehabilitation Institute of Chicago
ROI	return on investment
RVUs	relative value units
SEER	The Surveillance, Epidemiology, and End Results Program
SEZs	special economic zones
SIMC	Shanghai International Medical Center
SIO	Society for Integrative Oncology
SIT	special interest tourism
SNAP	Supplemental Nutrition Assistance Program
STDs	sexually transmitted diseases
SWOT	strengths, weaknesses, opportunities, and threats
TCM	traditional Chinese medicine
TSA	Transportation Security Administration
UNCTAD	United Nations Conference on Trade and Development
US	United States
WTTC	World Travel and Tourism Council

FOREWORD 1

Destinations and healthcare providers who have executed a clear and focused strategy got their marketing right, delivered an outstanding international patient experience, and tracked their return on investment (ROI).

This book highlights some of those successes and provides valuable insight that will inform your business decisions, guide your strategy, and drive forward your international patient business.

In this enlightening volume, Fred DeMicco and Ali Poorani have brought together some of the world's leading experts to share their advice and experience on those areas that have proved challenging for many health-care providers and destinations-determining a successful strategy, creating competitive advantage, marketing, and branding a medical tourism (MT) offering and meeting and exceeding the expectations of the international patient.

The authors encourage you to learn from the past but, of course, to look to a future when medical travel moves from the niche to the mainstream.

—Keith Pollard
Editor in Chief, International Medical Travel Journal,
www.imtj.com

FOREWORD 2

Since the 1970s, with the advent of managed care, diagnostic-related groups, and other social and policy forces, healthcare providers have faced reimbursement constraints and an increasingly competitive marketplace. These developments and the later advent of digital communications hastened the emergence of the patient as a consumer. Arguably, once patients became consumers, it was inevitable that they would eventually be seen as customers or guests. This paradigm shift fostered the now ubiquitous array of hospitality-style service enhancements throughout the healthcare industry.

Rising health care costs in some countries, long waiting times in some healthcare systems, the widespread availability of air travel, and the ability to digitally search the world almost instantaneously have all coalesced to foster the medical "tourism" industry. By searching the Web, prospective patients can find available health care from numerous sources. They can access quality data, compare costs, and interact with proprietors and caregivers before ever leaving their homes.

This has, of course, also increased competition in the healthcare industry. Whereas providers historically operated in relatively safe zones of local and regional "catchment areas," now medical centers vie for patients from places near and far. These centers increasingly seek a competitive advantage, not only by offering high-quality care but by providing patients and families with a range of differentiating hospitality services.

Indeed, healthcare leaders have come to appreciate that medical and healthful patient outcomes are most successful when integrated with a hospitality-driven process. The pre-planning of the medical journey leading up to the arrival at the medical/health/wellness destination begins a cascade of multiple hospitality bridging healthcare or H2H© services to create an entirely satisfying experience for the patient/guest, which can lead to positive health and loyalty outcomes, if performance in the entirety of the process is at the highest level. Hospitality and Healthcare services working in harmony and cadence will play a critical role for success.

In this second edition of their indispensable volume, *Medical Travel: Success Factors and Cases: What You Need to Know About Starting or Building a Medical Travel Brand*, by Frederick J. DeMicco and Ali A. Poorani have assembled an array of experts who provide invaluable guidance and insight illustrated with real-world case studies for any individual or firm interested in navigating the waters of this rapidly evolving industry.

From chapters on branding through to assessing accreditation and post-care quality metrics, Drs. DeMicco, Poorani, and their fellow contributors take the reader through the critical phases of the medical travel journey: pre-visit, travel, on-site care, discharge, and follow-up care. The authors address critical issues facing medical, health, and wellness travel from both macro and micro perspectives. Presentations of best practices and strategies demonstrate how some destinations have built, renewed, or engaged various stakeholders to construct or enhance their medical tourism destinations.

Medical Travel also showcases best practices and innovative ways of designing and operating a profitable and entrepreneurial practice. Quality issues, esthetics, and legal issues related to inbound and outbound medical tourism are also presented. The book explores the evolving nature of hospital design and the complex relationship between people and medicine as manifested in the relationship of hospital esthetics to patient satisfaction.

Besides medical travelers at the center of medical travel, there are three main players in the industry: government agencies, intermediaries, and health and wellness providers. Selected best practices, research, cases, innovative strategies, SWOT analysis, and toolkits address the aims of all stakeholders.

In short, Drs. DeMicco and Poorani have done a masterful job of presenting the full range of issues and opportunities in the field of medical travel. For the student hoping to enter this field, the entrepreneur about to make an investment, or the established professional firm planning a significant brand extension, this book is essential reading and should be the standard reference for anyone involved in the field of medical travel.

—John La Forgia, MBA
Emeritus Chief Officer of Marketing and Public Affairs, The Mayo Clinic;
Author of *Brand Aid: The Revolutionary Transformation of the Mayo Clinic Brand* (2019)

PREFACE

Medical travel is an emotional journey, and the effects of COVID-19 during the year 2020 have taken an emotional toll not only on medical travelers but also on healthcare providers, medical travel businesses, transportation, and everyone involved. Furthermore, it has presented serious economic challenges to those countries and destinations that are heavily dependent on tourism and medical travel.

When the time comes to resume safe travels, and some sort of normalcy prevails, we feel there will be a new normal, when mostly the fittest, those destinations that are proven safe, well-branded, and exceptionally equipped to create unique experiences will endure.

The events surrounding medical travel since the announcement of the COVID-19 pandemic by the World Health Organization (WHO) has further motivated us to bring together major players, renowned authors, practitioners, and researchers to create this book and prepare the medical tourism (MT) market to not only recover from the devastating effects of the pandemic but provide tools and cases that will help to structure successful destinations for medical travel.

The Coronavirus Pandemic has changed so many aspects of our lives. The summer of 2020 especially was a year without many vacation holidays, MT, and medical travel. The fear of COVID-19 coupled with the loss of income dissuaded many (medical) tourists from traveling. The fear of taking flight (flight phobia) dominated the travel market. In addition, some countries had to close borders, and many also implemented pre- and post-COVID quarantines.

Few events short of nuclear war could be as devastating to the global economy. Tourism is its heartbeat and a major contributor to jobs and careers. There were about 1.5 billion international trips in 2019. About 350 million hospitality and tourism industry jobs depended on travel-from the front desk agents to chefs to tour operators and cruise ship captains.

- **Facts:** In April 2019, nearly 7 million passengers arrived or departed through London Heathrow. In April 2020, the number was only about 200,000, a nearly 97% decline. Flights trans-Europe were down by about 85%. In the U.S., the Transportation Safety Administration screened only about 3 million passengers in its airports in May 2020, a decrease from about 70 million during the same time period in 2019. The majority of travelers who had confirmed air travel or hotel plans prior to COVID-19 had canceled those plans; and trips that require longer booking windows, such as cruises and group tours, were more likely to have been postponed for later in the year.

- **International Travel Stuck on the Tarmac:** In 2020, the United Nations World Tourism Organization projected a drop in international travel spending of about $1 trillion and a reduction in tourist arrivals by about 70%. Of the over 200 key tourism destinations modeled, more than 70% had closed their borders to international tourists. Generally, Europe receives over one-half of global travelers each year. The fear of COVID-19 kept these travelers closer to home, depressing the travel and hospitality industry throughout the continent.
- **Baby Steps Towards Opening:** While optimism for the gradual opening had increased for both business and leisure travel during the third quarter of the year, the second wave of pandemic began to brush many countries and regions in red towards the end of the 4th quarter of 2020, with many restrictions and limitation, resulting in actual trip cancellations and postponements (47.4%) in the United States (U.S.), according to Destination Analysts, with 62.8% expressing less likely to travel in the next 3 months.

While writing this script, during the last quarter of the year 2020, the news regarding the development and efficacy of vaccines have produced some glimpse of hope. Pfizer, with the early data showing its vaccine, is more than 90% effective. In addition, Moderna's vaccine is showing 94.5% effectiveness against coronavirus, according to CNN Health News. This news has made Americans more confident that they can travel safely in 2021, particularly during the third and 4th quarters.

Internationally, countries that heavily depend on tourism were working hard to restart this sector sooner. Italy and Greece, for example, wished to welcome global tourists during summer 2020. Cyprus was offering free medical care to anyone who contracted the virus while visiting, plus free hotel room, food, and drinks for their traveling companions. Geo-Connected Regional areas within which vacationers will be allowed to travel freely—Travel Bubbles—were being considered around the world, especially in Europe, Australia/New Zealand, China, and Thailand, Singapore, and AEU, India with 13 countries, and many more. Travel bubbles or travel corridors are an exclusive partnership between neighboring or nearby countries that have demonstrated considerable success in containing and combating the COVID-19 pandemic within their respective borders. Despite all of this, many would-be travelers had to stay in their own home countries last summer. Rising coronavirus cases throughout the world in the last quarter of 2020 have adversely affected the hopes. International bookings for future travel to the U.S. have plunged to (–63%).

- **Medical Tourism (MT):** The general state of tourism has created serious concerns and massive impacts regarding the future of medical and health tourism. According to a study of 70 industry participants and opinion leaders from across the globe conducted by the *International Medical Travel Journal* and Health Tourism Worldwide, COVID-19 has reduced their international travel by 50%, as reported by 60% of participants. In addition to the negative impact on the number of travelers, business operations, liquidity, and capital resources were also seriously impacted (impact score of 4.1 out of 5). Nearly 16% of respondents had reported closing their operations to international travel, and the majority also stated that they were changing their practices and operations to be ready for the new opening. Nearly 41% anticipated the impact to last 6–12 months, and 31% anticipated the impact to last more than 12 months.

  Among the other important findings regarding the future of MT were: (a) the time between the decision to book and the actual travel date will increase; (b) the patient concerns about safety will reduce the distance when travelers selecting a destination; (c) with safety the primary concern, patients more likely will select destinations that managed COVID-19 effectively; and (d) medical travelers will also choose well-known and established and branded healthcare facilities and packages.

  In this book, we hope to achieve how to design, renew, and manage well-established and branded medical travel facilities and destinations to accommodate the future needs and preferences of medical travelers.

PART I
Getting Started

CHAPTER 1

MEDICAL TOURISM AND WELL-BEING: TRENDS AND STRATEGIES

FREDERICK J. DEMICCO,[1] M. CETRON,[2] and O. DAVIES[2]

[1]*School of Hotel and Restaurnat Management, the W.A. Franke College of Management, Northern Arizona University Flagstaff, Arizona, USA*

[2]*Forecasting International, 22 Commerce Road, Newtown, CT 06470, USA*

ABSTRACT

For most of us, getting sick is a good way to ruin a vacation. However, for growing numbers of people, needing to see the doctor is the whole point of going abroad. When they require surgery or dental work, they combine it with a trip to the Taj Mahal, a photo safari on the African veldt, or a stay at a luxury hotel—or at a hospital that feels like one—all at bargain-basement prices. Simply walking across the bridge from San Ysidro, Texas, to Tijuana can cut the price of dental work in half or less, sometimes much less. This is medical tourism (MT), and it is one of the hottest niche markets in the hospitality industry. At least, we believe it is. Or will be.

1.1 MARKET PROJECTION OF MEDICAL TRAVEL

As you may have figured out already, reliable data about medical tourism (MT) is remarkably hard to come by. Observers disagree about the size and growth rate of this travel segment. Sometimes, they even disagree with themselves. One specialist in MT suggests that about 7 million people each year qualify as medical tourists (DeMicco, Cetron, and Davies, 2019).

On the other hand, the Organization for Economic Cooperation and Development (OECD)—which we trust in most of the fields it studies—estimates that nearly 50 million patients travel abroad for medical services each year.

In part, this variation results from disagreement about exactly what constitutes MT. Some authorities count not only visits to hospitals, clinics, and individual providers, but travel to spas and wellness services, and we tend to follow their lead throughout this chapter. However, other analysts do not. This difference alone could reduce the count of medical tourists by 90%. In addition, in some regions where undocumented migrants are common, it becomes difficult to count the number of people making the trip. This is particularly an issue in the market for medical services among travelers from the United States (U.S.) to Mexico.

Revenue estimates are just as variable. In 2016, one analyst placed the market for MT at only $5 billion a year. We believe this is considerably too low.

At the other end of the scale, in 2016, VISA and Oxford Economics carried out a major study of MT. The first version of their report put annual revenues at $349 billion worldwide. (One garbled account from no less than the American Medical Association transposed the digits and made it $439 billion!) A hasty revision brought the figure down to around $50 billion per year. In mid-2017, the latest version available online suggests that MT brings in global revenues of $100 billion annually. Call that the highest credible number.

This report also projects revenue growth of 25% per year, another number we believe marks the high end of credibility. Starting at $100 billion and expanding at that pace, the medical-tourism market could be worth about $600 billion in 2025. This seems very optimistic, but not totally impossible.

Unfortunately, the study also predicts that 3 to 4% of the world's population will go abroad for medical services in 2025. That is about 300 million people. We find this much harder to believe.

Tentatively, then, we will cast our vote with Patients Beyond Borders, an online clearinghouse of information about MT. They estimate that 12 to 14 million people travel internationally for care each year. Only one in 10 go for medical treatment, and 70% of them want cosmetic surgery or other elective procedures. The remainder seeks a variety of "wellness" services like spa visits, supervised exercise, and dieting. Medical tourists, the company estimated, would spend an average of $3,800 to $6,000 per visit that year, including direct costs for medical services, accommodations, and local and international transportation. This puts revenues between $40 billion and $60 billion per year. The firm's best guess is that the market is now growing by 15 to 25% annually.

For a quick analysis, let us accept that there were about 14 million medical tourists in 2016 and put the revenue and growth numbers in the middle of their ranges, at $50 billion and 20%. This multiplies out to about 60 million medical tourists in 2025, producing revenues of $260 billion, or about $4,333 per person. We find these numbers much easier to accept. Until better data becomes available, we will make this our official forecast: Through 2025, expect MT to grow by 20% per year, reaching 60 million people and a market value in the neighborhood of $260 billion.

Note that the exact numbers matter a lot. Raising revenues to an average of $5,000 per person adds $40 billion to the bottom line. Accepting the high-end estimate of $60 billion a year in 2016 revenues brings the 2025 forecast to nearly $310 billion. Thank you, compound interest!

1.2 WHY GET HEALTHCARE ELSEWHERE?

The U.S. has the best medical care in the world, or so we are told despite many statistics that say otherwise. Most countries in Europe offer universal healthcare, either paid for entirely through taxes or heavily subsidized and with relatively modest charges. In fact, 75 of the world's 192 countries have passed legislation mandating universal access to healthcare services independent of income, and in 58, from South Korea to Botswana, at least 90% of the population has effective access to necessary medical services. Why go elsewhere for them?

It turns out that medical tourists have several good reasons to seek care away from home, and sometimes far away, even when their countries provide generally adequate care at reasonable cost—as too many do not.

One motivation, particularly in the U.S., probably is general dissatisfaction. A study by the Commonwealth Fund rated healthcare systems in 11 prosperous countries according to satisfaction ratings by both patients and primary-care physicians. Britain, with its much-maligned National Health Service, was rated best in 9 of the 11 measures and first overall. In four categories and overall rank, the U.S. came in dead last. With a 5th-place rating for timeliness of care, the U.S. even landed below the UK, which came in third.

Another benefit of MT is comfort. A private room in a clinic or hospital with the amenities of a four-star hotel is a lot easier to stay in than the bare two- and four-bed cells typical of American hospitals.

Ordinary sightseeing also is a factor. Why spend time in a hospital at home when you can combine your procedure with a good vacation at lower cost? Almost 90% of patients or their companions take advantage of the standard attractions while in the country for care.

Yet, there are more compelling reasons to go abroad, especially for critical care. In some regions, state-of-the-art medical facilities are hard to come by, if they exist at all. For this reason, patients throughout the Middle East regularly travel to Jordan or Dubai, Asia, and even the U.S. for complicated surgery; close to 1 million Indonesians make a yearly trip to Malaysia or Singapore for routine matters as well as more challenging treatments; and 100,000 Russians annually find care in Israel, Turkey, and the U.S.

In other countries, the public health system is overburdened, and needed care can be delayed, particularly for difficult, uncommon, or expensive procedures. In Britain, regulations require the National Health Service to provide surgery no more than 18 weeks after diagnosis. In 2016, more than 193,000 patients waited longer every month. (At that, this is a substantial improvement over the experience reported in the first edition of this text, published in 2006). In Canada, the waiting list for neurosurgery in 2016 averaged 47 weeks after referral by a general practitioner. Esoteric procedures can take even longer. Waiting times of three months or more are common for stem-cell transplants needed for more than 80 diseases, including several blood cancers. In one well-publicized case from 2016, an 18-year-old girl with acute myeloid leukemia died while waiting for a transplant 8 months after learning that several matching donors were available. Until May 2017, cancer patients who had relapsed after chemotherapy were barred from stem-cell transplants that might have saved their lives.

Stem cells bring up another reason to leave home for care. Some procedures not yet approved or banned for nonmedical reasons in one country may be available in other countries for patients who want and can afford them. Canadians with multiple sclerosis, for example, must leave the country for treatment of abnormal blood drainage from the brain and spinal cord, which a few specialists believe is a factor in the disease. And in the U.S., stem cell transplants are accepted only for patients with certain cancers and a few other disorders. Despite this, nearly 600 clinics across the country offer them for purposes ranging from repair of spinal cord injuries to breast augmentation. However, insurance rarely covers these procedures, and costs of $25,000 or more can make it worth searching for less expensive providers in other countries.

China is another such case. New treatments there are not blocked by ethical or religious sensitivities, but bureaucratic inertia has the same effect.

Beijing estimates that half a million of its citizens traveled abroad for medical care in 2016, many of them for cancer and heart disease that required treatment not yet approved at home.

The delay between early research and clinical use of new therapies provides many such examples. In probably the most dramatic, scientists in the last few years have found that giving the elderly blood plasma from the young reverses many effects of aging. To date, these benefits have been clearly demonstrated only in experimental animals. Yet, at least one clinic in the U.S. already provides these treatments for human patients in the guise of research—at a cost of $8,000 per two-liter dose. Predicting that similar infusions will soon be available much less expensively in any number of developing countries is probably the easiest forecast we will ever make.

This of course brings us to the real attraction of MT for patients from countries where modern healthcare is readily available: price. In polls, nearly 80% of medical tourists say they left home to save money on care. They have good reason to do so. Patients Beyond Borders offers some ranges for expected savings on care in 10 popular destinations. Compared with costs in the U.S., the savings add up:

- **Brazil:** 20 to 30%;
- **Costa Rica:** 45 to 65%;
- **India:** 65 to 90%;
- **Malaysia:** 65 to 80%;
- **Mexico:** 40 to 65%;
- **Singapore:** 25 to 40%;
- **South Korea:** 30 to 45%;
- **Taiwan:** 40 to 55%;
- **Thailand:** 50 to 75%;
- **Turkey:** 50 to 65%.

For some procedures, the savings can be even greater. A heart-valve replacement that would cost $200,000 or more in the U.S. goes for as little as $5,000 in India. Call it $10,000 including round-trip airfare and a brief vacation.

A coronary artery bypass can cost anywhere from $70,000 to $200,000 in the U.S. and sometimes more. In India, the price is $7,000 to $9,000.

Dental implants are worth $4,500 or more per tooth in the States. They go for around $770 in Thailand, $600 in Mexico, and as little as $500 in Hungary.

A knee replacement that averages close to $60,000 in the U.S.—not counting physical therapy—costs as little as $5,000 in Thailand and $3,300 in India.

LASIK eye surgery averages about $2,500 per eye in the U.S. It can be had elsewhere for between $300 and $465.

1.3 QUALITY CONCERNS

Inferior medical care would not be worth having at any price, and there are still a few skeptics who warn that Third-World surgery cannot be as good as that available in the U.S. Many of them represent organizations representing American physicians, including the American Medical Association. In fact, there were cases of botched plastic surgery, particularly at some Mexican clinics in the days before anyone figured out what a gold mine cheap, high-quality care could be for practitioners in developing countries.

Yet, hospitals and clinics that cater to the tourist market often are among the best in the world. Many are staffed by physicians trained at major medical centers in the U.S. and Europe. Bangkok's Bumrungrad International Hospital has some 900 physicians, including about 220 surgeons who are board-certified in the U.S. One of Singapore's major hospitals is a branch of the prestigious Johns Hopkins University, which collaborates with 18 hospitals around the world. The entire healthcare system in Singapore does a better job than the U.S. in measures such as life expectancy, infant mortality, and maternal deaths in childbirth. More than 450 hospitals around the world are accredited by the nonprofit Joint Commission International (JCI), a branch of the organization that evaluates and certifies the quality of care provided by a wide variety of medical institutions in the U.S.

Another indicator that suggests quality care is the number of procedures a facility performs. In a field where experience is as important as technology, Fortis Escorts Heart Institute in New Delhi, carried out some 70,000 heart operations in 2016. Its death rate among patients during some heart operations is about that of most major hospitals in the U.S.—a little higher in some years, a little lower in some. In other heart procedures, the hospital's death rate is consistently less than at comparable American institutions.

It can help, too, that clinics specializing in MT in some countries are backed by sophisticated research infrastructures. India is one of the world's leading centers for biotechnology research, while both India and South Korea are pushing ahead with stem cell research at a level of sophistication at least equal to that anywhere else in the world.

Skilled doctors and state-of-the-art equipment are only two of many benefits these medical centers offer foreign patients. In many, the doctors are supported by more registered nurses per patient than any Western facility could afford. Some facilities provide a nurse for each patient 24 hours a day. Many assign patients a personal assistant for the post-hospital recovery period. Some Asian national airlines offer frequent-flyer miles to ease the cost of returning for follow-up care. Some offer special discounts for medical tourists visiting affiliated clinics.

Under the circumstances, it is no surprise that the MT market is growing rapidly. Around 10 years ago, it was hardly large enough to be noticed. Today, something over 250,000 patients per year visit Singapore alone; nearly half arrive from the Middle East. Perhaps half a million annually travel to India for medical care; in 2002, it was only 150,000. McKinsey, the consulting firm, estimated that MT could bring India as much as $2.2 billion per year by 2012. Argentina, Costa Rica, Cuba, Jamaica, South Africa, Jordan, Malaysia, Hungary, Latvia, and Estonia all have broken into this lucrative market or are trying to do so, and it seems that a few more countries join the list every year.

1.4 TOP MEDICAL TRAVEL DESTINATIONS

A number of countries entered the market for MT early on or worked hard to develop their offerings quickly, and they have benefited from low wages and other costs. As a result, they have become the most popular destinations in the market today. Some details about 11 at the top are discussed in subsections

1.4.1 COSTA RICA

Costa Rica has ecological wonders found in few other lands, from some of the largest, best-protected rain forests in Central America to the fire show of the Arenal volcano. For those with more urban tastes, the casinos of San Jose, Puntarenas, and Guanacaste provide all the action even a jaded Las Vegas regular could ever want.

But for North American patients, what Costa Rica really offers is inexpensive, high-quality medical care in their back yard. Two major hospitals are accredited by the JCI, while many others have specific accreditations for ambulatory health care or ambulatory surgery. In all, the World Health Organization rates Costa Rica's healthcare system as one of the three bests in Latin America. Costa Rica received about 2.6 million foreign visitors in 2015; about 130,000 went for medical care or wellness services.

Cosmetic surgery and dental work are the traditional specialties here, but just about any medical service that can be found in the U.S. can be had in San Jose. Many will be delivered by English-speaking specialists board-certified in the U.S. In the last few years, eye surgery and weight loss procedures have become common among medical tourists there. Wellness clinics and prevention services abound. There even are so-called “recovery retreats,” hotels with all the amenities but with nurses and interns on hand to provide medical services for post-surgery patients.

For most procedures, costs can be half those in the U.S.—not the prices they would find in India or Thailand, but a lot closer to home for medical vacationers with limited travel time or budgets. Get a facelift, and chill on the beach until the bruises go away, and the folks at home will never quite be sure why you suddenly look so good.

1.4.2 HUNGARY

Hungary may be the continent’s oldest destination for MT. The wealthy of Europe have been visiting the country’s spas for centuries. They still are, and patients are beginning to join them from North America, western Russia, and the Middle East.

Hungary’s healthcare system was modeled on the British National Health Service, so almost any form of care is readily available. Although cosmetic surgery is a growing market, this country is best known for its dentistry. No wonder. In Manhattan, there is one dentist for every 630 residents. In Prague, Czech Republic, the number is one in 1,200. Hungarian towns near the Austrian border, where dental tourism is at its busiest, average about one for every 200 residents.

This competition, combined with Hungary’s generally low prices, has the expected benefit for patients. A dental implant that would cost from $3,000 to $6,000 in the U.S. goes for as little as $450 in Hungary; the costliest we have heard of was only $1,200. Removing a tooth averages only $50. A root canal that would be up to $900 for a front tooth and $1,400 for a molar average just $99, with front teeth costing as little as $50. For anyone in Europe, it does not take much work to justify the cost of visiting Budapest.

Add to this spa towns fed by more than 1,000 hot springs, historic castles and palaces, the Danube, and four-star hotels that can cost as little as $85 per night in downtown Budapest, and Hungary becomes one of the more inviting destinations for the medical tourist.

1.4.3 *INDIA*

India has a two-tier medical system. In the vast rural areas, doctors are thin on the ground, and an estimated 700,000 who practice as physicians are unlicensed, and many are untrained. In the cities, the wealthy, at least, have access to state-of-the-art hospitals and clinics, and doctors are as skilled as any in the world. It is this second group that medical tourists get to see.

India is not the oldest "brand" in MT, but it has grown into one of the most inviting and successful. Medical tourists visit India in large numbers—estimates range from 250,000 to 500,000 of them each year—and the flow is believed to be growing by 20% annually.

Several factors contribute to this success. It helps a lot that English is one of the many native languages in India, one spoken by all educated Indians. Unlike some competing nations, India offers a high degree of transparency; visitors need not worry about unexpected problems with their funds or legal status. Some hospitals begin evaluating patients over the Internet, so they are ready to begin treatment as soon as the patient arrives. Some will even take care of travel arrangements from visa applications to making hotel reservations for the patient's family. And, of course, India is the only country in the world where medical tourists can visit the Taj Mahal while convalescing from a kidney transplant.

Yet, India's most obvious attraction is its cost of care, which can be as little as 15% of the price in the U.S. and for many procedures is the lowest in the world. Even trips for follow-up care are cheaper than for ordinary vacationers, because Air India subsidizes them with frequent flyer miles.

Another attraction is the range of high-level services available in such a large, technologically advanced country. There are top-notch centers for open-heart surgery, pediatric heart surgery, hip, and knee replacement, cosmetic surgery, dentistry, bone marrow transplants, cancer therapy, and just about any other specialty a patient could need.

Many of those centers are among the best in the world. Virtually all are equipped with the latest electronic and medical diagnostic equipment—and India, unlike some of its competitors in this market, has the technological sophistication and infrastructure to maintain it. Additionally, Indian pharmaceuticals meet the stringent requirements of the U.S. Food and Drug Administration. It helps, too, that most facilities specializing in MT provide accommodations that could be mistaken for five-star hotels.

Some Indian medical centers even provide services that are uncommon elsewhere. For example, instead of having the entire hip joint replaced, patients can undergo "hip resurfacing," in which damaged bone is scraped away and replaced with chrome alloy. The result is a smoothly functioning joint with less trauma and recovery time than total replacement, and at lower cost. The operation is now widely available in the U.S., but Indian tourist clinics offered the procedure long before it received FDA approval; this long practice has given Indian orthopedic clinics success rates better than those of most other institutions in the world. Other advanced services available in Indian clinics include radiotherapy and radiosurgery to destroy cancerous tumors from outside the body, robot-assisted surgery, and bone marrow transplants for genetic disorders yet to receive FDA approval in the U.S.

Of course, before surgery or after, India has a broad array of unique and exotic destinations for Western tourists. From attending one of the Dalai Lama's public talks at Dharamsala—you will need to plan well ahead—to a half-day safari in the White Tiger-Bandhavgarh National Park, shopping for handicrafts in the tribal villages of Orissa and Madhya Pradesh, or skin-diving in the Indian Ocean, this 4,000-year-old civilization has something to offer anyone who visits.

1.4.4 ISRAEL

These days probably is best known for its long-running conflict with its neighbors and a relatively high risk of terrorist attack. (Of course, "relatively" conceals important detail. Even in Israel, people are statistically more likely to be struck by lightning than by terrorist violence.) Yet, among medical tourists, this country is known for exemplary care, a high doctor-patient ratio, a high-tech manufacturing industry that supplies state-of-the-art equipment, and some of the best hospitals in the world.

These and other advantages bring Israel about 50,000 medical tourists per year, many of them from Russia and other parts of the former Soviet Union. English is spoken almost universally, so travel from the U.S. also is growing quickly.

If Israel has a medical specialty, it is probably in-vitro fertilization, which is readily available for about one-quarter of American prices and provides success rates higher than those of comparable clinics in the States. Several hospitals and clinics specialize in this field. Others are known for their skill with organ transplants, cancer care, and orthopedic surgery.

The country's other marketable asset for MT is the Dead Sea, where the saline waters—nearly 10 times as salty as the ocean—are said to have healing properties. Health and wellness travel to the Sea, on the borders of Jordan and Palestine, is growing quickly, and if the water will not cure psoriasis, as many believe, at least the heated mud baths temporarily ease the pain of arthritis. A side trip to the region's spas and clinics makes a pleasant follow-on to a trip for more serious care.

1.4.5 MALAYSIA

For many in the West, it seems a lot like staying at home. Although its fundamentally Asian core is always easy to see, the country was a British colony until 1957. It retains a thick veneer of English tradition, particularly in major cities such as Kuala Lumpur and Georgetown.

However, for MT its important asset is not culture but location. Four out of five medical tourists to Malaysia-call it 800,000 out of 1 million in 2016—arrive from Indonesia, just three miles across the Strait of Malacca. Substantial numbers also come from nearby Australia, and many travels from the Middle East; a moderate Islamic culture makes this country a comfortable destination for patients from other Muslim lands. It also helps that Malaysia is only a half-hour flight from Singapore, where prices for equivalent care are up to 50% higher. Compared with health costs in the U.S., prices are two-thirds or more lower.

Like most destinations these days, Malaysia offers a full range of care, with specific hospitals and clinics known for surgery, pain management, eye care, cancer treatment, dentistry, and weight control. More than one-third of medical tourists come for dental care, with cosmetic surgery and orthopedic treatments close behind. Malaysia also is well known for health screening, which has made it a favorite place for pre-employment physicals for companies throughout the region. For around $500, hospitals, and clinics provide an astonishing array of tests, many of which Western patients will never have heard of, much less undergone.

A few years ago, MT to Malaysia was growing by 35% per year. It continues to increase quickly, if not quite that fast, thanks substantially to the efforts of the Malaysia Healthcare Travel Council (MHTC), part of the Ministry of Health. MT is expected to expand by 30% per year, with revenues reaching $3.5 billion by 2024.

1.4.6 MEXICO

Mexico is the obvious destination for American medical tourists, and they travel there en masse. Although statistics are sketchy at best, an estimated

8 out of 10 Americans who seek health care abroad go one country to the south. Most travel relatively short distances to destinations just across the border from Arizona, California, Nevada, and Texas. Mexico, in turn, receives nearly all its medical tourists from the U.S. or Canada. One reasonable estimate puts the number of tourists visiting Mexican hospitals and clinics at around 1 million a year. However, many are native Mexicans, and an unknown number of them are undocumented migrants. All this makes the cross-border trade in medical services very difficult to quantify.

Like most destinations, Mexico sees the greatest demand for dentistry and cosmetic surgery. Tijuana alone has 20 dental clinics within three miles of the border, nine in the first mile, and 20 cosmetic-surgery clinics in the same neighborhood. Nuevo Laredo, just over the river from Laredo, TX, is home to 19 dental practices and 10 clinics specializing in cosmetic surgery. Juarez, next door to El Paso, is home to 20 dental clinics and 15 cosmetic surgeons, seven of them sharing office space with dentists. In Nogales, the numbers are 19 and 12. In Mexicali, we stopped counting at 40 and 22. Simply looking at a map tells us why American medical tourists head south.

1.4.7 SINGAPORE

Singapore is tiny, just 277 square miles. (For comparison, New York City is 309 square miles, London 607.) Yet, this wealthy city-state has one of the world's best medical systems. Around 14 of its 20 hospitals are accredited by the JCI. In most measures, its healthcare ranks better than healthcare in the U.S., Canada, and much of Europe. Singapore's rates of maternal and infant mortality are lower, its life expectancy higher. The medical school at the National University of Singapore, a joint venture with Duke University, is prestigious enough that would-be physicians come to study there from Oxford, Cambridge, and the American Ivy League universities.

This is far more medical care than a city of 5.5 million needs, so the government has put a good deal of work into attracting medical tourists to keep its doctors occupied. It has had considerable success. Nearly 600,000 people each year travel to Singapore for care. Most are from neighboring countries, but the number arriving from Europe and the U.S. is growing.

With healthcare so plentiful and skilled, there is little point in thinking about specialties for which medical travelers should consider Singapore. If doctors anywhere in the world can treat, heal, or cure it, their colleagues in Singapore can do it, too, and very possibly better.

However, most patients will find that care is no cheaper than it would be at home. In a private hospital, a heart bypass will cost most patients between $35,000 and $52,000, while replacing a hip will run from $14,000 to $25,000. These prices will seem cheap to patients from the U.S., where a bypass averages about $75,000 and hip replacement is around $40,000, but travel expenses will eat up much of the saving. For patients elsewhere, Singapore is an option mostly when the highest-quality care is worth the extra cost.

1.4.8 SOUTH KOREA

South Korea is not the cheapest destination for American medical tourists. Neither is it the closest. Savings typically range from 30% to 45% compared with prices in the U.S., and getting to South Korea from North America means spending 11 to 14 hours in the air. As a result, most American medical tourists to the country are expatriates or military personnel assigned to bases in the region. For the rest, at least Korean Air offers nonstop flights from many cities in the U.S. and Canada.

For certain procedures and for tourists within Asia, South Korea can be much more attractive. Patients needing a heart bypass, valve replacement, or artificial hip or knee can save over 80% of the price at American hospitals. For uninsured patients from Japan, care costs much less than it would at home, and travel expenses are nominal. China's National Tourism Administration Rates South Korea the most popular destination for its country's outbound medical tourists.

About 600,000 foreign patients visit the country each year. Like the government of Malaysia, Seoul is working hard to attract more medical tourists. Roughly 1 million are expected by 2020. Two years later, MT is forecast to bring the country $2 billion a year in revenues.

What tourists receive for their money is access to one of the most sophisticated, high-tech healthcare systems in the world. All the usual medical services are available, but South Korea is particularly well known for spinal surgery and care, robot-assisted surgery, and organ transplants. One transplant center achieves 96% survival rates one year after liver replacement, compared with 88% in the U.S.

By some accounts, Seoul is the plastic surgery capital of the world. However, this South Korean specialty has become controversial in the last few years. In 2014, the national health ministry reports, one in four Chinese patients came for cosmetic surgery. So did 28% of Russians and 23% of Americans. However, this is one specialty that medical authorities do not police, and unregulated clinics are reported to have caused problems. In China particularly, stories of patients who were overcharged or received poor treatment are common. The number of patients seeking cosmetic surgery in South Korea has declined, though estimates of how much vary widely.

Other attractions in South Korea include some spectacular seaside resorts, national parks rated among the world's most beautiful, the serenity of countless Buddhist temples, and a market for medicinal herbs that has been in operation since 1658. The country's mountainous geography, particularly in the north, also means that visitors will not face the heat and humidity found in Southeast Asia.

1.4.9 THAILAND

Thailand got its start in MT in 1997, when the economic crash that hammered much of Asia sent canny healthcare providers looking for new markets. Today, it is one of the world's largest and best-established destinations for foreign patients, particularly from Japan and the U.S. Around 53 JCI-accredited hospitals are distributed among four major cities: Bangkok, Pattaya, Phuket, and Chiang Mai. Nearly 3 million people reportedly visited Thailand for medical care in 2017, another 13 million for wellness treatments. In all, they accounted for more than half of tourist arrivals that year and contributed about $4.7 billion to the economy.

In 2018, Thailand expects to receive about 1 million tourists for medical care and more than 25 million for wellness treatments.

The largest centers for MT are Bangkok and Phuket. No fewer than six medical facilities in Bangkok have hospital accreditation from the U.S. Bumrungrad Hospital alone sees 850,000 patients per year, 40% of them from abroad. As in most tourist-oriented medical communities, the major attractions are cosmetic surgery and dental treatments. However, eye surgery, kidney dialysis, and organ transplantation are among the most common specialties sought by medical travelers in Thailand. When not pinned down by medical treatments or recovery, patients usually spend their time shopping or in local sightseeing.

Yet, for vacation possibilities, Phuket is clearly your destination, with some of the most spectacular beaches and shorefront scenery on the planet. For a few patients, Phuket has another attraction as well: In all the world, Bangkok Phuket Hospital probably is *the* place to go for sex-change surgery, one of the top 10 procedures for which patients visit Thailand.

1.4.10 TURKEY

Turkey may not be the first place Americans think of when considering care overseas. For patients in the Middle East, Russia, North Africa, and parts of Europe, it comes to mind much more readily. One reason is the country's enormous supply of public and private hospitals, more than 1,200, including 51 accredited by the JCI—more than in any other country in the world. Another is the Turkish government's hard work in putting the word out that medical tourists are welcome and will find what they need in Istanbul, Ankara, Izmir, and probably many smaller cities. It probably helps that Turkish Airlines offers a 50% rebate on fares for medical tourists.

These efforts have been remarkably successful. According to the Turkey Health Tourism Development Council, 727,000 international patients arrived in 2014. Other sources put the number closer to 425,000 that year. Yet another estimate suggests that 32% of patients seen in Turkey are medical tourists. Add perhaps half a million visitors to the country's hot springs, spas, and wellness centers, as some industry analysts do, and the number would be in the neighborhood of 1 million medical tourists a year. The Council hopes to attract 2 million actual patients a year by 2023.

For some reason, Turkish hospitals have become especially popular for eye operations, at least among medical tourists. LASIK surgery averages about $2,500 per eye in the U.S. and perhaps $2,000 in Britain, but around $1,250 in Turkey. Cataract removal is about $3,600 in the U.S. and $4,500 in the U.K. In Turkey, it is only $1,500. There are good reasons to make the journey, especially if a side trip to the Topkapi Palace, the Ayasofya Museum, or the ruins at Ephesus sounds like a good way to test your eyesight.

Yet, if a medical specialty or service exists, tourists will find it in Turkey, with a high standard of care and prices well below those at home. A coronary angioplasty that would cost an average of $30,000 in the U.S. or about half that in Britain is $5,000 or so in Turkey. A knee replacement that averages about $50,000 in the U.S. and around $17,000 in the U.K. is only $7,500 in Turkey. For patients in Europe, where travel costs to Turkey are low—a flight from London can be had for as little as £85, or about $110—these are obvious bargains.

1.4.11 UNITED STATES (U.S.)

Americans may find it ironic, dissatisfied as many are with their country's high-cost care, poor insurance coverage, and statistically mediocre results compared with other medical systems, but the U.S. is one of the world's most popular destinations for medical tourists. A look at some of the institutions that actively compete for overseas patients makes the reason obvious. They include the Mayo Clinic, Johns Hopkins, Memorial Sloan Kettering, the Cleveland Clinic, and MD Anderson Cancer Center. These are some of the most prestigious medical centers in the world, and their quality of care lives up to their reputations. For those who can afford their cost, they are the obvious places to go for treatment, especially when life is at stake.

As a result, some 800,000 patients from all over the world seek treatment from American medical centers each year. If the case is complicated or doctors at home have failed, a trip to the U.S. is often the last resort. For those wealthy enough that cost does not matter; it is often the first.

1.5 TOP TRENDS FOR MEDICAL TRAVEL

MT sounds like a specialty where health workers are the only essential personnel, and they do account for the great majority of employment in the field. However, opportunities are expanding almost as quickly as the market. Hospitals need planners, coordinators, and marketers to bring in patients and promote the institution's abilities to health travelers in other countries. Several companies serve as intermediaries between medical providers and employers needing low-cost care for their workers. A number of firms have arisen to coordinate MT for patients, making travel reservations much like traditional agents, but also helping to select the right institution for individual needs, arrange for care, set up accommodations for people accompanying the patient, and planning pre- and post-care vacations. Many more such opportunities are likely to appear in the decade ahead.

Like all these market-segment chapters, this one has given only a snapshot of its topic. Anyone working as a manager in this field will need to assemble a movie of its evolution to help them anticipate changes and develop effective strategies for dealing with them. This, of course, is where the trends come in. Here are five of the most important for this fast-growing field.

Travel (especially international) and tourism are growing fast, as they have done for many years. This trend, of course, is critical to all segments of the industry. Two statistics are significant for MT.

The UN World Tourism Organization believes that the number of international tourist arrivals will grow by 3.3% per year, on average, for the two decades ending in 2030. At that rate, there will be 1.8 billion at the end of the period. Note that travel to developing lands is expected to grow by 4.4% annually, compared with only 2.2% in the industrialized countries.

The World Travel and Tourism Council (WTTC) predicts that industry's direct contribution to global GDP will grow 4.0% per year, on average, from 2018 through 2027.

These numbers provide an effective floor for this segment. The number of medical tourists will grow at least as rapidly as the overall industry. So, will the value of the market.

Let us use the 4.4% growth estimate for the developing countries, as that group includes most destinations for MT. If we accept the narrowest definition, there will be about 8.3 million medical tourists in 2020 and about 10.3 million in 2025. If we use the broader definition favored by the OECD, the numbers come to about 59.4 million in 2020 and 73.7 million in 2025. We will be surprised if the number of medical tourists does not exceed these figures by at least a few million annually.

Market value in the years ahead is even less certain. Assume, as we did at the beginning of this chapter, that the current value is in the middle of the range calculated by Patients Beyond Borders, some $50 billion in 2016. Accepting the WTTC estimates, MT would generate nearly $52 billion in 2017, $58.4 billion in 2020, and $71 billion in 2025.

Again, these are the minimum figures we can expect for MT in the years ahead. This segment is much more likely to grow faster than the international travel market in general, reaching values close to those we calculated at the beginning of this chapter: some 60 million medical tourists in 2025 and worldwide revenues of $300 billion, plus or minus about 20%, depending on per-person spending during the trip. Consider these a reasonable baseline for your future calculations.

1.6 ECONOMIC GROWTH IS SLOWING IN MANY COUNTRIES

Not in all, however, and it turns out that both slowing and accelerating economies can be good for MT. It depends on the countries involved, both as tourist sources and as destinations.

China and India will be critical to the future of MT, and they will take most of our attention here. This is a function of their economic growth, at least as much as their billion-plus populations. Both economies are expanding faster than in any other large nation in the world. Even as China's slows to around 4% per year in the 2020s, the country will remain one of the two growth leaders among those with populations upward of 50 million. India, of course, will be the other.

As we saw in Chapter 6, the rapid growth of China's economy has raised hundreds of millions of people out of abject poverty and into the middle class. By 2020, more than 130 million of the country's 250 million households are expected to fit into this category, with half of them classified as affluent. That adds up to some 400 million people who might be able to afford medical care in another country.

"Middle class" in China does not mean quite what it does in the West. The Chinese government considers anyone with a household income of $7,250 to $62,500 to be in the middle class. McKinsey defines the Chinese middle class as those with an income of $11,500 to $43,000 per year. In the U.S., incomes at the lower end of these ranges would qualify as poverty. Yet, in China, they are enough to pay for food, clothing, and shelter with some disposable income left over. This explains why China has the world's fastest-growing fleet of automobiles. It also tells us that this country is destined to become one of the world's largest markets for MT in the 2020s.

The Indian economy is growing even faster, by upward of 7% in recent years. It will continue to expand rapidly when China's high-growth period is clearly behind it, averaging 6.4% per year through the 2020s. Again, this suggests that the Indian market for MT will grow at least that fast.

What constitutes the Indian middle class, and how many people fit that category? Definitions and estimates vary widely. At Mumbai University, economists include anyone spending from $2 to $10 per day, based on purchasing power parity (PPP.) By this definition, more than 600 billion people already rank in the middle class, upward of twice as many as in 2004. This seems too inclusive; $2 a day is a starvation-level income.

A more common definition considers the middle class to be those who can spend between $10.01 and $50 per day, again as PPP. Call it $3,650 to $18,250. This standard pares the middle class to around 50 million people. By 2020, this version of India's middle class will grow to 200 million people, according to Ernst and Young, and 475 million by 2030. Yet another source predicts that the middle class will number 575 million soon after 2025. No wonder outbound travel from India is growing by 25% each year.

In both China and India, healthcare is readily available. At least, that is the theory. In practice, there are large gaps.

About 95% of the Chinese population has at least minimal health insurance. On average, it covers about 70% of medical expenses, but less for severe or chronic ills. Cancer or heart disease can still leave Chinese families destitute.

Beijing hopes to provide basic healthcare at affordable prices for all its citizens by 2020. That might be expected to minimize the number of people who leave China for care. However, hospitals in China vary widely, with services in major cities generally much better and more modern than those in rural areas. And even in major cities, the best hospitals tend to be profit-making ventures run, or at least partially owned by, foreign healthcare companies. Bills at these hospitals can run 10 times those in rural facilities. Throughout the country, medications are expensive and, for some, unaffordable.

The situation is worse in India. Perhaps one in four families are covered by insurance, and even they pay 70% of medical expenses out of pocket. The government runs an extensive network of medical facilities. Yet, India has so many people that a community health center with four doctors and 30 beds must cover a population of 80,000 to 120,000. Care at public facilities is free, but shortages of staff and supplies mean that needed services often are unavailable. In a 2016 study by *The Lancet*, a prestigious medical journal, India's healthcare system rated 154th out of 195 countries. China came in 82nd. Bangladesh, with a per-capita GDP of around $4,000 compared with India's $6,700, was 52nd.

All this means that in China, and especially in India, the poor are stuck, unable to afford the price of sophisticated medical care or continuing treatment. In India, they also are likely to find that timely care is unavailable.

The Indian upper middle class and wealthy can buy care at the same high-quality hospitals medical tourists do, and for much less than the same services would cost in the West or in many other tourist destinations. However, Indian culture values travel, and as we have seen, destinations like Hungary and Turkey have developed reputations for excellence in particular specialties like cosmetic surgery and eye care. These factors both suggest that many well-off Indians are likely to go elsewhere for care even if strictly economic concerns do not justify it. MT, like most kinds of travel, will be seen as a luxury and proof of affluence. Even for those in the next tier down, MT may be an attractive option, especially for those who are going abroad for other purposes.

Their peers in China have another reason to travel for healthcare. It may be more expensive in other countries, but at least it is available. It will help a lot, also, that insurance has begun to cover travel expenses for those who cannot find care close to home.

The UN World Tourism Organization expects some 50 million Indians to go abroad in 2020. Around 150 million Chinese will join them. If even

1% of them seek medical care, they will add some 2 million patients to the medical-tourism market. We consider this very possible.

In fact, for Chinese travelers, this is a simple projection of the established record. They took an estimated 500,000 outbound medical trips in 2016. The greatest number went to Thailand. Bumrungrad Hospital alone saw 7,500 Chinese patients in 2015—compared with 9,000 for all the hospitals in Singapore—and has added a ward with a Chinese-speaking staff. We expect Indian medical tourists, too, to think of Thailand first when seeking any but the most specialized care elsewhere.

Now consider Europe and, to a lesser extent, the U.S. and Canada. They are well supplied with skilled doctors and high-quality hospitals, and patients in Europe and Canada are covered by government-supported medical systems. In the U.S., we believe the long-term trend leads to the establishment of a single-payer healthcare system, but only after further delay by politically powerful lobbies. Until then, care will remain expensive and insurance coverage spotty. In the UK and Canada especially, even urgently needed treatment can be delayed by shortages of capacity and other factors.

Economic growth in the developed lands has slowed from the heady pace seen in the late 20th century and the first decade of this one. Through the 2020s, we therefore expect people throughout Europe and North America to feel increasingly short of disposable income. They will be eager to save money on healthcare, especially for elective procedures like plastic surgery. It is at least possible, also, that some European countries will choose to cut back on the coverage provided by social programs. (See below.) The UK and Germany, where politicians facing economic problems automatically think first of austerity, seem particularly likely to follow this course.

All these developments will be especially welcome for Hungary and the Czech Republic, which draw most of their foreign patients from other parts of Europe, and for Turkey, where the number of patients from Europe and North America has begun to grow. They will be less significant for Thailand and India, which already receive the largest shares of Western MT and will be proportionally less affected.

1.7 DEMOGRAPHIC CHANGES IMPACTING MEDICAL TRAVEL

For MT, it makes sense to consider three trends simultaneously: the growth of the world's population to 9.6 billion in 2050, the global extension of life expectancy, and the dramatic growth of the elderly population in rich and poor countries alike. These trends all foretell rapid expansion of the travel-for-treatment market. Together, they are the largest single reason to expect growth in the medical tourism market that is even faster than many observers do.

The effect of simple population growth is obvious. If the planet is home to roughly 7 billion people and the number of medical tourists is about 14 million, then when the world's population reaches 9.6 billion, the number of medical tourists should be about 19.2 billion. The real number will, of course, be much higher.

The reason is the stretching of our life expectancy and the growing elderly cohort. As we saw in Chapter 6, better nutrition, new pharmaceuticals and medical technologies, and government health programs raise life expectancy with every generation. In 1950, life expectancy at birth in the U.S. was 68.2, and an 80-year-old American could look forward to another 6.5 years. Today, life expectancy at birth is up by more than 10 years, to 78.9, and the average 80-year-old will reach 89.6. In many other places, the average is even greater. In 25 countries, life expectancy is upward of 80 years. In Monaco, it is 89-plus. A life expectancy just under 79 puts the U.S. in 53rd place. By 2050, most of the world's countries are expected to have more people age 65 and above than younger than 15. In Japan, South Korea, and Germany, more than half of the population in 2050 will be over 50.

As a result, the world's senior population will more than double from 901 million in 2015 to 2.1 billion in 2050. The over-80 population will more than triple, reaching 434 million. Those in their traditional retirement years made up only 15% of the global population in 2000. Around 50 years later, they will be 27%.

This trend alone guarantees that the cost of medical care will rise quickly in the years ahead. In the U.S., healthcare for the average child costs about $3,550 per year. For working-age adults, it is around $6,600. However, among the elderly, it is nearly $19,000 per year, and much of those goes to the oldest old. Patients in their late 60s make up 26% of the Medicare population but account for only 15% of the program's cost. Those age 80 and above make up 24% of the population but account for one-third of Medicare spending. All these numbers change over time and from one country to another. However, the general trend is clear: The older the population, the higher their medical expenses.

This is mostly good news for individuals; in general, we all want to live longer. However, it is much less so for national economies, which will have to spend more and more on healthcare over the next few decades. And that, in turn, is great news for MT.

Exactly how these trends will play out depends on the region. The critical issue is the dependency ratio. This is the number of children and retirees compared with the population in their prime employment years, ages 15 to 64. The smaller a country's dependency ratio, the easier it will be for workers to support those who do not contribute to the economy.

High dependency ratios change the way economies perform. Savings translate to investments that promote growth and, because more money is available to those wishing to borrow, keep interest rates low. Retirees, on average, begin spending what they have saved, so interest rates go up and economic growth declines. Housing prices also go down; one study of 10 countries found that they had sunk, in real terms, by 0.2% per year as the ratio of seniors to working-age adults rose. One more likely effect is a shift in consumption from goods to healthcare services. This is particularly true when the elderly population is comparatively large, producing a high "old-age dependency ratio."

In most of the less developed countries and some of the developing, high birth rates and reduced childhood mortality in recent decades are enlarging the workforce, and dependency ratios are falling. In India, for example, it was 54 in 2010 and is expected to be 48 in 2050. In Nigeria, it is slipping from 88 to 69. These trends should lead to faster economic growth and improved standards of living.

In the developed lands, dependency ratios are climbing rapidly thanks to a combination of low birth rates and longer life expectancies. In the U.S., it was 49 in 2010 and will be 66 in 2050. In Germany, it is rising from 52 to 83. In Spain, the dependency ratio began the period at 47 and will end at 94. In Japan, it was 52 and is expected to reach 96!

In all these countries, the change is due primarily to the growing population age 65 and over. In the U.S., there were 19 per 100 people of working age in 2010. By 2050, there will be 36. In Germany, there were 32, and the number is expected to reach 60. In Spain, the increase is from 25 to 67. In Japan, the ratio will double, from 36 to 72. South Korea's age-dependency ratio will reach only 66, but that is more than four times its ratio of only 15 in 2010.

China is a special case, thanks to the one-child policy introduced in 1979 to stem out-of-control population growth and not replaced—by a two-child rule—until 2016. Population growth has slowed as intended, but as a result, more people are now reaching retirement than are entering the workforce. The working-age population is declining, from a peak of 925 million in 2011 to only 700 million expected in 2050. The old-age dependency ratio is more than tripling from 11 in 2010 to 39 in 2050.

China is not the only country with a declining workforce, nor even the one with the greatest percentage ahead. Germany's workforce will shrink by about 17% over the same period by 2060. Japan's is due to shrink by 12.4%, but its high life expectancy means that by 2060 some 40% of the population will be age 65 or older.

However, Germany and most other countries of the European Union provide ample programs for retirement pensions and old-age medical care. Japan's are not as generous, but the government is working to expand them before need too badly outstrips availability. China is much less ready for the future. If its social programs are not in crisis already, they are approaching it much faster than they can be improved.

Beijing has attempted to prepare for this future by establishing a variety of social programs. In 2016, some 890 million people had government-funded pension accounts, 750 were covered by medical insurance, 220 million by workplace injury insurance.

Unfortunately, these numbers sound more impressive than the reality they represent. Expenses at the five largest government-funded insurance programs are growing about one-third faster than the money available to pay them. Many former military officers already report that they have not received pensions, jobs, or insurance since retiring; some have been waiting 20 years. By 2020, the number of elderlies collecting pensions will exceed the working-age population. And when the elderly need geriatric services, few are available. China has about half as many beds per 1,000 seniors as the developed countries. In the West, between 4% and 8% of seniors live in residential care facilities. In China, the number is 1 to 2%.

Consider one more factor, the growing shortage of medical personnel in much of the world. In the early 2020s, the U.S. is expected to have some 90,000 too few doctors and more than 1 million fewer nurses than it needs. Japan will have 200,000 too few nurses by 2020 and 380,000 too few by 2025. China had 500,000 fewer doctors than it needed as early as 2015.

All these factors point in the same direction. In China, Japan, and South Korea; in Western Europe; in Canada and, to a lesser extent, the U.S., demand for medical care will grow at least as fast as the supply, and in most cases faster. At the same time, people in many parts of the world will be watching their disposable income shrink even as their expenses grow. People will need much more care in the years ahead, and many of them will be unable to find it at home. Those who can afford the cost of travel will seek it in other countries. Demographic trends guarantee a prosperous future for companies that serve the world's medical tourists.

1.8 CONCLUSION AND IMPORTANT MEDICAL ADVANCES WILL CONTINUE TO APPEAR ALMOST DAILY

We really do not need to make an exhaustive list here. For that, see Appendix A, Trend 34, and every website that publishes medical news. In general, our medical knowledge doubles every 4 to 5 years, and faster in some specialties. Clinical knowledge, the kind doctors use in caring for patients, doubles every 18 months. Half of what medical students learn in their freshman year about the cutting edge of medical science and technology is obsolete, revised, or taken for granted before their junior year is over.

Many of the most important discoveries over the next decades will emerge from genetics. Some 10,000 human diseases arise from defects in a single gene. In 2017, researchers are already beginning to conduct small human trials of gene-editing for the treatment of lung, bladder, prostate, and kidney cancer. Other disorders likely to be treated by genetic therapy, and perhaps even eliminated from our species include heart disease, HIV/AIDS, and Alzheimer's.

Other advances will come from new developments in immunotherapy. Experimental results against some forms of cancer are particularly promising.

Work in regeneration medicine is not yet quite so advanced, but it seems likely that in the 2030s, it will be possible to heal paralysis by restoring damaged spinal nerves and someday even regrow lost limbs.

By 2025, work on nanotechnology is likely to produce the first cell-sized robotic machines that can travel through the bloodstream, scouring out whatever plaques of artery-clogging cholesterol other therapies have not eliminated and killing cancer cells before they have a chance to form a tumor.

As wonderful as all these developments promise to be, their impact on medical costs may be less welcome. New medical treatments and technologies, however, they operate and whatever they cure, have one thing in common: They tend to be expensive, often horrifyingly so. On balance and over time, this latest generation of medical advances seems likely to improve our health in old age enough to bring down the overall cost of care. However, during the transition period—say, the next 20 years—medical expenses are likely to rise sharply.

This is, of course, one more trend that will help grow the market for MT throughout the careers of today's hospitality students. In the healthy, fast-growing market for hospitality and tourism services, this should be one of the most vibrant segments for at least the next 20 years.

KEYWORDS

- **Joint Commission International**
- **medical tourists**
- **Organisation for Economic Co-operation and Development**
- **Patients Beyond Borders**
- **wellness services**
- **World Travel & Tourism Council**

REFERENCE

DeMicco, F. J., Cetron, M., & Davies, O., (2019). *Strategic Planning Today and Tomorrow*. In: Kendall-Hunt Book. Iowa, USA.

PART II

Designing Medical Travel Programs (Cases)

CHAPTER 2

A NEW PERSPECTIVE INTO AFFORDABLE, QUALITY HEALTHCARE: THE CASE OF PRONTO CARE

ADEL ELDIN[1] and FREDERICK J. DEMICCO[2]

[1] *Managing Director, ProntoCare Cardiologist Tampa, Florida, USA*

[2] *School of Hotel and Restaurnat Management, the W.A. Franke College of Management, Northern Arizona University Flagstaff, Arizona, USA*

ABSTRACT

There has been an exponential increase in the demand for healthcare in America due to many reasons, including an aging population, increased incidence of people with chronic disease affecting almost 100 million Americans, which constitute 85% of total healthcare expenditure. The baby boomers who are going to Medicare age being older than 65 years having increase demand for medical services. To add to the so many millions of Americans who are either uninsured or underinsured.

On the contrary, this increased demand is not matched with increased supply coupled with the ever-rising cost of healthcare and more out-of-pocket expenses being shifted to consumers of healthcare. Many problems face the American healthcare system, including being the most expensive in the world but rank 38 when evaluated with measured healthcare endpoints. Inefficient administration of healthcare and wasted spending could also be leading factors in the ever-declining healthcare climate. Thus, this cannot be ignored anymore.

2.1 INTRODUCTION: PRONTO CARE

It is a beautiful story that I would like to share with the world; hopefully, this will demonstrate that each one of us can have a positive contribution to make, which will ultimately turn challenges into opportunities that could potentially do so much good.

Specifically, here, I mean the chronic, complicated, and seemingly unsolvable healthcare crisis. I have witnessed a lot of devastation to both my patients and to my fellow physicians. Patients are suffering from increasingly uncontrollable out-of-pocket expenses (cost is being shifted to consumer responsibility) in the form of co-payments to office visits, procedures, medications, hospitalization, and rising deductible costs which has reached unsustainable levels exceeding the rate of inflation for many years now!

Patients cannot afford to eat, let alone support their families, because a big chunk of their earnings goes to pay for healthcare bills. It is a well-known fact that higher costs do not necessarily translate into better care. As a matter of fact, the U.S. is ranked the world's highest spender on healthcare but is ranked number 37 when measured with health outcome data!

This effort has picked up substantially after the world economic crisis started in 2008 when I knew then that the current healthcare model is unsustainable, as healthcare cost is breaking the bank and forcing many people into bankruptcy.

Also, the doctors got hit with denied, delayed or ridiculous payments after spending more than 38% of their time dealing with insurance bureaucracy, paperwork, escalating overhead costs dealing with more regulations (including expensive electronic medical records) and in some cases, doubling working load and hours to make ends meet!

So many small practices were forced to shut down as acquisition mania started and continued since 2010 which has led that for the first time in America. The independent physicians who were once the majority are surpassed by the employed physicians who sold their practices to bigger healthcare systems such as hospitals and big physician practice groups. This was the only choice left for physicians to continue to practice medicine and earn a living!

This was further complicated with the affordable care act (ACA) with decreased reimbursements, high overhead, and new coding systems when (ICD9 changed into ICD10) adding thousands of new codes made it very difficult for physicians to get paid unless these physicians are in the employed category. Non-insurance payment led to the closure of many small rural hospitals as well as small practice offices.

Small independent pharmacies also suffered with ACA as patients were forced to go to the big pharmacies to get their medications, not the ones they have been established with for years, making billions of dollars in profit!

Now, the new healthcare plan under the new administration will not only stop the expansion of ACA but would most likely lead to its demise, and thus, beginning of 2019, the individual mandate will be lifted. Even Medicaid is being proposed to be replaced with block grants which would not be enough to give enough coverage for millions of Americans. Skinny Insurance, which is being used as a temporary fix is lacking essential coverage and ultimately costs more!

Due to physicians losing autonomy and professional gratitude and the fact that they are being replaced with automation and lower-paid staff (Nurse Practitioners), the rising problem of physician burnout (up to 66%), depression, and even suicide cannot be ignored anymore!

Thus, I have developed a direct contracting for healthcare model with combined internal medicine and cardiology services with incredibly affordable pricing with transparency, great trusted care and impeccable credibility which qualified our program to be offered by the City of Tampa for Non-Medicare Retirees for three consecutive years. This has made our Affordable Quality Healthcare Program a reality that many wanted to join, but many were reluctant to do so because of a concern about individual mandate, tax penalty, and the fact that they could not get out of their employer health plan.

Our constant campaign of community outreach has been a priority: educating fellow Americans about the importance of wellness, early disease detection, and management to help patients stay healthy and cut down their healthcare costs substantially. Many trips were taken to the State Capitol to meet and educate lawmakers as well about the model. We even went as far to make presentations in front of the healthcare innovation subcommittee, which required long early travel in the dark in the cold winter hours!

The HB37 Bill did not pass for many years as it was blocked by lobbyists, but it is official now as Governor Scott signed into law on March 23rd, 2018. It has been an exciting journey, and now is only the beginning of a new phase as we are able to reach out to help so many patients (consumers) and providers (doctors). As we build our network with prontocare.co to help fight disease and hunger, we donate 5% back to the community as a part of our social responsibility. Also, we provide a dignified model for the practice of medicine and stabilize the immune system with natural immune supplements.

2.1.1 BACKGROUND FOR PRONTO CARE

1. Traditional Insurance has been failing for many years with rising premiums and out-of-pocket costs to consumers.

Similarly, doctors and hospitals are not getting paid for services forcing many places to shut down even in areas where medical services are needed most.

2. Patients and doctors go through daily hassle that stressing both, spend 37% on paperwork and bureaucracy not on patient care.
3. Very little attention to Wellness and disease prevention, especially when it comes to chronic disease management, which consumes 85% of total healthcare cost.
4. Saves money to patients, employers, and governments and delivering better care for less cost.
5. Average savings 5 to 10 k per person per year and you have testimonials.
6. Strengthening the immune system naturally with black seed to prevent disease under daily stresses.
7. Make all appropriate referrals to subspecialists for discounted cash pricing through a transparent mechanism so as no hidden cost to consumers or employers.
8. Most dignified way to practice Medicine as it prevents physician burnout and improve patient satisfaction as each practice will be maxed between 600 and 1,000 patient per provider with no pressure to run the patients through the mill.
9. Will protect consumers from many incidents of confidential data breach which took place under traditional insurance (identity theft and fraud).
10. Pronto Care will connect you with all your needs (pharmacy, labs, diagnostics, and more at great savings to help our patients and advocate for them (Affordable Quality Healthcare Program; www.prontocare.co).

2.2 CONCEPT OF CONCIERGE MEDICINE

Concierge medicine is becoming more and more a viable option to traditional medicine for both patients and physicians (https://www.linkedin.com/pulse/concierge-medicine-makes-sense-adel-eldin-md-facc-facp/). Establishing direct care with the patient and getting rid of the hassle and focus on the delivery of quality healthcare for a monthly or a yearly charge with 24/7 access to the physician, same day or next day appointment with more time spent with the patient addressing his or her concerns, making care for patients much easier and with positive experience during and after the visit. We have concierge medicine programs serving patients from 20 to 100 years of age in Tampa, Florida, with very transparent services and fee schedule as follows:

- Age 20–30 years old $50 per month, with 10% off for the annual enrollment.
- Age 30–40 years of age $100 per month with 10% off for annual enrollment.
- Age 40–65 years of age $150 per month with 10% off for annual enrollment.
- Age 65–100 years of age $75 per month with 10% off for annual enrollment as a wellness plan.

In addition, we donate 5% back to the community mostly to fight hunger! No denial for any preexisting condition, no copayments, no pre-authorization, no referral needed, just come, and be treated with dignity and respect you deserve! These unique combined internal medicine (primary care) and cardiovascular disease all together makes it a one-stop health care which so many patients like and appreciate. When patients see the real value in these services which would save them additional thousands of dollars in healthcare by aborting hospitalization or emergency visits in more than 95% of cases as they were managed successfully on an outpatient basis on top of saving so much on medications, blood work and diagnostic testing, so patient really appreciates the services and the savings and you can see our testimonials on Facebook (www.brooksvillecariology.com; www.facebook.com/pages/Heart-Health-and-Wellness-Program/278124625560440).

So, when patients do the math, one visit to the emergency room average cost about 5,000 dollars which will be equivalent to 10 years of enrollment for someone between 20 and 30 years of age! And one-day hospitalization average cost 15 to 20,000 dollars which will equal to 40 years, yes 40 years of enrollment in my plan for the same age group, and you can do the math for the rest of the various age groups, so when the patients are educated, the decision which they will make during this tough economy will be so easy. This goes to support that the Concierge Medicine program that I developed was really developed to help the middle-class hard-working Americans and dispel the myth the concierge medicine is for the rich! (not true) and when you put a few dollars a day less than $5/day to invest in your health would make a great investment compared with what you have to spend elsewhere with emphasis on disease prevention, comprehensive wellness, and stress management.

As Healthcare is going to continue to be driven by consumerism with the current impact of social media that makes patients (consumers) more educated about choices in healthcare, including concierge medicine. A good example of that is the positive impact of ACA on the growth of concierge medicine as many big corporates such Walmart, CVS, and soon Target will also provide concierge medicine in retail clinics as needs grows!

Concierge medicine allows doctors to concentrate on the practice of medicine and provide a personalized quality services with their patients as they are NOT pressured compared with traditional practice (usually half number of patients seen per day), so it is good for the patient and the doctor as well who makes a fair compensation for years of education and training!

In brief, I think there will be further growth in concierge medicine in 2015, It makes sense and that is why some health insurance carriers like Cigna made concierge medicine available for employees and made it complaint with ACA as it made direct primary care (DPC) (concierge medicine) bundled with catastrophic coverage Policy. I welcome any input or opportunity to cooperate with our local community and beyond as there is so much need out there and we have a good model to share (Have a Happy and Healthy 2015! https://www.linkedin.com/pulse/most-valuable-commodity-adel-eldin-md-facc-facp/).

The search continues for the next big thing to sell, after iPhones, iPads, and all the long list of high technology, is going to be the most valuable commodity, that is health! Simply without it, you can never enjoy any of your wealth; no matter how much money you have when you are sick, you will never enjoy it and if you are in fact going to lose most or all of it on healthcare bills. As a lot of my patients share with me that they actually feel sicker when they see those bills coming in the mailbox!

As it is reported that number one cause for bankruptcy is the unpaid medical bills as it is too expensive. But spending so much money in healthcare does not necessarily translate to better health outcome. This is evidenced by the fact that U.S. is the biggest spender in the world on healthcare but ranked number 37 when measured with clinical outcome data (which is the bottom line)!

So, there has been a lot of research about (cost-effectiveness) in so many top universities such as Harvard and others to calculate the cost of healthcare dollars for various diagnoses. For example, the cost for hemodialysis care for end-stage kidney failure patients, cost of implanting a cardiac defibrillator to shock the patient out of a lethal heart rhythm called ventricular tachycardia or ventricular fibrillation and help jump-start the heart, cancer care, cost of open-heart bypass surgery or a coronary stent, all very costly, and develop appropriate guidelines for the management of patients without breaking the bank! One important factor is being proactive in disease prevention and wellness. Only doing a spot fix or acute disease management without addressing the entire head to toe care has caused so much waste in the overall

U.S. healthcare dollars. As a matter of fact, according to the AARP article in November 2012, there is estimated a total of $765 billion of wasted spending in healthcare divided as follows:

- $55 billion wasted as missed prevention opportunities;
- $65 billions wasted as medical fraud;
- $105 billions wasted on excessive prices;
- $130 billions wasted as preventable errors/mistakes;
- $190 billions wasted as insurance and bureaucratic costs;
- $210 billions wasted as Unnecessary services.

So, after full diagnosis of the current unhealthy healthcare system, that should be followed with effective treatment strategy, as everybody is hurting, including doctors, providers, patients, and the entire community!

That is why four years ago we started our low-cost quality healthcare program to restore the direct doctor-patient relationship without third party interference or hassle and with mutual trust and respect working together to achieve the best health results, save patients thousands of dollars, decrease stress, abort, or limit costly hospitalization without compromising the patient healthcare. We are very proud to share with you outstanding results for so many of our patients who enrolled in our low-cost quality healthcare program, all available on our website: www.brooksvillecardiology.com.

So, now is the time to switch from an old ineffective, costly cycle of healthcare model and join in our program of disease prevention, comprehensive wellness, and Stress management. As you start 2015, think about investing in your health between 2 and 5 dollars a day by joining as the rate of return on your minimal investment is phenomenal and guaranteed for you to be able to get up, go to work, take of your family and be a healthy productive member of the society. So, do invest and protect your most valuable commodity, your health!

2.3 BUSINESS OF HEALTHCARE (DIRECT CONTRACTING)

Whether you are a big employer such as Boeing with thousands of workers, or a small, medium-sized business, individual or a hospital. They all win with direct contracting. When companies and providers have direct relationships via clear and transparent agreement with regards to pricing, healthcare services and quality of care delivered. The bottom-line is that there is significant cost saving without compromising that quality of care. As primary care or combined primary care /cardiovascular sub-specialty services in our program could provide up to 95% of the patient care for a flat rate which is only a fraction compared to the traditional coverage (https://www.linkedin.com/pulse/direct-contracting-healthcare-adel-eldin-md-facc-facp/).

As all businesses are looking to save money on healthcare costs, they will be considering the alternative model (self-funding for healthcare by direct contracting). The growth of direct contacting explained in part by saving money, but also gets rid of the stressful relationship between providers and Insurance companies as while delivering the care, providers complain that they do not get paid or too little, too late which will strain their operating cash flow and eventually force them to shut down. Direct contacting is also good for the consumer looking for a break.

Employer direct contracting may also include an innovative approach to cut costs to include (medical travel or domestic medical tourism) where employees will travel out of town or state to another part of the country to have their elective surgery done (travel surgery program).

Direct contracting with medical providers already proved to help self-insured folks and is a viable alternative to ever-escalating costs with the insurance-based model. It is simple, when services are delivered directly to businesses/consumers, this will help control unnecessary expenditures and minimize the administrative cost associated with traditional insurance.

Putting emphasis on wellness, disease prevention, even chronic disease or acute illness management will translate to savings to consumers and employers alike. We have zero co-payment for any visit or testing done and people like that as this saves them big!

Adding on-site clinics will help to save costs and minimize loss of productivity when employees have to take time off work to see medical care. The data is positive and very encouraging proving lower cost, improving care and satisfaction.

Now, the focus would be to reach out to the community and educate small, medium businesses, big employers and individuals that direct contacting is the way to go as a good strategy which they should seriously consider choosing for their healthcare (https://www.linkedin.com/pulse/retail-clinic-growth-its-impact-healthcare-adel-eldin-md-facc-facp/).

retail clinics have had a substantial growth over the past few years and are expected to have an even faster growth rate in the upcoming years.

retail clinics are booming as consumers seek new ways to get convenient, cost effective healthcare, which constitutes a formidable challenge to traditional practice. As a matter of fact, some insurers like Kaiser Permanente partnered with Target to staff retail clinics in some California stores.

Up to 60% of patients seen in retail clinics do not have established primarycCare providers with more than 10 million patient visits to retail clinics every year mostly utilizing low acuity care for self-limiting conditions. But with convenience and reassurance, most consumers find those clinics to be a much cheaper alternative than being seen in the emergency room.

The ACA, which added millions of patients into the already strained primary care capacity, created great opportunity for the retail clinics. Most of these Clinics are concentrated in the metropolitan areas, but tele medicine is proposed to provide care for the rural communities in America, giving 46,000 current physician shortages, which will hit 94,000 by 2025!

Drive-Thru style of healthcare offered by retail clinics in Shopping Centers, Malls, and Supermarkets was actually preferred by millennials compared to senior patients. The majority of these clinics are staffed by nurse practitioners, not physicians, and because of the possible delay of few days to be seen by the physician due to shortage and scheduling issues, consumers now choose to go to retail clinics as walk-in and be seen the same day.

In addition to physician shortage, the ACA, and higher out of pocket costs are all factors fueling the growth of retail clinics. So, patients prefer low-cost visits (average $75) per encounter rather than very high out of pocket cost when seen in the emergency room. Patients are also getting to be more educated consumers of healthcare as they shop around and compare prices for various healthcare services. Some retail clinics are taking advantage of managing chronic diseases as they constitute the biggest burden of health-care cost (up to 86% of total money spent) and are getting the electronic Medical records connected with other health-system affiliate to capture the chronic disease management market and filling this gap in healthcare delivery. Our concierge medicine model offers patients another viable option to deliver cost-effective quality healthcare with convenience and personal-ized continued Care. Thus, we can count on more and more retail clinics growth but hope to have a positive impact on lowering overall healthcare cost and working in harmony with traditional physician offices, concierge medicine providers, and hospitals to best serve our patients! (https://www.linkedin.com/pulse/support-small-business-adel-eldin-md-facc-facp/).

Every business was once small and then grew and became a big business. This is the natural healthy sequence of events. Small business is usually referred to as the engine of the economy as 60–80% of the new jobs come from small businesses. Small businesses employ 57% of the country's private workforce and pay 44% of the U.S. Payroll. Unfortunately, only 50% of these small businesses survive the first 5 years, increasing the burden on small businesses with minimal or no reserve. For example, in 2009, and average small business went bankrupt every 8 minutes as 70% of small businesses are/were owned and operated by a single person.

Recent challenges to the small businesses have sharply increased during the current phase of surge in merges and consolidation.

For example, there are now four major financial institutions (Citibank, Wells Fargo, Chase, Bank of America) out of the previous 37 financial institutions only two decades ago. Similarly, big insurance companies are merging, so there are maybe only three or four choices for patients to choose from. The same would apply for retails shops, and local restaurants, which are now being replaced by big businesses that invested in technology which makes it easier for people to order online. And since only big businesses can afford these technological advances, people usually flock towards them due to the extra services. Now also a trend towards limiting number in outpatient surgery centers as they compete with hospitals and big medical centers.

Recent articles have echoed the looming death of small business in America and found small business optimism is crashing as all businesses need working capital which may not be easy to access with limited options and prohibitive cost.

On the bright side, some realize what is going on and now are trying to give support back to small businesses such as Shop Small on Saturday and others to keep the traditional engine of the economy working to keep the overall economy alive and well. Small businesses are going through transformation to adapt to the changing economic climate and challenges through innovation, entrepreneurship, and adopting new technologies and reaching out through social media.

There is always a niche for small business where consumers find satisfaction in quality customer service and personalized care.

We have adopted these concepts to make small business have a big impact on people lives through improved health outcome, saving big for individual patients, and small and medium-sized businesses through establishing the (Direct care Medicine) as we can very much share the value of affordable quality healthcare that does not break the bank. This has been proved over the past few years in our program and being replicated all over the nation.

This holds true for all independent physicians who have been running their small businesses for a long time but now realize the importance of unity, cooperation, and transparency as they owe it to their customers (patients). This will be the healthiest response to thrive during changing economic climate (https://www.linkedin.com/pulse/unsustainable-rise-healthcare-cost-adel-eldin-md-facc-facp/).

Healthcare cost is rising at an alarming rate and disproportionately to income earned. Most Insurances have sharply raised their premiums, and similarly, the pharmaceutical companies did the same even with the (generic drugs) which prices have been raised up to 25 and 75 times compared to what they were only 2 years ago. Additionally, in many cases, these drugs are frequently denied even when attempting to get preauthorization for critical medications which patients need as the insurance and pharmaceutical companies control them jointly.

What has become clear now, people realize that the threshold of sustainability of this traditional model has already been surpassed, and people cannot continue to pay more for their healthcare simply because they cannot afford these escalating costs. The middle class in America has shrunk into half and has sadly moved to poverty level over the past decade.

So, the mandatory insurance could be passing only the cost of millions of dollars in taxes and fees to finance the expansion of the healthcare coverage to the slipping and economically strained middle-class Americans.

Then comes another issue of (Risk Adjustment) which would require insurers and healthier members to make payments to companies with (sicker members) to share risks and cost of insuring people with many medical problems to secure their profitability. Similarly, health plans for individuals and small businesses were raised with rates ranging between 4.5% and 16.5% with more out of pocket responsibility shifted and dumped on patients!

According to recent New York Times Article (http://www.nytimes.com/2016/01/04/us/many-see-irs-fines-as-more-affordable-than-insurance.html?), many would be happy to pay the I.R.S penalty by intentionally skip enrolling in the ACA which will be much less than the cost of insurance. For some healthy people, the combined cost of premiums and deductibles would exceed $12,000 per year.

That is why so many dropped insurance all together as their insurance did not cover most or all their medical expenses, as they had to pay out of pocket in addition to paying insurance. So, many people are taking the risk of not being insured simply because they cannot afford it. They would adopt healthy lifestyles, eat healthy food, exercise regularly, and try to minimize stress as they educate themselves by going to the Internet to learn about reducing their risk and save money (Do It Yourself Style).

We believe as we enter the 6th year of our direct care medicine program is the way to go by achieving the goals of disease prevention and total wellness. With patient education, advocacy, and saving thousands per person every year through discounted medications, lab work, and diagnostic testing and aborting costly ER visits, hospitalization, and procedures. Yes, quality healthcare can be truly affordable, effective, and hassle-free through direct physician-patient relationship without third party control or unsustainable rise of healthcare costs, which have devastating effects on the entire community (https://www.linkedin.com/pulse/risk-sharing-versus-shifting-adel-eldin-md-facc-facp/).

Are you ready and prepared yet? The risk is being shifted and being (dumped) on the consumer, that is what is happening and became official. It is seen in many cases whether with the new credit card machines, if you do not update your machine, the risk will be shifted from the financial institution literally to become laid on your shoulder, similarly if you do not (code it right) referring to the ICD 10 coding system, you will be delayed and denied payment for your professional medical services even in some cases with hiring professional billers that are well trained in this transition, turning again all the responsibility on the provider who are most likely will have most of the denials and delays in payment!

Similarly, with subprime mortgage crisis, when someone who may not qualify brought home or property and then the bubble burst, and the economy went into a recession, then millions of Americans lost their homes and properties with foreclosures that devastated millions of families and businesses with effects of which still being felt throughout the nation!

So many people struggle to make ends meet and good luck jumping those hoops to make it on a day-by-day basis, with a significant crunch in the cash flow cycle which is going to impact so many industries, for example the medical offices give business to many suppliers of medical equipment and supplies, hires staff, pay taxes, and contribute to overall economy. But those small entities that contribute up to 84% of the total engine of the economy, may come to a full stop as the cycle of business is not moving and forcing so many to early retirement or bankruptcy! On the contrary, the bigger providers are merging together (for example, Anthem merger with Cigna and Humana merger with Aetna insurance companies) to get stronger and richer with stronger control of the market, leaving consumers with fewer options and more risk-shifting on the patients with more out of pocket expenses, copayments for office visits, testing, deductibles to the point that people are refusing the coverage simply when they add up what they have to pay on top of their insurance, they will not have enough left to feed their families or pay bills.

The rising trend of risk shifting is not good for consumers as there is no balance and sense of security left anymore. Everyone is afraid to lose their shirt if they get sick! This is unsustainable and is bound to fail. We have to restore a well-balanced model when we all enter into a fair and transparent partnership with all our providers and go back to fair risk-sharing rather than devastating and stiffing risk shifting to the consumer with more options. That is why we advocate direct care model to cut unnecessary cost of third-party involvement and advocate on patient behalf to put patient first rather than send them to the cleaners! (https://www.linkedin.com/pulse/amazon-going-healthcare-adel-eldin-md-facc-facp/).

Our unhealthy Healthcare system is getting now new and aggressive players, as American sickness is becoming a huge Multi-Trillion Dollar Business. Now Amazon is going into healthcare in direct competition with CVS and walgreen. Also, walmart, and Uber are moving in. Walmart has closed 154 stores in the U.S. as its sales were adversely affected by Amazon online shopping from home!

Because somebody did their homework of healthcare market analysis, the results showed as technology is getting better and better every day, you can have up to 40% of house calls done via telemedicine. Similarly, on-demand services with delivery of Flu-Shots successfully, for example in 35 cities by Uber health was tried in a pilot phase of 2000 people who received Flu-shot at home rather than going to doctor office or clinic, which used to be a frequent reason to make the trip to the clinic.

Similarly, Wal-Mart has built 100 on-site retail clinics (like CVS minute clinic staffed by a nurse practitioner, not a doctor to cut down overhead and maximize profit. Now, if you go to Amazon.com and check out the health care categories. You will see as if you are shopping in a virtual pharmacy with first aid, eyecare, feminine care, pregnancy test, baby supplies, skin products, pain relief, anti-reflux medications and the impressive list goes on. Not only that, but nutrition supplies, medical equipment, health, and wellness, essential oils up to 400 pages to shop from!

Medical equipment include splints, mobility aids, wound care, respiratory, urological care, ostomy supplies, fitness devices, occupational, and physical therapy aids. Sexual health with all associated supplies and even funeral services!

Yes, caskets, cremation supplies for human ashes, and more all are included. Thus, there is an Amazon superstore for medical supplies in Florida where large populations of senior citizens and retirees live!

Medications as well as Blood work services also included with delivery even with drones to remote locations and on long weekends and holidays. diagnostic testing for colon cancer screening (non-invasive) with cologuard for those 50 years or older done as well. It has so many advantages over conventional colonoscopy as it is done at home, no procedure-associated risks such as perforation, bleeding needing surgery or blood transfusion, no special prep, do not have to take time off work with high accuracy up to 92% of colon cancer and can detect 69% of pre-cancerous lesions. It is FDA approved to test Stool DNA for colon cancer. All related educational material and Videos are also available for sale. Similarly, Diabetes monitoring supplies, blood pressure monitoring and related educational material. Diagnostic Imaging with CT scan or MRI or Ultrasound study is available.

Amazon also carries various catheters, infusion pumps, IV bags, Infusion tubing, sutures, and hospital beds. Now, you start to see that consumerism is going to be the main driving force shaping the Healthcare market. Amazon understands that aging population will generate more healthcare services and opportunities. It is also driving mergers and consolidation in the Medical Supply industry with recent 24-billion-dollar deal as Becton-Dickinson acquiring C.R. Brad!

Now, Amazon, and Microsoft going into fierce competition not just in the cloud, but also in the healthcare market. Amazon is working with Electronic Medical Records, virtual medical visits, and applications for devices such as Amazon Echocardiogram!

Microsoft is doing the same with health vault and specific focus on proactive wellness and big data analysis to guide decisions. Amazon is going global as well with videos /music expanding to middle east buying souq.com which is online market place exactly as souq in Arabic means market place. Through its e-commerce platform, Amazon has more than 130 million visitors to its site every month!

Well, now there is a big concern about Amazon expressed by President Trump in a Tweet saying (Amazon is doing great damage to tax paying retailers, towns, cities, and States throughout the U.S. are being hurt, many jobs are being lost) and replaced with automation and digitally-run e-Commerce Business.

Well, fasten your seat belts as we are going through tough rides and watch events shaping the future of healthcare unfold!

2.4 PATIENT EDUCATION AND ADVOCACY

It sounds like everyday conversation with my patients regarding various choices of therapeutic options, including medications for various medical conditions, procedures, or surgeries. The discussion that usually takes place about the established guidelines regarding the specific issue with its indications, safety profile explaining side effects which could be potentially experienced as well as the expected benefits of being complaint with intake of this medication (https://www.linkedin.com/pulse/help-patients-take-charge-health-adel-eldin-md-facc-facp/).

After all, a well-informed patient is much more likely to be complaint with diet, exercise, intake of medications at the prescribed times and the route of administration. He or she will have a better outcome with less cost. The opposite is true as well; patients who do not take charge of their health are mostly non-compliant with more adverse clinical outcome, which is obviously very costly as well.

Simply, patient engagement is a key to better healthcare and can make a big difference from being stable and doing well medically to going back to hospital with its subsequent high cost. So, patients are encouraged to shop for good doctors, good hospitals and surgery centers to compare services, prices, and customer service because they are the customers, and without them, there will be no healthcare industry.

So, this is how it should work, doctors will share their expertise and knowledge with their patients, informing them about what needs to be done (proper nutrition, lifestyle changes, such as stop smoking and avoid excessive drinking, regular exercise and take medications when indicated? Certainly, patients have to do their homework for the plan of care to work. This is why (direct care medicine or concierge medicine) is the best way to achieve the desired health goals, spend the needed time without rushing patients, answer their questions and give them reassurance and psychological support which they need as well.

I found throughout my medical career if you show your patients genuine concern about their health, patients could feel and appreciate that concern and respond in a very positive way. For example, some patients have been smoking for 1–2 packs per day for 40–50 years as I never give up on any one and take the time to explain the devastating health effects due to smoking and its associated cost. After a candid conversation with the patients, we had as high as 99% success rate in getting people to quit smoking and stay off smoking.

Those patients when seen in follow up visits, feel so good about their accomplishment and show a desire to share their experience with others. Technology can play an important role (for example sending reminders) to patients to review their Medical Notes (Open Notes) which helped patient engagement, compliance, and outcome.

With Direct Care Medicine, there is more time spent on actual patient care rather than Computer care checking many boxes to justify payments. Thus, empowering patients to be active participants in their care and encourage open communications channels with the healthcare team through technology with Tele Medicine will certainly lead to achieving the goal of better healthcare delivery with less cost! (https://www.linkedin.com/pulse/empathy-what-we-need-adel-eldin-md-facc-facp/).

Empathy is defined as identifying with and understanding of another person's situation, feelings, and motives. In Medicine, patients seek empathy from their doctors. Thus, the whole point of empathy is to focus on the patient, build trust, which will help patients to manage or deal with (bad news) and improve outcomes. Listening more attentively and showing genuine concern to what the patient is saying and expressing what they are going through will certainly help successful diagnosis, treatment, and outcome and cut health care costs.

This saving comes from limiting waste like repeated unwarranted testing, risk management and minimize burnout and litigation, which can adversely affect not only the care-giver but also the institutions they work for as frustrated patients seeking help and understanding never got either or both. This need is seen in the teachings at universities such as Duke and others that teach doctors about how important but overlooked skill, Empathy.

Teaching Clinical Empathy which is the ability to communicate understanding of the patient situation and willingness to help. A healthy doctor-patient relationship is becoming more important in boosting patient compliance to treatment plan as a good doctor is defined as the one who understands people and communicate well with them. It is all about patient satisfaction and experience that is closely linked to better outcome, incentives, and reduce physician burn out as they are more satisfied professionally as well. This make Concierge Medicine the ideal model for the practice of medicine as 90% of healthcare needs can be delivered on an outpatient basis, no time-pressure, concentrating on quality personalized care rather than volume-based care where patients feel no body has time to listen to them, as they are being rushed in and out without having the opportunity to communicate with their care providers.

Empathy is needed in all aspects of service-based professions such as banking, law, insurance, etc., no wonder most of big companies have included Empathy in their corporate training manual to improve business operations by enhancing customer service and cut costs from potential lawsuits. So, automation, and technology which is growing very fast, will not replace the human touch and effect as Empathy involves the human interaction and cannot replace with automation. That is why Clinical Empathy is being taught to both providers and medical students as only medical Knowledge and technical skills are not enough to fulfill patient needs and have a positive impact on the overall healthcare system (https://www.linkedin.com/pulse/do-you-practice-honest-medicine-adel-eldin-md-facc-facp).

Do you practice evidence-based Medicine? Or you feel you were put in a situation where the patient best interest is not being looked at as first priority. But rather as revenue-generating for job security especially in the employed physician category!

To put it all in perspective, here are real examples to illustrate this challenge. A 58 year old female patient with persistent dry cough, a tickle in the throat who has seen multiple consultants including Ear, Nose, and Throat specialist, then allergy specialist, then a Gastroenterologist (stomach doctor) who saw the patient and performed upper Endoscopy and gave patient anti-reflux medication, but cough continued and a 4th consultant was added being a lung doctor, He did workup including CT scan of the chest which was negative and so was the multiple allergy testing done was negative.

Then I saw the patient and went over carefully her medication list that she was currently taking and Bingo, it was clear that these symptoms were obviously due to side effects of one of the medications she was taking, causing her dry cough and tickle in the throat symptoms. This medication belonged to a class of drugs called ACEI, or Angiotensin-Converting Enzyme Inhibitors, which after being switched to a different class of medications called ARB or angiotensin receptor blockers that does the same therapeutic benefit without the side effects. Wow, the patient, after the switch of medications, dry cough, throat irritation, and the need to clear her throat every couple of minutes, miraculously went! Only, if this was done earlier, could have saved tens of thousands of dollars of unnecessary healthcare money and procedures, and achieving better outcome for only few dollars spent instead!

The same can be seen in cholesterol-lowering therapy that prevents progression of atherosclerotic plaques or hardening and clogging of the arteries which if left untreated will certainly lead to subsequent clinical events such as Angina or chest pain first on exertion and if left untreated will lead to Unstable angina and acute heart attacks, strokes, and peripheral vascular disease leading to gangrene and amputation. That is why we all should address these issues seriously and aim to achieve target numbers for cholesterol and triglyceride lowering by diet, exercise, life style modification with smoking cessation, limit alcohol intake and medications to prevent close to 80% of cardiovascular morbidity and mortality affecting millions of Americans and costing billions of dollars.

Around 86% of every healthcare dollar spent goes to chronic disease management, which is largely preventable and achievable on outpatient ambulatory basis with empowering patients through education and advocacy.

The fast changes dynamics of healthcare will dictate the adoption of evidence-based Medicine for cost-saving while improving outcomes. Similar examples including unnecessary IV therapy, including vitamins such as B12 and others, iron, fluids, antibiotics, etc., which can be given orally effectively unless than patient has a malabsorption syndrome and cannot hold anything down due to uncontrolled vomiting or diarrhea which is the only justifiable indication then. This will help the patient to avoid unnecessary complications such as IV site infection with the risk of introducing Staph infection into the blood stream, especially in cases where the patient has hardware like pacemakers, defibrillators, artificial heart valves, or joints. This risk increases with every time the integrity of the skin is compromised with needle and catheter insertion plus the so many trips that patient make with some of whom are elderly, frail, and cannot drive. Adding enormous stress on patients making these trips to justify billing the third parties for these services. And in many cases, the patients are stuck with the bill for unnecessary IV therapy or its complications putting additional stress on the already stressed patient. Believe me, those patients tell me their unfortunate stories.

In brief, patients are fed up with the non-sense and they need and appreciate their honest doctor who will tell the truth and support them no matter what!

2.5 COMMUNITY-BASED HEALTHCARE SOLUTION

Investing in the community is one of the best things anyone can do. Being useful, helpful, a problem solver, an uplifter, and a lifesaver all together make a difference in people's lives which is badly needed in our community these days. So, going out and giving free healthcare to the less fortunate, giving food, mentoring young students and future doctors, mending fences among friends all will change the prevailing culture of doom and gloom into hope and happiness. The fact that you wake up in the morning in good health, safe in your home and have your sustenance for that particular day then you qualify to be on top of the world. Around 17 years ago, coming to Florida from New York City, pioneering the community outreach program, which started with going off to different businesses, churches, and local chamber of commerce, homeowner associations and have many educational programs for the community. And build strong relationships with many wonderful diverse groups and organizations, this is giving back to community programs such as going back to school, Youth Educational Services (YES Program), Thanksgiving Food Basket program, free health screening and in turn over the years these programs have helped develop direct relationships with my patients (https://www.linkedin.com/pulse/community-outreach-program-adel-eldin-md-facc-facp/).

It is a very encouraging sign that yes, patients are mostly no longer passive, and they are searching for both (trust) factor as well as the most economical way to fulfill their healthcare needs. Patients do feel sincere care from the heart rather than (production line medicine) which has built relationships that have overcome many barriers, including language and culture. So, the local story is being now replicated globally as patients (consumers or clients) are searching for quality of care, compassion, and transparency as we are now connected via the Internet and social media everywhere. We certainly hope that our local community outreach program will impact positively on global healthcare delivery. It is the same local message however, now it is being circulated across the globe (https://www.linkedin.com/pulse/bringing-heart-back-medicine-adel-eldin-md-facc-facp/).

As we observe the month of February is known as the Heart month, as well as Black History month, and even known for Valentine's Day. However, also in in February this year, we celebrate the 4th anniversary of our Low-Cost Quality Healthcare with its initial vision, to "Bring the Heart back in Medicine." We have been continuously working hard over the course of 4 years to fix the ills of unhealthy healthcare delivery system. It should be driven by doing what is right by patients, address their needs, but sadly the main driver for the healthcare system is (Profit)!

Doctors who dedicate their life to being trained to take care of patients are supposed to be the ones who control 100% of health care dollars as no patient can get out of bed unless there is an order written by the doctor. Similarly, tests, procedures, and even type of meals are only done according to doctor's orders. However, due to the change in the healthcare system, doctors have been pushed to the bottom of the chain of the healthcare industry. This ended up in harming patients in many ways.

For example, when acquisition mania started in 2008 ended up in 50% of independent physicians to be forced to work for corporate. This in turn caused consumers to pay for the rising cost of health care. In addition, this also created an environment where the employee has to serve the employer's interest in order to produce and help gain profit. However, the employee ultimately gets reimbursed based on the famous formula of relative value units (RVUs) they produced.

For example, for Emergency Room Colleagues, there is pressure to admit patients to help generate revenue until recently where when CMS (Center for Medicare Services) started to impose penalties for patient readmission to the hospital. In an instance where a patient suffers from Congestive Heart Failure, they would have to pay up to $2,500 per readmission. This is the penalty for a patient being readmitted within the duration of 30 days for another episode of Congestive Heart Failure. In 2015, the penalty is expected to increase even more because of a strained budget. This model has not proved to be effective, as people have begun to realize that 'you cannot keep kicking the can down the road any further.' Serious steps must be taken to face the ever-increasing healthcare cost that is running individuals to bankruptcy and potentially our current healthcare system as well!

According to a recently published article in The New York Times, the importance of disease prevention and outpatient delivery of most healthcare services will constitute significant saving up to 30 billion dollars simply when the same services are done in an outpatient setting. This will get rid of the big gap in pricing between in-patient and outpatient for the same test as usually the price difference is being dumped on the patient to pay out of pocket.

In the Internet Era, most patients or a member of their family have already done some sort of a quick Google search about a medical condition they might be experiencing. They read about the required testing and therapeutic options, after that, they will shop around for the best price and will question any additional cost because of the economic pressures that so many are facing nowadays. Patients will go to 'Dr. Google' to help them with their condition as seen in an instance where a patient will simply type in their symptoms into a search bar and will receive all information needed to relieve the symptoms; however, this did not work, and usually patients will ultimately be admitted.

Now with digital technology, there are applications that will help physicians take care of patients. For example, there are applications that pick up on abnormal heart rhythm and can differentiate between a serious heart rhythm and an insignificant rhythm thus helping in patient triage.

So, I believe the middle of the road approach which will best serve the patient (consumer) and change the current culture of looking only at the bottom line, will be a hybrid approach which will allow patients/ consumers to get most of their healthcare needs 90–95% on outpatient basis with emphasis on Disease prevention, Comprehensive Wellness and Stress Management, all elements needed by everyone and keep the catastrophic coverage to cover any emergencies that could happen!

The key elements of future healthcare are Low Cost which we have proven in our program over the past 4 years that is real and achievable with so many testimonials on social media to share their experience with our program, but never the less, all this was achieved without compromising the quality of healthcare!

Patients do appreciate such programs as they shop around for the best prices for caring affordable healthcare provider that will serve them, save them plenty, educate them and advocate for them as well. Also, as we look into those, we served we noted how we made a positive impact in their life, took the added stress away, and helped give back to our community fight hunger especially during this cold winter days as there is so much need for it out there!

It is a dignified model for the patient, provider, and the entire community as we strive daily to 'Bring the Heart back in Medicine,' not just during the Heart month, but rather all year round!

2.6 NUTRITIONAL ASPECTS OF HEALTHCARE

Food deserts were first described in the United Kingdom in the 1990s as low-income neighborhoods with no access to nutritional food. If you happen to be living where local residents do not have access to a supermarket or a store with fresh fruits and vegetables within one mile, you might be living in a food desert. Thus, individuals are forced to either eat at fast-food restaurants or a convenience store with soda, chips, high carbohydrate meals leading to high incidences of heart disease, diabetes, obesity, and increased adverse cardiovascular events such as acute heart attacks, strokes, and mortality (https://www.linkedin.com/pulse/more-compelling-reasons-end-food-deserts-adel-eldin-md-facc-facp/).

Food deserts affect more than 23 million Americans, with a majority of low-income status individuals. So, geographic location as well as economic status determines where food deserts are; these food deserts create eating habits that end up with the consumption of food that causes inflammation in the body, which mediates all disease. Thus, the importance of an anti-inflammatory diet comprises of fresh fruits and vegetables, in addition to no red meat, no processed food or chemicals. The best anti-inflammatory diet is the Mediterranean diet, which is rich in Omega 3 Fish Oil and other anti-oxidants well.

The list of anti-inflammatory food would include green-leafy vegetables, Chinese cabbage, beets, broccoli, blueberries, pineapple, salmon, walnuts, flax seeds, chia seeds, turmeric, and ginger. Food deserts are characterized by residents who have little to no access to transportation, communities of color, low-income communities, communities that are highly saturated with liquor stores and fast-food outlets which serve high sugar, high salt, and/or high carbohydrate and processed food(s).

Unfortunately, instead of dedicating more funding for food security, there have been budget cuts of more than $150 billion dollars in programs such as SNAP (supplemental nutrition assistance program), Meals on Wheels, and Medicaid. Other factors such as social determinants of health (such as low income) heavily co-exist and affect access to healthy diets in the Food deserts!

A fresh approach to this critical problem of food deserts could help save billions of dollars in preventable healthcare!

Solutions would include making nutritious food available even in smaller outlets and institutions to prevent many chronic diseases related to poor nutrition in food deserts. At the same time, minimizing the huge amounts of wasted food in America as to create less disparity in different socioeconomic communities. According to the *Guardian Report*, roughly 50% of all produce in America is thrown away, which translates to 60 million tons a year, or 150,000 tons of food a day! In other words, 30–40% of the entire U.S. food supply is being wasted, reaching a value of $160 billion dollars in wasted food! The American family of four would average $1,600 worth of discarded food products.

Additionally, this enormous amount of wasted food is contributing to global nitrogen pollution, thus having a direct relationship to global warming, and its impact on the environment. Bacteria in organic waste products produce methane gas which is 25–72 times more potent than carbon dioxide and contribute to global warming. Thus, minimizing the wasted food would not only help solve the food desert but eliminate hunger in more than 50 million in America, and save our environment. This will decrease nitrogen emission from wasted food products which contributes to floods, hurricanes, and wide spread destruction with increased frequency here in the U.S. and around the world.

Simple solutions of minimizing food waste should be implemented to fight hunger and acute or chronic diseases resulting from poor nutrition. Having a number of large refrigerators to keep food fresh and safe for consumption, have available transportation to access the healthy food (especially in the food deserts), is pertinent in order to provide access to high-risk communities. This also will increase food security and boost our national security by keeping our country safe from devastating consequences related to climate changes such as hurricanes and coastal flooding.

After eliminating hunger in America, statistically, there would be still so much food left, which can be donated to food-starved places around the world like Africa, for example, as it is estimated that there are 795 million people in the world who are undernourished. This kind gesture can avoid so many conflicts, uprising, and instability which are caused by people that are hungry and cannot find food to eat. It would be a win-win strategy.

Successful companies in Europe that handle the food waste recovery use mechanical separation technology as food items can be also utilized for feeding animals as well. Some innovative companies in Europe are using food waste in energy companies which use thermo-mechanical technology to process food waste and packaging into dry stable bio-fuel!

There is an increasing trend in the hospitality industry to have Green Restaurant certification to be environmentally friendly with food waste recycling policies. We can work together implementing these solutions, ending hunger, preventing disease, saving billions of dollars in food, using technology for food waste management to feed more people and animals, reducing risks of devastating environmental and geopolitical changes, and turning food waste management into a great business strategy here locally and globally! (https://www.linkedin.com/pulse/eat-healthy-stay-adel-eldin-md-facc-facp).

Very interesting recent article published in JAMA (Journal of the American Medical Association) found a link between consumption of high calorie subsidized food and the subsequent development of the obesity and many related Cardiovascular and metabolic diseases and their complications.

The study used the seven subsidized food items in a (Subsidy Score) including corn, soybeans, wheat, rice, sorghum, dairy, and livestock as a percentage of total caloric intake. Then, various clinical parameters were used including body mass index (BMI) which is a simple ratio of waist circumference to height, C-reactive protein (a marker of inflammation) blood pressure, HgA1C which is an indicator of blood sugar control, and (non-HDL) cholesterol which is basically the bad cholesterol. Also, avoiding any other possible contributing risk factors such as smoking, physical inactivity, poverty, and food insecurity.

The study showed with higher consumption of high calorie subsidized food items; there was a higher incidence of heart disease and diabetes, all related to obesity resulting from consuming those subsidized food items. The current cost of treating those diseases and related complications would reach $300 billion dollars annually, which would exceed the money spent on subsidizing these high calorie nutritional items.

This study supports the current nutritional recommendations to limit carbohydrate intake, use fresh fruits and vegetables. Also use unsaturated fat to replace saturated fats and trans-fats. After all, you are what you eat. So, if you eat healthy food, you will stay healthy and if you eat unhealthily, then will lead to many chronic illnesses, exactly like the car to run well, it needs good gas!

In our Concierge Medicine Program, we put big emphasis on proper nutrition, use food as medicine to strengthen the immune system, thus preventing disease, provide energy and improve quality of life (https://www.linkedin.com/pulse/you-what-eat-adel-eldin-md-facc-facp/).

Some people used to say, "It doesn't matter what you eat…" but now almost everybody knows that it does actually matter. It can make a big difference-eating healthy or eating unhealthy shows that it affects your overall health.

Because it costs so much money to be sick, especially if you ended up being hospitalized and having procedures done will now make a dent in your financial health and could adversely affect you now and in the future, simply because of consuming wrong food. It is like using the wrong gas for your car!

The evidence is overwhelming with the impact of diet and proper food on your health. For example, using the dash diet (low-salt diet that helps controls blood pressure), or the Mediterranean diet, which showed a positive effect in lowering blood pressure and the rate/risk of strokes. Simple components include extra virgin olive oil, omega-3 fish oil, fish that is barbecued or sautéed (not fired!), beans, hummus. They are starting to become popular in supermarkets as the word is getting around because it affects close to 70 million Americans with hypertension (which leads to heart disease, heart failure, strokes, and renal failure requiring dialysis-especially among the African-American population). The good news is, that we and do something about it. We can change the natural history of these diseases and improve both morbidity and mortality, resulting from hypertension.

Similarly, there are about 20 million American with hyperlipidemia (elevated cholesterol and triglycerides) that is related to clogging of the arterial tree from head to toe. A process of the hardening of the arteries called atherosclerosis, which can clog the carotid arteries leading to strokes, clogging coronary arteries leading to heart attacks, and clogging kidney arteries leading to kidney failure, and clogging of the arteries of the lower extremities causing leg pain, claudication, and ischemic ulcers (that would progress to gangrene) which would lead to amputation.

Also, poor nutrition leads to other diseases entities like diabetes, obesity, cardiac arrhythmia, arthritis, sleep apnea, and many cancers (which would become chronic diseases claiming more than 85% of healthcare dollars spent).

Luckily, there is a cultural change of learning about healthy nutrition, adopting conservative lines of therapy that are less expensive and non-invasive, without compromising the quality of healthcare delivered with measured clinical endpoints.

For example, "Courage Trial," that was published in *New England Journal of Medicine* have shown for stable coronary artery disease that would have equally effective therapy; one of them being optimal medical therapy and lifestyle modification with exercise and diet was found to be equally effective as expensive, and invasive coronary angioplasty with similar clinical outcome data.

Also, the adoption of clinical guidelines that will have evidence-based medicine for various therapies and interventions only to be reimbursed by insurance companies if they are in accordance with the guidelines. Not only that, but evolving technology, just like every aspect in our lives, also, in medicine have new technology like CT angiography which is noninvasive to actually reconstruct 3D images of the coronary arteries and which have replaced conventional invasive cardiac catheterization in many instances. Also, MRI is evolving and improving to replace the need for many invasive surgeries as well.

Even guidelines established for the intermediate blockages in the coronary arteries to be studied first with physiological flow study before putting a stent as a knee-jerk reflex as it used to be done in the past!

So cost is becoming an increasingly significant factor on which patient gets a specific therapy for a similar outcome, and the choice will go with the less cost and less invasive therapy. And, if someone decides to do procedures that do not match the established guidelines in addition to not getting reimbursed may also get flagged.

Over the past 5 years, we have demonstrated better outcomes for less cost to the patient who enrolled in our Health and Wellness program in addition to patient education andand empowerment as a part of a comprehensive approach to healthcare delivery. Your health depends on your decision of what goes into your body.

2.6.1 WELLNESS

For more than three decades of being involved in delivering Healthcare in the USA, it has been always my passion to introduce real solutions to our unhealthy Healthcare system. Here are basic facts to start with (https://www.linkedin.com/pulse/all-wellness-folks-adel-eldin-md-facc-facp/).

We spend the most money on healthcare but ranked number 37th in the World according to healthcare outcome!

We face many challenges relating to implementing the needed change of culture into Health and Wellness. The main elements for culture change to take place are Clinical, Business, and policy elements, and I have been fortunate to be involved in all three elements to varying degrees. Hopefully, this will add a critical element in providing real healthcare solutions to provide better care for less cost.

I believe, we need to promote concepts of Free Healthcare Market including (direct contracting) between patient and doctor without third party hassle and abuse to both.

The cash flow is what keeps any business viable and would not thrive without positive cash flow (more money coming in than going out) to also allow future growth and operational development. This is a case of Concierge Medicine which brings back the direct relationship between Patient and Doctor for better health outcomes and less money spent. Promoting the concept of early risk factor detection and intervention to prevent future health problems, which are certainly getting more expensive every day. Simply, Wellness is great business for Employers, Employees, individuals, and the entire society.

It also helps to keep the small businesses alive, (traditionally) small businesses are considered the main engine of the economy, contributing to 84%, and Corporate contribute to 16% of the total economy.

Unfortunately, during the past 10 years this ratio flipped and thus many small practices (businesses) have shot down due to nonpayment by the insurance companies or get very little that does not cover the operational overhead forcing many physicians to go out of business, early retirement, declare bankruptcy, or change to non-clinical positions at a time that our Nation is suffering from critical physician shortage!

Or simply have physicians consumed with paperwork adding to the physician burnout which is affecting more than 60% of physicians nationwide. This is clearly adding harm to our unhealthy Healthcare system.

It costs an average of 1 million dollars to recruit and train a replacement physician and costs an average of 1 million dollars to settle a malpractice claim originating from burnout physician. This will add financial strain due to added costs.

The issue of Automation of Medicine and increase in rate of Physician extenders including Nurse practitioners and Physician Assistants can potentially help to ease the work load on physicians in a supportive and collaborative way rather than replacing physicians with lesser pay providers which would affect bottom-line dollar spent!

This is important to point out that doctors would always be needed for delivering quality care and meet the demands of patients who, for the most part, request to see doctors only!

The proper use of healthcare Information Technology to facilitate, co-ordinate the patient care cannot be over stressed. Technology can help reduce healthcare disparity, promote patient education, engagement, improve patient safety, reducing errors and help implement the practice of evidence-based medicine.

The current model of healthcare is better described as (Sick Care) with quick fix and neglecting to address the root cause of the problem till it comes again or other diseases arise.

Thus, we must switch our attention to Wellness and Disease Prevention by early detection and Intervention. This has been proven to be economically successful across the board to patients, employers, and the whole Nation as healthcare cost is consuming about 20% of the GDP.

Thus, smoking cessation, staying physically active with regular exercise, limiting alcohol intake, and proper nutrition are critical goals in achieving better health outcomes with billions of dollars saved in cost!

This will prevent many chronic diseases such as high blood pressure, high cholesterol, diabetes, obesity, arthritis, cancer, and chronic pain with associated depression from the disease process and the high cost involved leading to further economic devastation resulting in many cases in Bankruptcy.

The status of our healthcare and economy are closely correlated. So, restoring the healthy status of both which will decrease the added stress on low and medium class of hard-working Americans. Help their respective employers become more productive economically by keeping them healthy on the job.

Ultimately, consumerism will play a key role in the choice of healthcare providers, products, and services. Consumer-driven healthcare would also increase demand for healthcare advocacy to help patients (consumers) to navigate through a very complicated system with complicated choices of coverage and providers as more consumers take charge of their health management through active education and engagement to empower patients (consumers).

Consumerism would also drive completion towards better-personalized care, price transparency, and overall experience, and hopefully the consumer will be the winner at the end of the day.

The goal is to help alleviate the pain of many Americans resulting from the ongoing healthcare crisis. By helping patients, doctors, economy, and the entire community by implementing successful Health and Wellness program through Direct Contracting (https://www.linkedin.com/pulse/promote-culture-prevention-adel-eldin-md-facc-facp/).

The concept of prevention is not practiced in so many aspects of our day-to-day life for which a real culture change is needed.

For example, viewing the most recent tragic mass shooting in Parkland High School is a continuation of similar tragic events that have taken place. Since December 2012, when Sandy Hook Elementary School shooting killing 20 children, there has been 1,607 mass shootings killing at least 1,864 and wounding 6,459 fellow Americans.

According to a study done at Harvard University, published in the (American Journal of Public Health) which showed strong relationship between violet homicides and fire arms ownership, similarly with high rate of suicide. This strong relationship is true of firearm ownership and homicide even among law enforcement officers (American Journal of Public Health, 2015, Vol. 105, pp. 2042–2048).

Preventive measures which proven effectiveness in reducing gun-related death/violence include Universal background checks for high-risk individuals such as felons, those with mental disorders and violent behavior, and drug abusers. Banning high-capacity magazines, bump stock and limiting purchase of a firearm for only those above 21 years of age as Technology can easily accomplish that effectively.

Thus, decreasing the number of guns (a study in 2013, found 357 million guns in America, which has increased since then, surpassing the total number of U.S. population!). It is clear now that gun violence is a major public health risk affecting everybody, not just the kids who are getting scared going to their schools to learn in a safe environment.

The value of prevention cannot be overstated. Preventing disease for example comes with eating healthy food, drinking clean water, regular exercise, avoid harmful agents such as tobacco, high salt, sugar intake, excess alcohol as well as management of stress through Wellness and Relaxation Programs This will save lives, money, and improve quality of life.

A good example of a successful public health preventive strategy is wearing seat belt while driving and vaccinations to prevent the spread of diseases.

Prevention and management of stress would lead to a better outcome and enhanced productivity which will prevent depression and suicide. Prevention and management of pain by avoiding obesity, overweight, doing regular exercise, balanced diet, physical therapy, acupuncture, massage therapy, cupping, and non-pharmacologic therapy which will be a great alternative to opioid therapy which have been way overprescribed for obvious financial gains killing thousands of Americans annually and costing billions of dollars.

Prevention of poverty by improving education, job training as there is a strong association between low socioeconomic status and drug abuse, violence.

Prevention of STDs (sexually-transmitted diseases) and their complications by avoiding high risk behavior and promote abstinence.

Prevention can help avoid most of the 76 million cases of food poisoning in the U.S. annually, leading to 60,000 hospitalizations and 1,800 deaths. Proper sanitary food handlings, thorough washing of fruits and vegetables, proper packaging, cooking, and thorough hand washing of all food handlers would dramatically reduce food poisoning.

Prevention of fake news at it has become another public health risk, since false news spread much faster than real true news, thus adversely affecting health, businesses, and even our democracy itself. Solution is promoting truth and honesty in all aspects of our life.

Prevention of hospital-acquired serious infections including Penicillin-resistant Staphylococcus Aureus or MRSA), Catheter-related infections, ventilator-associated pneumonia, catheter-associated urinary tract infections by following sterile techniques and early administration of effective therapics and measures to contain and control infections.

Fall prevention among the elderly who are at high risk for falling, especially those with advanced degenerative joint disease or those with unsteady gait by taking precautionary measures to minimize falls and its adverse impact on social and economic well-being.

Preventing wars by engaging in active diplomacy and dialog aiming at peaceful resolution of conflicts rather than engaging in escalation.

The same message can go on to reinforce a culture of prevention with early effective intervention strategy that works well, thus will save lives and money by maintaining the well-being of the entire community.

Ignoring our problems by not implementing the appropriate preventative measures will only result in additional adverse outcomes measured by every metric out there! (https://www.linkedin.com/pulse/invest-your-health-adel-eldin-md-facc-facp/).

Everyone agrees that without Health, any amount of Wealth has no value. Also, everyone hopefully agrees that Physical Health is closely linked to

Financial Health. So, adopting healthy lifestyles at homework and everywhere have been proven to be a good investment to combat stress, lower incidence and complications of chronic diseases, improve productivity and minimize waste. This also means job security with good health versus losing the job over being too sick or unfit for the job health-wise!

Physical activity incorporated in the daily schedule decreased incidence of many obesity-related diseases such as heart disease, arthritis, diabetes, high lipids, Sleep Apnea..., etc. By Limiting time spent watching TV, as it was found that more than 3 hours daily watching, was associated with increased incidence of premature death (could be related to snacking while watching). Similarly, sedentary office work was found to be associated with 5.3 million deaths globally higher than 5.1 million deaths globally due to smoking!

Research from Eastern Kentucky University showed companies spend more than $300 billion dollars on healthcare and lost productivity due to missed work as a consequence of workplace stress. According to the survey used, 73% of those surveyed reported psychological symptoms such as being irritable, angry, nervous, lack of energy and emotional outbursts. Also, 77% of those surveyed encounter physical symptoms such as fatigue, headache, upset stomach, feeling muscle aches.

Many companies adopted solutions such as personal Resilience Program, Yoga Classes in addition to having regular social events to help manage workplace stress. The good thing here is that no one can go on ignoring stress and the importance of reducing it as no one can get rid of 100% of stress!

Even when people approach their retirement age, they are often reminded to plan for covering medical expenses and put aside saving to cover health expenses calculated an average person above 65 years of age to reach about $200,000 dollars.

We believe in the Concept of Comprehensive Health and Wellness being physically, emotionally, socially, and spiritually will be the ultimate investment. That is why we implemented those Wellness Components in our Concierge Medicine Program which has kept enrolled members healthy, saved each 1,000 of dollars (between 7k and 10k per person enrolled per year). The Program kept them away from going to Hospitals as none of these members was admitted to the hospital over the past 7 years and patients feel great, especially for those who used to be admitted to the hospital two to 4 times a year on an average every year, not anymore. It is simple, we work with our patients to improve nutrition, have a regular exercise routine, get rid of bad habits such as smoking and excessive alcohol consumption, limit salt and sugar intake, and have a positive approach to daily challenges which will ultimately minimize stress. Having direct contact between patients and doctors for trusted care will simplify life, cut costs, get rid of unnecessary hassle causing stress during challenging economic times for so many hard-working middle-class Americans. The patients pay so much out of pocket with deductibles, copayments, and added (patient responsibility portion of the bill on top of their insurance premiums). Just about all patients complain of the unbearable costs. So, the advice here to all my patients and everyone reading this important blog is to pay cash and save so much money and headache. This is our solution for healthcare crises for individuals, small, and medium size-Businesses to get relief felt immediately after making a small investment in your health. Our Wellness program is time tested, with proven efficacy and recession-proof. You, the consumer now will make an informed decision about your investment.

2.7 LEADERSHIP ROLE IN HEALTHCARE

February is the American Heart Month; physicians are at the Heart of Health-care, and they should be the brains leading Healthcare delivery as well. One major element in the current Healthcare crisis is the almost complete absence of the physician leadership in healthcare! Physicians sitting on the top of the pyramid as they are in charge of 100% of the healthcare dollars, as there is no hospital admission unless the patient is seen by a doctor or have approved and attested the note if seen by a non-physician staff being a nurse practitioner or a physician assistant. There is no surgery; procedure or a treatment plan is valid unless approved by a physician. Any decision for a specific device or therapy, discharge planning, follow up plan is worth-less unless approved by a physician. Any medical facility cannot get paid for medical services delivered unless it has the physician signature. Even if not physically on-site, the physician will be set up with a remote access to sign any of the forms or consults with (E-sign). The physician gives orders when to give or stop a certain medication. The physician writes orders if patient can eat or kept NPO (Nothing by Mouth) and what to eat being clear liquid diet, soft mechanical diet, or regular diet with diabetic (low calorie) or cardiac restriction (low salt, low cholesterol diet) or when a patient has renal failure (low protein, low potassium diet). In some cases, physicians would allow only parental nutrition through a feeding tube or Intravenous line in certain cases when the patient cannot have food by mouth. Doctors make decisions if the patient is allowed to have any activity as to being allowed out of bed or to have activities restricted. You now see what I mean, the physician approval and signature to all orders, including verbal ones. The physicians are the one who discusses with the patient and family the medical condition prognosis and explain adverse events to help family in making the best decisions regarding healthcare and decide on the code status in terminal and critical cases. Everyone else job is to help the physician delivering the best care possible to the patient. That is why the physician will lead a team that consists of (nurses, physician assistants, pharmacy staff, nutritionists, social workers, infection control, quality, and risk management staff and occasionally students) (https://www.linkedin.com/pulse/physicians-must-lead-healthcare-adel-eldin-md-facc-facp/).

Unfortunately, due to lack of physician leadership, physicians have suffered major losses in their independence, autonomy, and ability to make the proper decisions for the patient care. When majority of physicians who became employed workers and became totally helpless as they are frequently reminded, they are just employees and must do what they were told to do even if not in the patient's best interest as it is all about the bottom line. Physicians are frequently reminded that they could be replaced anytime and for any reason (such as not generating enough revenues for example) and in some cases terminated without cause!

This has resulted in an alarming high incidence of physician burnout that is threatening the quality of healthcare delivery to patients and being able to serve them well. Physician must unite to regain the natural role of Healthcare Leadership after many years of training, testing, and mentoring. This could be accomplished by simply being you. Do not be consumed with what the others are doing, excel in what you know and do the best job possible. Physicians have to adapt to changes through a continuous self-improvement and education to stay competitive. Stop complaining and always think solutions and turning problems into opportunities to do well. Be flexible by accepting a new niche, consider switching into the Wellness World, Disease Prevention, and changing the current sick care or spot-fix model of healthcare to get better outcome with less dollars spent!

A good clinical example is focusing on the (Mechanical) approach to treat Coronary Artery Disease by using Balloon Angioplasty or implanting Coronary Stents as a spot-fix and focusing on the mechanical part such as choosing the proper wire type to cross the stenosis (blockage), which guiding catheter to use, which device to use to de-bulk the lesion to make room for the stent (metal mesh used as a scaffold to keep artery patent) to be implanted and prevent the arterial re-narrowing (recoil) which happens frequently after using POBA (Plain Old Balloon Angioplasty). Then the choice of stent mostly now drug-coated with a polymer that prevents cellular replication (endothelial hyperplasia), which is another mechanism leading to re-narrowing of the artery at the stented segment with recurrence of symptoms of Angina (Effort-induced Chest Pain, relieved with rest.) That is mostly done without addressing the entire arterial tree as if stents were inappropriately implanted, this will create a new disease which may require additional stents leading to eventual need for Coronary Artery Bypass Surgery after failed (full-metal jacket) of multiple stents placed. This can also lead to repeated Angina attacks, heart attacks, and subsequent Congestive Heart Failure, adding to already high-cost healthcare bill. Even after Bypass surgery, the bypass grafts can get clogged with further stents placed, which may fail leading to a redo Coronary Artery Bypass Surgery in few years due to progression of coronary artery disease!

Evidence from clinical trials showed aggressive risk factors modifications to control hypertension (High Blood Pressure), High Cholesterol, Diabetes, obesity, regular exercise, which could be as simple as walking, improving diet to limit carbohydrate intake and increase fresh vegetables and fruit intake. Abstaining from Tobacco and alcohol abuse, illicit drugs, pork consumption and processed food will have a profound effect on overall health, cardiovascular, and cancer mortality as well.

A new risk factor recently reported to be associated with increased cardiovascular mortality is the degree of Economic Health! Yes, Income Volatility and the incidence of cardiovascular mortality direct relationship is well established!

Physician leadership is critical in educating the public about Wellness to avoid added financial stress to the physical or mental illness, empower patients and certainly make a healthy partnership to involve patients in healthcare choices. To have their skin in the game by making them aware of significant potential savings for them when adopting Wellness strategy rather the old model of (sick-care strategy) as to see the doctor only when you get sick. Patients will listen to the doctor they trust and the one who shows sincere empathy and compassion. Working together with physician leaders, we can bring the Heart Back in Medicine!

Physicians are considered to be the ideal leaders in healthcare as they have been through tough training for many years, and they are in charge of 100% of the healthcare dollars spent. They know medicine, but unfortunately, they lack management skills such as strategic planning, negotiation, budget, and financial decisions (https://www.linkedin.com/pulse/critical-role-physician-ceo-adel-eldin-md-facc-facp/).

There is a big need for these skills to be taught and doctors to be trained to fill this gap. For better healthcare management, physicians have to be involved and engage more with extra mentoring. Only few physicians take on the leadership positions and become successful. Those physician leaders will communicate, coordinate, advise, and lead so as to achieve the target goals of their particular institution.

In addition to their clinical skills, they need Information Technology skills and conflict management skills which they will be better suited for. As the non-physician CEO may sometimes be treating doctors as their adversaries rather than a physician CEO who will treat fellow physicians as partners alleviating a lot of friction especially when doctors feel that have lost autonomy as they are constantly told to increase (productivity) by increasing the number of patients seen during the workday with some doctors do not get to have a lunch break or even go to the bathroom because of the increasing workload and tasks assigned to them. Then comes other intrusive issues as they are told what medications to order (the cheaper ones even if patients cannot tolerate them, what tests to order, and even which subspecialists to refer to!

A real story comes to mind about a very well-respected colleague who had a very big private practice for about 24 years, and when the acquisition mania started few years back, hospitals started buying small and medium-sized medical practices, and those doctors became hospital employees. Once this senior physician who became a hospital employee and requested a physician-to-physician communication within the company regarding issues related to patient care. He was intercepted by a young administrator (younger than the physician son) who found about it and came reprimanding the senior physician that he should not have bypassed him, and the doctor has to go through him (the administrator) first!

When the senior physician explained that this is a physician-to-physician discussion and that the administrator is not a doctor, then the administrator simply fired the good doctor on the spot and the doctor was given only 15 minutes to clear his desk and take his belongings, but he was not allowed to have access to his patient charts from his own private practice that he spent his life building as they became hospital owned and was the good doctor was left out in the cold!

Needless to say, he suffered a very severe psychological trauma, and he now had to start from scratch all over again more than 25 years later. Similar stories have been reported with some physicians resigned and went out to start their private practice one more time. Some, with Concierge Medicine to escape stress and mental torture of being controlled by a non-physician CEO who may not understand or appreciate what doctors have to go through on a daily basis taking care of their patients.

Physician CEOs in addition would appreciate autonomy that fellow physicians need to make decisions that would be good for patients and good for business most of the time. They will be better qualified as trained leaders to lead towards evidence-based, cost-effective healthcare, which we are all need now more than ever.

2.7.1 *WORKING WITH FLORIDA LAW MAKERS*

It is about time for good news for Florida; with so many people in the Sunshine State who are either uninsured, non-insured or cannot afford current health insurance anymore, the DPC bill passed in the Florida House by 97-10 Vote. This is really great for small and medium-sized businesses as they cannot retain good employees or attract new hires without offering healthcare coverage. The current costly insurance premiums make it almost impossible for small business owners to afford employees, let alone offer other benefits to employees. This will certainly break the bank for small businesses struggling to thrive! (https://www.linkedin.com/pulse/good-news-florida-adel-eldin-md-facc-facp/).

Now, there is another option to choose without going through the insurance bureaucracy, on top of the hassle and ever so rising out of pocket expenses added with paying high premiums already. Direct contracting is great for physicians and their staff avoiding the daily stresses with insurance companies' denials, paperwork, filling forms (which can consume about 35% of the time that could be spent on taking care of patients). Direct contracting offers personalized service to keep patients healthy so that patients will not go broke with so many expensive bills. Today's system can become a further nuisance, especially when patients get sick and end up in the Emergency Room or even get admitted to hospital. That is where there is an exponential increase in healthcare costs which will add further stress on patients who are already economically strained.

Governor Scott finally signed DPC Bill into law on 3/23/2018, which comes after many years of perseverance and hard work. The overwhelming majority of lawmakers agree on its importance and positive impact on all Floridians. The Bill got additional traction after the shrinkage of ACA (Obama Care) and "skinny insurance" that does not cover much as an emplacement.

2.7.1.1 SUPPORT BILL HR 37! | ADEL ELDIN, MD FACC FACP | PULSE | LINKEDIN

Some individuals were inspired and took leadership role to sponsor the bill even after many previous failed attempts to pass it. I am personally very proud to be a part of this effort for years; I have made many trips to Tallahassee and talked to several lawmakers to explain the rationale and benefits. I gave an example of a model that we started 8 years ago with emphasis on early detection of risk factors and management to prevent future adverse events and save thousands of dollars to patient, millions of dollars to the local economy, and billions of dollars to the nation.

2.7.1.2 PATIENT TESTIMONIAL

The signed HB37 by Governor Rick Scott was to amend the Florida Insurance Code to clarify that the DPC arrangement between patients and physicians does NOT constitute the making of insurance and consequently will NOT be regulated under insurance laws (https://www.facebook.com/Heart-Health-and-Wellness-Program-278124625560440/?fref=ts).

Thus, the direct agreement between patients and doctors, while describing the services offered outside the traditional insurance model which is currently going from patient to insurance to provider. However, the direct agreement makes it patient to provider directly and keeps the insurance out of it.

2.7.1.3 A DIFFERENT KIND OF DOCTOR'S OFFICE: PATIENTS PAY DIRECTLY, KEEPING

This will help to bring back the healthy relationship between patients and doctors, with less cost to patient, better outcomes, and less hassle for both parties. It broadens the access to healthcare by empowering consumers to choose the best service for the best price. Improving accessibility and affordability through direct contracting will enhance the medical experience for both sides; The doctor(s) will spend more time with the patient(s), rather than pushing the patient(s) out under many pressures emanating mainly from insurance-related pressures, and thus creating a more friendly environment at doctor(s) office, as patients pay directly for services they know they are getting with transparent pricing coming directly from the doctor and their staff (https://www.facebook.

com/hearthealthandwellnessprogram/videos/10210909771772302/UzpfSTI3ODEyNDYyNTU2MDQ0MDoxOTk1NTc2NjUwNDgxODg3/?fref=ts).

Additionally, DPC will help management of acute and chronic disease with reduced need for costly and repeated emergency room visits, and costly hospitalization admittance and its related procedures. DPC model is certainly very innovative in reducing cost, which is becoming a critical element in healthcare. The rising rate of healthcare cost is much faster than the rate of inflation, which would compel local and state governments to adopt such a model for cost containment. Therefore, DPC is good for all stakeholders, not just patients and doctors. We would certainly welcome the opportunity to collaborate with those who are interested in starting or adopting such a great model for dignified practice of medicine to doctors and patients (http://brooksvillecardiology.com/?page=ARTICLES/Preventative%20care).

We also have used this model to help promote patient total wellness through education, active engagement, empowerment, and advocacy. That is the way to go Florida! (https://www.linkedin.com/pulse/support-bill-hr-37-adel-eldin-md-facc-facp/).

Four days ago, on January 14th, Bill HR 37 was presented in Tallahassee by its main Sponsor Rep. Castillo Known as DPC, and was approved by the House and now moves to the State Senate for approval. Everyone realizes that there is a Healthcare crisis all over America and certainly in Florida as it costs so much, and many millions are still without affordable coverage. That is why we move away from acute care to proactive care with emphasis on Wellness and Disease Prevention and Stress Management to save money.

As for the growing trend, moving towards open Healthcare Market delivering fair, transparent, competitive market with regards to price of services.

So, this is accomplished by empowering the patients with tools and advocate for patients to better care for themselves to reduce the main cost driver in healthcare being the (chronic diseases) by early detection and intervention.

Looking forward to joining other 13 States which approved Legislation which defines DPC agreements or Services as outside the Scope of State Regulation (Washington, West Virginia, Oregon, Utah, Arizona, Louisiana, Michigan, Mississippi, Idaho, Oklahoma, Kansas, Missouri, Texas). Florida having more than a million-plus uninsured and so many providers are unable to take of patients as they are overwhelmed with bureaucracy, paperwork, overhead includes costly billing services. With DPC, the doctors will focus on delivering quality care to their patients and spend more time with them, thus achieving higher rates of customer sanctification without the third-party control or hassle. Thus, putting value instead of volume to the Healthcare.

Join me and so many great providers in the State of Florida to urge the Senate President and leadership to pass HR 37 and do not repeat in 2016 what happened in 2015 as the Bill stopped at the Senate. We want to be able to provide the service that so many Floridians who need at a price they can afford, save patient and the State so much of healthcare dollars and improving outcome. We know because we have the model to prove that and have been already active in 13 other states already!

KEYWORDS

- **Affordable Care Act**
- **body mass index**
- **Centers for Medicare & Medicaid Services**
- **direct primary care**
- **Journal of the American Medical Association**
- **relative value units**
- **Supplemental Nutrition Assistance Program**

REFERENCES

Adel, E., (2011). *Wesley Chapel Chamber of Commerce Wellness Health Fair [Video file].* Retrieved from: https://www.youtube.com/watch?v=TrmcrFa4QmI (accessed on 14 July 2021).

Adel, E., (2012). *Health Heart and Wellness Seminar at the Palace Grand Part 01 [Video file].* Retrieved from: https://www.youtube.com/watch?v=xxyKa0ubmzI (accessed on 14 July 2021).

Adel, E., (2018). *Pronto Care is Here to Serve You! [Video file].* Retrieved from: https://www.youtube.com/watch?v=GHZLvFqBLGI (accessed on 14 July 2021).

Adel, E., (2018). *Pronto Care Testimonial [Video file].* Retrieved from: https://www.youtube.com/watch?v=1k3LB6M4SHU (accessed on 14 July 2021).

Burns, J., (2017). *With Direct Contracting Boeing Cuts Out the Middleman.* Retrieved from: https://www.managedcaremag.com/archives/2017/11/direct-contracting-boeing-cuts-out-middleman (accessed on 14 July 2021).

Byers, J., et al., (2018). *CMS Pushes Hospitals to Post Prices Online.* Retrieved from: https://www.healthcaredive.com/news/cms-IPPS-2019-proposed-rule/522089/ (accessed on 14 July 2021).

Eldin, A., (2015). *Black Seed, Global Natural Therapy.* Retrieved from: https://www.linkedin.com/pulse/black-seed-global-natural-therapy-adel-eldin-md-facc-facp/ (accessed on 14 July 2021).

Eldin, A., (2015). *Bringing the Heart Back in Medicine*. Retrieved from: https://www.linkedin.com/pulse/bringing-heart-back-medicine-adel-eldin-md-facc-facp/ (accessed on 14 July 2021).
Eldin, A., (2015). *More Hospital Closing in America?* Retrieved from: https://www.linkedin.com/pulse/more-hospital-closing-america-adel-eldin-md-facc-facp/ (accessed on 14 July 2021).
Eldin, A., (2016). *Direct Contracting for Healthcare*. Retrieved from: https://www.linkedin.com/pulse/direct-contracting-healthcare-adel-eldin-md-facc-facp/ (accessed on 14 July 2021).
Eldin, A., (2016). *Support Small Businesses*. Retrieved from: https://www.linkedin.com/pulse/support-small-business-adel-eldin-md-facc-facp/ (accessed on 14 July 2021).
Employer Health Care Congress, (2018). *Employer Direct Contracting & Bundled Payments Summit*. Retrieved from: https://www.employerhealthcarecongress.com/employer-direct-contracting-summit/ (accessed on 14 July 2021).
Ferguson, J., (2017). *Rising Health Care Costs Cause Some Tucsonans to Skip Meals, Doctors Visits*. Retrieved from: https://tucson.com/news/local/rising-health-care-costs-cause-some-tucsonans-to-skip-meals/article_c043a37e-4b38-5f0e-98de-5cbc38fddd12.html (accessed on 14 July 2021).
Frieden, J., (2018). *'We Have Failed' at Giving Diet Advice, Says Former FDA Chief*. Retrieved from: https://www.medpagetoday.com/primarycare/dietnutrition/73631 (accessed on 14 July 2021).
Haberrichter, E., (2017). *Direct Contracting and Bundling: Remedies for Rising Health Care Costs*. Retrieved from: http://www.ifebp.org/inforequest/ifebp/0200792.pdf (accessed on 14 July 2021).
Herschman, G., et al., (2016). *The Future of Healthcare: Direct Contracting Between Healthcare Providers and Self-Insured Employers*. Retrieved from: https://www.ebglaw.com/news/the-future-of-healthcare-direct-contracting-between-healthcare-providers-and-self-lnsured-employers/ (accessed on 14 July 2021).
Livingston, S., (2018). *Direct Contracting with Employers is Getting Easier for Health Systems*. Retrieved from: http://www.modernhealthcare.com/article/20180127/NEWS/180129920 (accessed on 14 July 2021).
Livingston, S., (2018). *Left Out of the Game: Health Systems offer Direct-to-Employer Contracting to Eliminate Insurers*. Retrieved from: http://www.modernhealthcare.com/article/20180127/NEWS/180129919 (accessed on 14 July 2021).
Masterson, L., (2018). *Medicare Part A Funding Will Dry up by 2026, KFF Finds*. Retrieved from: https://www.healthcaredive.com/news/medicare-part-a-funding-will-dry-up-by-2026-kff-finds/526409/ (accessed on 14 July 2021).
Masterson, L., (2018). *More Employers go Direct to Providers, Sidestepping Payers*. Retrieved from: https://www.healthcaredive.com/news/more-employers-go-direct-to-providers-sidestepping-payers/518269/ (accessed on 14 July 2021).
Medical Tourism Magazine, (2015). *How do Direct Contracts Work*? Retrieved from: https://www.medicaltourismmag.com/employer-direct-contracting-in-medical-tourism/ (accessed on 14 July 2021).
Muchmore, S., (2018). *AHIP18: Payers Face Threats from All Directions*. Retrieved from: https://www.healthcaredive.com/news/ahip18-payers-face-threats-from-all-directions/526368/ (accessed on 14 July 2021).
Olsen, C., (2018). *Ditching PPOs for Direct Contracting*. Retrieved from: https://www.employeebenefitadviser.com/news/ditching-ppos-for-direct-contracting (accessed on 14 July 2021).
Prontocare, (2018). Retrieved from: prontocare.co (accessed on 14 July 2021).

CHAPTER 3

AN EXAMINATION OF AN INNOVATIVE SELF-INSURED MEDICAL CENTER: THE CASE OF ROSEN HOTELS AND RESORTS

FREDERICK J. DEMICCO[1] and ABRAHAM PIZAM[2]

[1]*Northern Arizona University, Flagstaff, Arizona, USA*

[2]*The University of Central Florida, Orlando, Florida, USA*

ABSTRACT

Since its inception in 1991, the RosenCare® employee health and wellness program has saved Rosen Hotels and Resorts over $340 million. These savings have been invested back into employee education, improving the quality of life of underserved local communities, and optimizing the RosenCare® healthcare facility for the future. The RosenCare® state-of-the-art medical center provides employees and their dependents access to affordable and effective routine healthcare services. The success of RosenCare® is widely recognized as a viable model for similar businesses. It is conceivable that a widespread adoption of this model will result in an overall reduction in the costs of medical services and ultimately make the U.S. more competitive in the international medical tourism (MT) industry.

3.1 INTRODUCTION

To begin, here are some statements that demonstrate the multitude of problems and challenges that the healthcare system faces in the United States (U.S.). Some experts suggest that this system is dysfunctional, inequitable, controlled by myopic traditional business practices and values, full of inbred greed, and fiscally and financially out of control.

Wasteful healthcare spending consumes 79% of household income, leaving just 21% for everything else (Auerbach and Kellermann, 2011). This, more than nearly anything else, is destroying the American dream (Klepper, in Chase, 2017).

How is the U.S. spending $3.4 trillion on a healthcare system that is harming so many people? (Chase, 2017). The problems with the current healthcare system that damages the country and its residents include:

- The lack of accountability. Data shows that clinicians are not held accountable for their quality of care, with an epidemic misdiagnosis rate of 20 to 40%. (Chase, 2017).
- The power and status of a lobbying machine whose purpose is to protect the status quo of an enormous healthcare industry that is sustained with $3.4 trillion expenditures per year which equates to 17.5% of the total GDP.
- The myth of American healthcare as the gold standard of exceptionalism. In fact, U.S. healthcare is twice as expensive as, and delivers significantly worse results than, that of many other countries (Chase, 2017).
- Widespread conflicts of interest among benefit managers of business firms (insuring their employees) who hire benefit consultants who are paid by health insurers and providers.
- Pharmaceutical companies that, through their pervasive lobbying efforts, carry a lot of weight and manage to influence healthcare and insurance policies. The U.S. has some of the highest prices for prescription medicines in the world. This forces many American patients to buy U.S.-made drugs from Canada or Mexico, where they are sold for a fraction of the price that they are sold in the U.S.

The U.S. middle class has gone backwards financially in the past 20 years. One major cause is a costly and out-of-control healthcare system. These healthcare costs nearly doubled in the 10 years between 1999 and 2009. Families' spending on healthcare has increased 25% since 2007, while at the same time other spending on basic needs such as clothing, housing, transportation, and food at home and away has decreased (Wall Street Journal, 2016). Reasons for this astounding runaway-train increase in healthcare costs include overburdened and risk-averse human resources (HR) executives and CEO's who want to keep their employees happy and avoid lawsuits. The insurance industry is aware of this situation and takes advantage of it by increasing the premiums by about 11 to 14% each year, then often negotiating a somewhat lower rate—albeit significantly above the rate of inflation (Chase, 2017). Another healthcare-induced event is illness and medical bills contributing to 62% of all bankruptcies in the U.S. in 2007. State-level data demonstrates that healthcare costs are crippling other state budget line items such as education and social services (Chase, 2017). Given all of this gloom and doom of escalating healthcare costs in the U.S., the question before us now is, can we turn this around?

Fortunately, there are American entrepreneurs such as Harris Rosen, developer of the innovative RosenCare®, who have built successful and cost-saving healthcare systems born of evidence, quality outcomes, transparency, compassion, and accountability.

3.2 THE ROSENCARE® STORY

Harris Rosen, philanthropist, the founder and innovative entrepreneur of Rosen Hotels and Resorts in Orlando, Florida, U.S., is a firm believer in taking care of his employees so they will take great care of the Rosen Hotel guests. Rosen Hotels and Resorts is the largest independent hotel operator in Central Florida, with nine hotels there. In 1991, Rosen Hotels and Resorts started operating their own primary care health facility for their associates and their dependents. CNN, *Forbes Magazine,* Fox News, CBS, and other experts in healthcare have praised RosenCare® as an innovative solution to our national healthcare crisis—a privately run and managed healthcare success that can be instituted as a model by businesses around the U.S. A *CEO World* magazine described Harris Rosen as:

"The modest and unassuming president and founder of Rosen Hotels and Resorts in Orlando, Florida. His name has never appeared in *Medical Economics, Modern Healthcare* or in any other medical or health policy journal. Yet his company's health plan has been the subject of personal visits by members of Congress, an Army Surgeon General and top executive from Disney World to Whole Foods."

What is so special about RosenCare®? To begin with, the company's health plan is comprehensive. It covers almost everything. Out-of-pocket costs to Rosen's 2787 covered employees are minimal. The most they pay for a prescription, for example, is $30 and most of the time prescription drugs are free. Rosen Health care is also very accessible. Same-day appointments are available and transportation to the clinic is free. Doctor visits are on company time, so no one loses pay.

Under these circumstances, one would expect Rosen's health care costs to be sky high. Instead, on a per capita basis they are about 40% less than the national average. Even though Rosen employees are different from the U.S. general population in respect to advanced age and poorer health conditions, they are treated well for their ailments and experience a significant improvement in their health (Goodman, 2017).

Since its inception in 1991, RosenCare® has saved Rosen Hotels and Resorts over $340 million. These savings were invested back into the hotels for employee education, the medical center, and various philanthropic projects. Rosen's innovative approach to healthcare has brought international recognition and honors, including three awards by the World Health Care Congress in 2018 (Diamond Award: Lifetime Achievement Health Benefits Innovation; Diamond Award: Lifetime Achievement Public Leadership; Silver Award: Large Employer) (Health Value Awards, 2018).

3.3 SOME FACTS ABOUT ROSENCARE®

RosenCare®, "a patient-centered onsite medical home model, is based on five tenets: Access, Quality, Service, Cost, and innovation. Each provides solutions to common healthcare pain points businesses and organizations face. Those pain points, born of the turbulent healthcare landscape in the U.S., have serious consequences on our workforce and their families. The RosenCare® solution is straightforward and has profound, long lasting benefits. Protect your most valuable investment—your employees, your extended family—with common sense, affordable healthcare that has no deductibles or co-insurances." (RosenCare®.com, 2018).

RosenCare® has a state-of-the-art, comprehensive onsite 12,000-square-foot medical center that provides employees and their dependents access to routine healthcare services. The center is equipped with used but functionally modern medical equipment, purchased for a fraction of the cost of new equipment (Chase, 2017). Employees are able to visit the health center "on the clock," which removes a major hurdle for employees seeking care. In 2018, RosenCare®'s cost per covered life was approximately $5,500 for its 5,700 employees. In the same year, the national average per covered life was over $11,000. This results in substantial cost savings on healthcare. At RosenCare®, wellness check-ups have a zero-dollar co-pay, and co-pays for specialists are only $20. For any hospital admissions over the year, RosenCare® participants pay a $750 co-pay for the first admission in the year, a second $750 co-pay if admitted again in the same year, and then a zero co-pay for any additional admissions within the same year. About 90% of pharmaceuticals are available for a zero co-pay. Rosen employees get these at no cost to them. Rosen employees/dependents fill their generic prescriptions at Wal-Mart. In addition, RosenCare® premiums include dental coverage for $16.17 per week for the majority of individual associates and $53.05 per week for the majority of families. This is in contrast to the following story told to us by Harris Rosen: "A (non-Rosen-employed) patient was in the hospital and needed aspirin. The cost to the patient was $40. They gave the patient two aspirins, charged the patient for the entire bottle of aspirin, and … the hospital kept the bottle!"

RosenCare® hires nurses to serve as navigators throughout the medical journey. Rosen company events serve food and beverages that are approved by nutritionists and the director of health services. They also provide cooking classes. Wellness classes include Zumba, walking, Boot Camp, Tie-Chi, Yoga circuit training and spinning classes.

RosenCare® provides free health screenings for diabetes, breast cancer, colon cancer, sexually transmitted diseases, and high blood pressure. A Registered Dietitian is available for healthful eating. Rosen Hotels has zero tolerance for tobacco use and conducts random drug tests for its employees. Rosen Hotels has a mandatory stretching program for housekeeping staff and other employees who may have a higher propensity to musculoskeletal problems. These injuries were reduced by 25%. The Rosen associates dining area provides discounts for consuming healthier fare to reduce the consumption of less healthy foods.

3.4 THE ROSEN MEDICAL CENTER ENTERPRISE

The RosenCare® medical facility has a current staff of four full-time physicians, one part-time physician, three nurse practitioners (Noctors), two physician assistants, and support staff who see and treat about 160 to 180 patients a day. The patient care center pays special attention to the management of chronic health conditions, with particular emphasis on preventative health and wellness. In the U.S., 75% of healthcare spending is for people with chronic conditions. Therefore, RosenCare®'s focus on these chronic conditions results in substantial cost savings for healthcare. For example, their medication compliance rate is over 92% for cardiovascular medications (includes hypertension) and 96% for diabetes, compared to national averages of 46% and 50%, respectively.

RosenCare® also includes personalized and/or group counseling for smoking cessation and weight loss (Weight Watchers program which is subsidized by Rosen Hotels), in addition to physical therapy, chiropractic care, and wellness. Around 56% of their pregnancies are high risk as a result of high rates of advanced maternal age, hypertension, diabetes, and other diseases. RosenCare® is very proactive, assisting employees in managing pregnancies (as a premature birth can cost approximately $500,000!). RosenCare® has a premature delivery rate of 7% vs. the Orlando average of 10.5%. This focus on management of chronic conditions and pregnancy results in a drop in the development of new chronic conditions among associates.

3.5 THE RIPPLE EFFECTS OF ROSENCARE®

The ripple effects of RosenCare® are both within and outside the company's walls. These include healthier and happier associates and a lower employee turnover. In the hotel industry, annual employee turnover is approximately 60%, but at Rosen Hotels and Resorts, it is in the low teens. With the millions of dollars saved on healthcare, Rosen Hotels and Resorts pays for full-time associates' college tuition after 5 years of employment. The company also offers a full college scholarship to dependents after 3 years of fulltime employment along with an employee college scholarship after 5 years of fulltime employment.

Harris Rosen, the healthcare visionary, and philanthropist, has used the funds saved to invest in various underserved community projects around Orlando in the domains of education, healthcare, and crime prevention. The Tangelo Park Program is in one of these underserved communities. It is a three-fold educational community service initiative to benefit children and families living in the Tangelo Park neighborhood in Orlando, FL. The project provides: (i) Free preschool for every two-, three-, and four-year-old child living in Tangelo Park; (ii) Full college or vocational school scholarships (including tuition, room, and board, and books) for every graduating high school senior from the Tangelo Park area; and (iii) a Family Resource Center where parents can take parenting courses, obtain counseling, and access other resources to help them become positive role models. In addition to the above, Harris Rosen also provided funding for a new YMCA facility that includes a gym, weight lifting room, and basketball court.

The cost over 24 years of the Tangelo Park Rosen Community investment has been $12 million, which is about the amount RosenCare® saves in healthcare costs in a year. The results of the program have been extremely positive: crime was reduced by 63% and student graduation from high school increased from 45% to 100%. "Of those students in Tangelo Park who go on to 4-year schools, either directly or through community college, 77% achieve a degree—far above the expectation for a community of this socioeconomic level. Of those who go on to graduate school, 83% complete their programs. Tangelo students who enter community college complete their associate degree at the rate of 32%—again above the national average. High school graduates entering vocational programs complete at a rate of 83%" (Tangelo Park Program, 2017). For Harris Rosen, the approach is simple: Get involved, care for your people and those who live in underserved communities.

3.6 PLANS FOR EXTENDING ROSENCARE®

As mentioned earlier, RosenCare® has achieved national attention from the media, business leaders, and politicians. Several large companies like Disney have already created similar programs and others are considering its emulation. Because of the high interest in this venture, Harris Rosen has recently started a consulting company that will advise private and public sector organizations how to set up programs similar to RosenCare®. At one point, he was thinking about building a small hospital equipped with high-quality used medical equipment that would cost only a fraction of the $1.0 million per bed that most hospitals pay today. He strongly believes that the current method of charging for services in hospitals leads to exorbitant prices and sticker shocks when patients and insurance companies get the bill. He proposes that rather than charging for services, such as X-Ray, MRI, etc., hospitals should produce a "menu" of items with their associated *fixed* costs based on diagnosis. For example, the cost of an appendectomy = $X, bariatric surgery = $Y, etc. This way, the patients and their insurers will know their financial obligations in advance, and most importantly, this will lead to competition in the market. This hospital could serve not only his employees, but also his hotel guests and out-of-town visitors. However, at the time of this writing, Harris Rosen felt that he had too much on his plate and could not devote the time and effort to embark upon such a huge venture.

3.7 THE RELEVANCE OF THE ROSENCARE® MODEL TO MEDICAL TOURISM (MT)

We believe that if the RosenCare® model and the Rosen method of hospital billing spread throughout not only the U.S. hospitality and tourism industries, but also other industries, inbound medical tourism (MT) to the U.S. will increase and outbound MT will decrease for the following reasons:

- The costs of medical services in the U.S. will decrease. As a result, the U.S. will be more competitive for inbound medical tourists, who will come here for advanced medical procedures and treatments such as: coronary disease, cancer, cerebrovascular disorders, etc. It will also be more attractive to international tourists who seek cosmetic and dental surgeries. After all, the U.S. has the best medical clinicians, who practice in the most advanced medical facilities in the world. What kept the inbound MT industry from flourishing until now were its high costs compared to other countries.
- When inbound tourism to the U.S. reaches significant levels, many hotel companies will be interested in building accommodation facilities on hospital campuses that cater specifically to international visitors and their families. At present, there are already examples of top brand hotels built on hospital campuses, e.g., Hilton/Baptist Hospital; Cleveland Clinic/Rosewood Hotel (Hospital Hotels, 2018).
- It is also possible that either Harris Rosen or some other hotel owners will eventually decide to extend the RosenCare® model to their onsite guests, especially for non-complex or selective treatments.
- It is estimated that at present 900,000 Americans travel each year outside their own country for medical treatments (Patients Beyond Borders, 2018). Those who travel abroad for medical treatments do so for two reasons. The first and by far the most important is price differentials. The second is alternative treatments that are not available in

the U.S. When prices for medical service come down, the American healthcare industry will be more competitive with the healthcare industries of Malaysia, India, Turkey, Brazil, and Thailand, which are the top five MT destinations that compete with the U.S. (Top 5 Countries for Medical Tourism, 2018).

- At present, the majority of U.S. healthcare organizations provide healthcare services to their own employees through insurance companies that have specific programs tailor-made to the healthcare industry. If the RosenCare® model and the Rosen method of hospital billing are adopted, organizations in the healthcare industry might decide to be self-insured and provide medical services for their own employees. That, in turn, will increase the employees' satisfaction, reduce labor turnover, and ultimately increase the quality of service to their patients/employees. But most importantly, by adopting the RosenCare® model, these organizations will incur significant savings that can be used to reduce the fees for medical services to medical tourists.

3.8 CONCLUSION

Harris Rosen, Central Florida's largest independent hotelier and the founder of Rosen Hotels and Resorts, is widely recognized as an entrepreneur and passionate philanthropist. In addition, he has also become a healthcare innovator who founded RosenCare®, a superior self-insured healthcare model that has saved the company $340 million. The program has garnered the interest of members of U.S. Congress and top executives from very successful companies. If the entire public and private sector implemented a RosenCare model, we would anticipate annual savings of approximately $935 billion, simply by preventing unnecessary illnesses and diseases and by emphasizing the importance of a healthy lifestyle.

The financial benefits of a healthcare model like RosenCare® extend beyond the company's bottom line and well into the local community. With the multi-million dollars in savings in healthcare, Harris Rosen has been able to create numerous charitable initiatives. These include providing hundreds of college scholarships to his employees and their dependents, as well as to hundreds of young men and women in underserved Orlando neighborhoods. In addition, it has enabled him to build the Rosen College of Hospitality Management at the University of Central Florida—now ranked in the top five hospitality colleges in the world; to build the Jack and Lee Rosen Jewish Community Center; and, recently, to build a new school and rebuild more than 100 hurricane-devastated homes in Chantal, Haiti. In our opinion, a widespread adoption of this model will result in an overall reduction in the costs of medical services and ultimately make the U.S. more competitive in the international MT industry.

The RosenCare® model is good for employees, good for companies, and good for the community "*We owe it to ourselves to explore its viability on a national level.*" (Quote from Harris Rosen).

KEYWORDS

- **cardiovascular medications**
- **healthcare system**
- **human resources**
- **Medical Center Enterprise**
- **medical tourism**
- **RosenCare model**

REFERENCES

Auerbach, D. I., & Kellermann, A. L., (2011). *How Does Growth in Health Care Costs Affect the American Family?* Santa Monica, CA: RAND Corporation. Retrieved from: https://www.rand.org/pubs/research_briefs/RB9605.html (accessed on 14 July 2021).

Chase, D., (2017). *The CEO's Guide to Restoring the American Dream: How to Deliver World Class Healthcare to Your Employees at Half the Cost.* Health Rosetta Media Publisher, Seattle, Washington.

Goodman, J. C., (2016). *Welcome to the World of RosenCare: This CEO Solved His Company's Healthcare Problems without Washington.* CeoWorld Magazine.

Hospital Hotels, (2018). Retrieved from: www.imtj.com/articles/why-growth-medical-tourism-being-limited-lack-appropriate-hotels/ (accessed on 14 July 2021).

Patients Beyond Borders, (2018). Retrieved from: www.patientsbeyondborders.com (accessed on 14 July 2021).

RosenCare, (2018). Retrieved from: www.RosenCare.com (accessed on 14 July 2021).

Tangelo Park Program, (2017). Retrieved from: www.tangeloparkprogram.com/about/faqs/ (accessed on 14 July 2021).

Top 5 Countries for Medical Tourism, (2018). Retrieved from: www.Clements.com (accessed on 14 July 2021).

Wall Street Journal, (2016). Burden of healthcare costs moves to the middle class. *Wall Street Journal.*

CHAPTER 4

AN INNOVATIVE APPROACH TO GLOBAL MEDICAL AND WELLNESS TOURISM: A CASE OF FLORIDA MEDICAL TOURISM

ADEL ELDIN

Managing Director, ProntoCare Cardiologist Tampa, Florida, USA

ABSTRACT

Florida Medical Tourism (FMT) offers global services in medical care, as it is designed to offer special medical services to national and global customers (www.floridamedicaltourism.com/). This culturally sensitive medical tourism (MT) is a very important concept that FMT has adapted to. FMT aims to respect various cultures and to offer the best services, while also combining it with specific cultural tastes, so that every consumer feels comfortable, at home.

4.1 INTRODUCTION

Florida Medical Tourism (FMT) offers unique, global services in medical care, as it is designed to offer special medical services to national and global customers (www.floridamedicaltourism.com/).

The target market is mostly Middle Eastern and/or Arabic-speaking countries. This culturally sensitive medical tourism (MT) is a very important concept that FMT has adapted to. FMT aims to respect various cultures and to offer the best services, while also combining it with specific cultural tastes, so that every consumer feels comfortable, at home. FMT aims to understand every aspect of the consumer including language, traditions, and everything associated with their own culture. For example, if a Hispanic patient requests someone who also speaks Spanish, FMT can provide them with not only a Spanish speaker, but someone who also understands the culture, and the background of that patient. This experience will help create a more holistic treatment for the patient and will increase in customer satisfaction as well, which will certainly be critical for future customers and services.

4.2 MEDICAL TOURISM (MT) IN FLORIDA BY THE NUMBERS

Florida is one of the highest visited states; there are almost 100 million visitors that come to Florida annually, and the number is increasing over the years, because of the warm climate, the welcoming tourist sites, and the amazing scenery. The visitors tend to increase over the winter months due to the warm climate; these visitors are called snowbirds. Visitors are domestic or international, that migrate from areas that are colder in order to enjoy Florida's less harsh winter. In America, this mostly applies to Americans that live in the North: states such as Illinois, Wisconsin, Michigan, etc. Snowbirds usually stay in Florida for about five to six months, starting in September, and ending in April, in order to escape the brutal winters of the North, and then they end up going back to their home states for the spring and summer seasons. In some cases, some people opt to stay in Florida permanently.

FMT also offers international inbound patients/clients that come from neighboring countries such as Canada, and also across the Atlantic, including the UK, and other European countries. Florida's legislature passed a law called "Florida medical tourism (FMT)," the same name as our company, to encourage domestic medical tourism (MT) for those who would come to Florida anytime of the year, to have their elective surgery done. It has been known by research that sunshine would impact not only post-operative recovery, but would also help with preventing post-operative depression, which is more likely to happen in states that lack sunshine. The very unique services of FMT provide a culturally sensitive and specific courses of treatment catered to the individual. We not only offer medical services, but "A to Z" services; for example, we help with airport-to-airport services, transportation, stay, etc. This is provided for both inbound and outbound, domestic, and international patients.

The patients will have multiple services. First, we established a unique phase of a wellness tourism program and we introduced this program here in Tampa, Florida in 2012, shortly after FMT was founded, by Dr. Adel Eldin, to help create an avenue for so many people that would be in need for culturally sensitive MT services. The innovative work of the wellness tourism package was started by designing a very unique plan that could be purchased online, and it is called the "999 Plan." This plan encourages clients to come and spend time with FMT Headquarters in Tampa, but this could be replicated in any area within Florida, or anywhere in the world, for that matter. This plan fuses client entertainment and health; clients can spend some time and enjoy Florida, while also getting a health analysis as well. This concept has been replicated many times, through many models across the world (in many parts of Europe, parts of the Middle East, and parts of the Americas), since the founding of FMT. The program of the wellness tourism, would involve initial labs, which will have a basic lipid panel, a CBC, electrolytes, thyroid function, hemoglobin A1C (to screen for possible diabetes, through monitoring sugar intake in the blood) and a PSA for men, if they are above 45 years of age, and other services as well. This will certainly be a great opportunity, once patients have their blood work, to engage in a complete medical history and physical exam, as well as a complete cardiovascular evaluation, by an echocardiogram, and all the data will be uploaded on a flash drive, which the patient will be able take back home and show to their primary physician. Also, for female clients above the age of 45, FMT does offer mammogram screening for those who would be candidates for it (if it is clinically indicated).

During the stay, FMT does offer many places that the client can go to such as amusement parks, beaches, and museums, etc. Disney World seems to be one of Florida's main attraction, and it is in close proximity to our headquarters in Tampa, Florida. FMT's main goal is to provide medical services, while also incorporating tourism and wellness, in order to create a more comprehensive experience for the clients. Also, the traditional MT services include arrangements for various surgeries, ranging from orthopedic surgery, heart surgery, cosmetic surgery, weight loss surgery, and so on. This could be arranged with affiliate hospitals that specialize in these specific surgeries.

FMT has certainly been on the forefront of bringing the concept of culturally sensitive medicine. There is a definite influx of customers coming to the U.S. as inbound patients from all over the world, such as the Middle East, Central America, South America, and Europe. This certainly has increased traffic and has helped a lot of individuals to be a part of the benefit of the free health care market, when it also come to medical therapy and tourism. Certainly, the tourism is becoming the number one source of income for the state of Florida, rather than agriculture which is now the second-highest source of income for Florida. Florida has a lot to offer when it comes to unique courses of treatment and specialized surgeries; for example, in Tampa, Florida, one of the most well-known cancer institutions, Moffit

Cancer Center, is established in this area, alongside other institutions such as the Children's Hospital (specializes in pediatric management of patients) and Tampa General Hospital. There has been changes in the baby boom market (individuals born between 1946 and 1964), and now with the globality of health care, as well as the changes that has been happening with the reduction in the size and possible demise of Obamacare, that a lot of folks, especially middle-class Americans, will be turning to other health care options that are affordable for their procedures and wellness tourism.

Currently, there are about 2 million Americans that travel globally to get adequate health care, but this number is very likely to increase because of the changes in healthcare, mostly Obamacare decreasing in popularity, as well as replacing Medicaid with block grants. There will be a definite opportunity to serve all inbound and outbound domestic markets by specific tailored programs that will fit the needs and also the budget of various socioeconomic classes within the population, specifically middle-class relief for inbound and outbound individuals.

The MT market has been growing steadily and it is estimated that it is going to be having a steady growth over the next decade, and FMT is certainly going to be a critical and affordable source to serve many customers, whether in the outbound or inbound, domestic, or international tourism, for those who are uninsured or underinsured, and those who are looking to combine medical wellness, as well as leisure into one experience.

The concept of the culturally sensitive MT will attract so many people to this program. FMT always takes into account many different aspects of the individual; for example, the dietary restrictions of the patient, the area they live in, their language, their living arrangements, and so on. For example, a Muslim patient's dietary restrictions, which comprises of "halal" food and restraining from pork and alcohol, or a person of the Jewish faith, who primarily eats "kosher" products, will be considered, and prioritized when it comes to the overall experience of the individual.

Due to the ease of global communications, and also instant interactions through social media, and other various applications, will allow patients to be connected with their treating doctors. This will allow an easier flow of medical records to make patients have the opportunity to discuss with their treating physicians all the questions related to their surgeries or procedures, and to discuss all related questions related to the success rates, complication rates, and any other information that is deemed suitable between the doctor and the patient, so both parties have an open line of communication, and the patient is in the loop about the doctor's course of treatment. FMT will ensure the highest quality and the highest standards of both ethical and professional conduct to give the best service to the patients. FMT is committed to offering transparent pricing with excellent services and also give the option to the patients if they would like to have special programs to recover, such as non-medical healing, which involves going to tourist attractions or doing tourist activities.

FMT is also on the forefront of making new local and global needs for healthcare, and this particular program will be specifically tailored towards those who have chronic pain syndromes, those who have arthritis or degenerative joint disease, or those who have skin disease(s). This program offers 8 days, 7 nights package that people open the door that encourages the cessation of opioid use/dependence (which contributes to 60,000 deaths annually in the U.S.), through non-medical treatments. This program is combined with non-pharmacological therapy that would involve natural spring water treatment/therapy, black sand bath/mud bath, cupping, acupuncture, as well as a nice balanced leisure program that the patient will get exposure to hydration and other natural remedies that will help pain control and get people off the opioid addiction, and prevent future devastation to a lot of patients and their families. FMT would like to change the culture of pain management that includes little to no amounts of opioid use.

4.3 CONCLUSION

FMT is playing a critical role in creating solutions to both the MT needs for inbound traffic to the U.S., particularly to Florida, and also for outbound patients, especially those who fall in the middle class who for the most part, cannot afford American healthcare, especially with the demise of Obamacare. Not only is Obamacare decreasing in popularity, but many programs are being removed that have to do with subsidies, such as Medicaid, that will be replaced by block grants, leaving many millions of Americans who are either uninsured or underinsured looking for other programs that will cater to their health-related needs. I believe that FMT, through its unique services including the wellness tourism as well as various therapies, surgeries, and procedures, along with telemedicine that allows for second opinions will be impactful in so many areas to provide the best culturally sensitive MT services in the way(s) customers understand and feel comfortable with.

We take pride in providing a comprehensive experience for the patient while taking into account and having respect for their likes/dislikes, their background, their medical condition, and their needs. Through its global outreach, I believe that FMT and its various partners and affiliates will certainly have an increasing role in delivering very unique services to all individuals involved, whether it is recreative, therapeutic, spiritual, and, of course, medical. I believe that FMT will certainly fill a void for a huge need for all clients of various cultures and backgrounds, while also maintaining ethical and moral values that the MT has established through compassionate care. In summary, whatever you happen to be in need of, in terms of procedure-based needs, MT needs, or other related needs, FMT is dedicated to providing the best quality of all these factors whether it applies to inbound or outbound patients. FMT welcomes any collaborations to make an ecosystem that will have all "A to Z" services, which includes transportation, diet, hospitality, entertainment, medical needs, and so on, and so forth. We aim to please the customers first, and all these aspects will certainly have an impact on the patient's overall experience. It is our goal to provide the best services, the best prices, the best transparency policies, and the best experience for all parties involved, and to create an everlasting relationship with the customers, and create a new way of providing health care.

KEYWORDS

- **comprehensive experience**
- **domestic medical tourism**
- **Florida Medical Tourism**
- **non-medical healing**
- **transportation**
- **warm climate**

REFERENCES

Agency for Health Care Administration, (2017). Retrieved from: http://ahca.myflorida.com/schs/DataD/DataD.shtml (accessed on 14 July 2021).

Armstrong, J., (2015). *Florida Surging as Medical Tourism Destination.* Retrieved from: https://www.medicaltourismmag.com/florida-surging-as-medical-tourism-destination/ (accessed on 14 July 2021).

Florida Medical Tourism, (2018). Retrieved from: http://www.floridamedicaltourism.com/ (accessed on 14 July 2021).

Florida Tax Watch, (2014). *Medical Tourism in Florida*. Retrieved from: http://floridataxwatch.org/resources/pdf/MedicalTourismFINAL.pdf (accessed on 14 July 2021).

IndUSHealth, (2017). *Global Healthcare Options*. Retrieved from: http://www.indushealth.com/ (accessed on 14 July 2021).

Levey-Baker, C., (2017). *Sarasota's Medical Tourism Sector Faces Opportunities and Challenges*. Retrieved from: https://www.sarasotamagazine.com/articles/2017/11/15/sarasota-s-medical-tourism-sector-faces-opportunities-and-challenges (accessed on 14 July 2021).

Lopez, G., (2018). *2017 was the Worst Year Ever for Drug Overdose Deaths in America*. Retrieved from: https://www.vox.com/science-and-health/2018/8/16/17698204/opioid-epidemic-overdose-deaths-2017 (accessed on 14 July 2021).

MDG Advertising, (2018). *Marketing Medical Tourism Improves Lives and Revenues*. Retrieved from: https://www.mdgadvertising.com/marketing-insights/marketing-medical-tourism-improves-lives-and-revenue/ (accessed on 14 July 2021).

NASA (Neurosciences and Spine Associates, P.I.), (2017). *Experienced Physicians in a Top International Health Destination*. Retrieved from: https://nasamri.com/medical-tourism (accessed on 14 July 2021).

Smith, R., (2016). *Foreign Patients Flying to Florida for Medical Treatment, Tampa Bay Seeing Financial Benefit*. Retrieved from: https://www.abcactionnews.com/news/region-pinellas/foreign-patients-flying-to-florida-seeking-medical-treatment-tampa-bay-seeing-financial-benefit (accessed on 14 July 2021).

Williams, R., (2016). *Florida has $6B Medical Tourism Market with Opportunities to Grow Study Finds*. Retrieved from: https://www.bizjournals.com/orlando/news/2016/01/25/florida-has-6b-medical-tourism-market-with.html (accessed on 14 July 2021).

PART III

Destination Competitiveness Factors and Cases

CHAPTER 5

SUCCESS FACTORS OF MEDICAL TOURISM SPECIAL ECONOMIC ZONES

ALI A. POORANI and YUAN FANG

Alfred Lerner College of Business and Economics, Department of Hospitality Business Management, University of Delaware, 14 W., Main Street, Newark, DE–19716, USA, E-mail: poorani@udel.edu (Ali A. Poorani)

ABSTRACT

Medical tourism special economic zones' (MTSEZs) track records have been mixed. Therefore, this study attempts to redefine success factors for MTSEZs. As such, this exploratory study investigated the opinions of experts involved in medical tourism (MT) to identify attributes important for the success of MTSEZs, as well as the challenges and pitfalls to avoid. A model was built based on three perspectives: (i) special economic zone factors; (ii) tourism destination factors; and (iii) medical treatment factors. This model presents an inventory of success factors for MTSEZs to assist developers, investors, local, and regional authorities, medical communities, media, and all other stakeholders to make informed decisions about designing or reconfiguring a MT economic free zone (MTSEZ). In the final part of the chapter, we present the most important challenges that MTSEZs face, and pitfalls to avoid when commencing MTSEZs.

5.1 INTRODUCTION

In 2019, there were 5,400 special economic zones (SEZs) worldwide, with 1,000 established in the last 5 years. This enthusiasm is fueled by the successes of many SEZs. However, according to the United Nations Conference on Trade and Development (UNCTAD, 2019), nearly 1/2 of SEZs, especially in developing and latecomer countries are somewhat underutilized or heavily underutilized, and many without substantial spillover benefits to the rest of the economy. Therefore, a clear understanding of success factors, challenges, and pitfalls of medical tourism (MT) SEZSs and policy objectives require a renewed attention.

The special quest for MT SEZs is due to the globalization of healthcare that has impacted tourism sector. MT is among one of the fastest growing tourism markets worldwide (Adabi, 2017; MacReady, 2007; Jones and Keith, 2006); also see Chapters 1 and 2. According to Stephano (2015), MT, together with health tourism is poised to reach a trillion-dollar business. It is important that we briefly distinguish between MT and health tourism. Though sometimes used interchangeably, MT and health tourism differ. MT refers to the practice of traveling to another country with the purpose of obtaining health care. While, health tourism generally refers to visiting spas, homeopathy treatments or traditional therapies for preventive purposes (Derrien, 2019; Smith, Alvare, and Chanda, 2011).

There are many benefits to investment in MT, such as increased income, improved health services, foreign exchange earnings, a more favorable balance of trade, and income generated from tourism (Crooks, 2015; Ramirez de Arellano, 2007). To facilitate these economic opportunities, many countries like the Philippines, South Korea, India, and others have embarked on establishing medical tourism special economic zones (MTSEZs) to accelerate MT; and the trend is growing (Cohen, 2014; Hodges et al., 2012; Soundarapandian, 2012; Reisman, 2010; Singh, 2008; Connell, 2006). Though such national efforts can produce results, the performance of SEZs in general has been spotty, debatable, and inconclusive (Wang, 2013; Aggarwal, 2006).

According to our review of the existing literature, there are no studies evaluating the success factors particular to MTSEZs, suggesting the need for an investigation. In this chapter, we explore specific success factors for MTSEZs from the perspective of experts. It is hoped that this framework will help policy objectives in countries or regions that intent to initiate MTSEZs or implement changes.

5.2 SPECIAL ECONOMIC ZONES (SEZS)

Farole (2011) broadly defined special economic zones (SEZs) as demarcated geographic areas contained within a country's national boundaries where the rules of business are different from those that prevail in the national territory. These differential rules principally deal with investment conditions, international trade and customs, taxation, and the regulatory environment; whereby the zone is given a business environment that is intended to be more liberal from a policy perspective and more effective from an administrative perspective than that of the national territory.

Beyond those benefits outlined above, according to UNCTAD (2019), three specific features have increased the popularity of SEZs. First, relative ease, particularly in countries that governments are weak and nationwide reforms are relatively difficult. Second, low costs, obviously building a particular piece of land versus the infrastructure of the entire country is less costly. Third, competing, and attracting foreign direct investment (FDI) by offering lucrative packages. Within the last 5 years, the levels of FDI have decreased. To attract domestic and foreign investment, countries have chosen to create innovative, specialized zones with unique, focus, and value propositioning, such as high tech, green zones, tourism zones, or MT zones.

Increased interest in developing MTSEZs could be attributed to the success of many SEZs. Many countries, particularly in the Asia region, have adopted SEZs as a popular means for fostering and stimulating economic development (Ji, 2015; Osborne, 1986; Wong and Chu, 1985). China, for example, was the first to create SEZs as a trial for the process of industrialization. Since then, these SEZs have played a significant role in China's economic success. The SEZs in China has successfully tested the market economy and new institutions and established role models for the rest of the country (Ji, 2016; Zeng, 2012). Encouraged by this practice, a number of other countries including India (Shiji, 2016) have adopted the SEZ-model for improving their export potential (Roy Choudhury, 2010).

Most SEZs are multi-activity zones. MT SEZs is different from SEZs in that MTSEZs demand a number of specific components, namely tourism destination factors and medical treatment attributes. We believe combing the three could produce a comprehensive model for MTSEZs that experts could identify the most relevant items for the success of this particular segment. Further sections explain these determinants in more details.

5.3 FACTORS RELEVANT TO SEZS

According to a number of studies, the success of a SEZ depends on its unique factors (Crane, 2018), such as location advantage; macroeconomics of the country; industrial investment support; investment costs; skilled labor; management and services; government policies; laws and regulations; stability and consistency of the government (Murray, 2013; Monga, 2011;

Farole, 2008). In addition, Marrakech (2009) in the study of best practice for the development of economic zones, recommended that policymakers interested in designing new or modifying existing zone programs, should focus on six elements: (i) the type of zone to be developed; (ii) the policy framework; (iii) incentive framework; (iv) regulatory framework; (v) institutional framework; and (vii) physical development and management of zones. Thus, aside from all 5 necessary elements, development of MTSEZs must first meet the item (i) mentioned above, *the type of the zone,* which requires identifying specific assets supporting tourism activities and medical assets. In the next two sections, we will focus on those two elements specifically.

5.3.1 TOURISM DESTINATION COMPONENTS

Demand for health services has become an important niche (Béland, 2018) for tourism. The drivers behind this demand included: Longer life expectancies, stronger health orientation and the need for self-development (Haderlein et al., 2007). MT is often seen as 'adding medical services to common tourism' (FSCAP, 2009). Thus, MTSEZs should be treated not only as places for medical treatment, but also attractive tourism destinations. In the study of Determinants of Health Tourism Competitiveness: An Alpine Case Study, Schalber, and Peters (2012) concluded that health tourism is moving towards more specialized but narrow markets, such as medical wellness aimed at boosting the destination competitiveness by adding more value to its services.

Beyond national competitiveness (Smeral, 1996), the concept of destination competitiveness and its measurement have been widely utilized to evaluate the sustainability and development of a tourism destination. Evaluating tourism destination competitiveness is useful (Carayannis, 2018) for industry and government as they attempt to identify destination strengths and weaknesses from a visitor's perspective in order to highlight opportunities for tourism development (Dwyer et al., 2004).

Lee and King (2009) proposed a framework of destination competitiveness where they combined the concept of internal and external resources. These resources are listed below:

1. **Internal Factors:** Include Core resources and attractors, such as natural resources; cultural resources; special events; entertainment activities, supporting resources: infrastructure; facilitating resources; and, hospitality.

2. **External Factors:** Includes issues relating to destination management, such as marketing; human resource development; quality of service; environmental management, etc.
3. **Other Factors:** Include items related to location; safety/security; price competitiveness; awareness of destination; destination image.

Destination competitiveness is important (Ege and Uslu, 2018) because it is often conditioned or limited by external environmental forces over which the authorities exercise minimal control (Kotler et al., 1993). In addition to the above factors, destination strategies are used to match internal tourism resources and destination attractions with external environments (Crouch and Ritchie, 1999).

5.3.2 MEDICAL TREATMENT COMPONENTS

Medical treatment quality can affect tourists' decision on choosing a medical destination. Wang (2012), in the study of *Value as Medical Tourism Driver* concluded that perceived value, which is influenced by components, such as perceived **medical quality,** service **quality,** and enjoyment were key predictors of customer intentions. In the study of *Interrelations of Service **Quality** and Service Loyalty Dimensions in **Medical Tourism***, Debata et al. (2015) found that the **treatment** satisfaction dimension of service **quality** has a positive and significant impact on MT service loyalty. Lunt et al. (2011) suggested that a common regulatory platform and reporting system would serve as the basis of an assessment of comparative quality of care using a range of performance indicators as facilitated by international accreditation and certification.

Naidu (2009) argued, though difficult to measure patient satisfaction and healthcare service quality, a multi-disciplinary approach that combines patient inputs as well as expert judgments could provide an effective approach. Medical delivery quality has been complemented by the Joint Commission International (JCI) to accredit hospitals worldwide (York, 2008), adding to the credibility of destination healthcare providers.

Gan and Frederick (2011) found that patients value their services based on the entire process of pre-and post-care treatments. Though patient follow-ups by providers are very important, studies have shown that insufficient management of complications and a lack of post-operative care were not uncommon (Terzi et al., 2008). Guiry and Vequist (2011) suggested that MT providers should focus on improving service quality, particularly in the areas of reliability (i.e., billing accuracy, employee sympathy and reassurance, as well as providing services at the time promised) and assurance (i.e., employee knowledge, employees receiving adequate support from the health care facility to do their jobs well, feeling safe when interacting with employees, and employee politeness).

Based on the analysis of literature it was compelling to combine the three areas: (i) factors related to general economics and public sector support; (ii) factors related to tourism destination competitiveness; and (iii) factors related to components of the medical treatment process and propose a comprehensive model for MTSEZs. Base on this assumption, we have combined all necessary items and have asked that expert identify the most relevant items for the success of MTSEZs in order to design a working model (Wongkit, 2016) for professionals to modify their existing practices or develop new MTSEZs.

5.4 METHODOLOGY

This exploratory study investigated the opinions of a number of experts involved in MT to identify attributes important for the success of MTSEZs. The model was built based on three perspectives:

- Special economic zone factors;
- Tourism destination factors;
- Medical treatment factors.

5.4.1 QUESTIONNAIRE DESIGN AND ADMINISTRATION

We have chosen an exploratory method hence there were no established research regarding MTSEZs. Based on the literature review, we developed a list of determinants from three major perspectives explained earlier in the chapter: Factors important to successful SEZs; factors important for a charismatic tourism destination; and, elements significant to MT. In the initial phase, we submitted the questionnaire to three MT SEZs for their input to verify the validity and thoroughness of proposed determinants. The respondents were provided options to write-in additional factors, as they determined essential. Based on the responses provided, a more comprehensive and final list was emerged. Table 5.1 lists all the relevant attributes.

TABLE 5.1 Factors Important for MT SEZs

Special Economic Zones Factors	Tourism Destination Factors	Medical Treatment Factors
Location	Nature resources	Treatment quality
Accessibility	Culture and heritage	Safety and risk
Macroeconomics of the country	Distinctive core attractors	Clinical outcomes
Industrial investment support	Shopping	Continuity of care
Investment cost	Outdoor clusters	Infection rate
Labor pool	Hospitality service	Hospital facilities
Management and service	Destination management	
Government policies	Safety and security	
Laws and regulation	Price competitiveness	
Stability and consistency of the government	Destination image and brand	
	Tourism awareness	
Collaboration	Multilingual services	
Weather	Tour operators	
Currency exchange	Information services	
Environmental sustainability	Customer services	
Education/training programs	Retirement living	

We then submitted the table to 37 invited participants in the 8th World MT and Global Healthcare Congress in September 2015. The invited participants for this special Summit were experts in the MT industry from various regions of the world (N = 12 from Europe, N = 4 from USA, N = 3 from South America, and N = 18 from Asia and Africa). The questionnaire was distributed to participants before the start of the Summit and collected immediately after they were completed. We believe this method provided a more effective exploratory finding than a general survey method, inquiring input from non-experts.

To further consolidate the variables listed in Table 5.1, we asked the experts to rate each variable for their importance, on a 5-Point Likert Scale (1= Very important and 5= Not important). To further consolidate the list of variables, we established that those variables that reached a *mean score of 3 or less,* will be designated as essential factors, since being weighted more important than other variables by the experts.

Important Note: We express that all items listed in Table 5.1 deserve attention when making decisions, since no two destinations are identical. However, the experts gave more prudence to items selected as important to very important attributes to a MTSEZ's success.

We also added two additional open-ended questions: (i) challenges that MTSEZs are facing; and (ii) pitfalls to avoid when planning for a new destination or revitalizing an existing MTSEZ.

5.5 FINDINGS

5.5.1 *SPECIAL ECONOMIC ZONE FACTORS*

Experts in this study identified 10 variables significant to the to an Economic Free Zone. Those are listed on Table 5.2.

TABLE 5.2 Success Factors for MTSEZs

	Success Factors for Medical Tourism SEZs
Special Economic Zones Factors	Location
	Accessibility
	Macroeconomics of the country
	Industrial investment support
	Investment cost
	Labor pool
	Management and service
	Government Policies
	Laws and regulation
	Stability and consistency of the government

5.5.2 *TOURISM DESTINATION FACTORS*

Table 5.3 lists 10 variables important and relevant to establishing and branding the destination for travel and tourism. As seen in the list, natural, cultural, and distinctive core attractions are essential assets for establishing a tourism destination, followed by other services offered.

5.5.3 *MEDICAL TREATMENT FACTORS*

Obviously, people fundamentally care about their health. It is not a surprise that quality of care, safety/risk, and clinical outcomes top the list, as shown in Table 5.4.

TABLE 5.3 Success Factors for Tourism Destination

	Success Factors for Medical Tourism SEZs
Tourism Destination Factors	Nature-based resources
	Culture and heritage
	Distinctive core attractors
	Shopping
	Outdoor clusters
	Hospitality services
	Destination management
	Safety and security
	Price competitiveness
	Image and brand

TABLE 5.4 Success Factors for Medical Treatment

	Success Factors for Medical Tourism SEZs
Medical Treatment Factors	Treatment quality
	Safety and risk
	Clinical outcomes
	Continuity of care
	Infection rate
	Hospital facilities
	Accreditation

We suggest that public and private sectors to incorporate these important factors and utilize a detailed strategic planning as provided in Chapter 13.

5.5.4 MOST IMPORTANT CHALLENGES THAT MTSEZS FACE

We also asked the expert to list the most important challenges that MTSEZs face. Table 5.5 lists the challenges in three categories: (i) issues related to government and politics; (ii) issues relevant to business; and (iii) cultural matters.

5.5.5 PITFALLS TO AVOID WHEN ENTERING MTSEZS

We thought awareness of some of the significant pitfalls are important for entities entering this business. The following list was developed by the experts in no particular order (Table 5.6).

TABLE 5.5 Most Important Challenges That MTSEZs Face

Most Important Challenges That MTSEZs Face	
Government and Politics	Transparency
	Political challenges, stability, and consistent government policies
	Terrorism
Business	Approaches to business relations
	Finding specialists and educated labor
Culture	Tourist friendly culture
	Informed society and trained professionals in various disciplines
	Language
	Community feels participant in medical tourism

TABLE 5.6 Pitfalls to Avoid Entering MTSEZ

Pitfalls to Avoid Entering MTSEZ
1. When players involved in the business are driven by the profit, instead by the scope of the business
2. Embarking on a medical tourism strategy because it is nice to do
3. Only start medical tourism when your medical to your own citizens are well established
4. Regions where there is a lack of education, communication, safety, and quality in medical services
5. Avoid partnership with government and be selective in affiliation with an academic institution
6. Lack of accessibility and availability of patient registration translators
7. Regions with unregulated marketing and promotion policies and guidelines

5.6 RECOMMENDATIONS

Based on the expert opinion and their experiences, a comprehensive list of factors related to the success of MTSEZs was presented. This comprehensive list constitutes a comprehensive inventory of success factors for destination who wish to embark on MT business and MTSEZs. The inventory can assist developers, investors, local and regional authorities, medical communities, media, and all other stakeholders to make informed decisions. We also expressed that not all of the items identified by the experts are applicable to all destinations. Careful strategic planning is essential to identify strengths, weaknesses, opportunities, and threats for a particular destination. The inventory can also be used as an additional source of information while developing relevant strategies.

In this study we also itemized all the factors that experts identified as important or very important to MTSEZs. They ranked each factor in the order of importance. We summarized their rankings in three distinct categories: Factors related to SEZs; factors important for a tourism destination; and factors significant to medical and health reasons. These distinctions are significant as a destination may decide factors that they want to emphasize in branding. A strong linkage between all segments in consultation with local communities is essential for building or reforming a destination.

We also presented important challenges the experts identified in relation to the success of a MTSEZ. We categorized them in three distinct groupings: Challenges involving government and political issues; business related issues; and, culture.

Finally, we asked the experts about their opinions regarding pitfalls to avoid when embarking on MTSEZs. Among the topics were raised were: Caution involving with entities who wish to enter this business purely for profit; or jumping on the bandwagon because it is a nice thing to do; the region is not fundamentally armed with educated and trained workforce, including medical and other specialties; the region lacks consistent rules and regulations about doing business, such as marketing and promotion; and when the region overlooking the medical needs of local population.

KEYWORDS

- **foreign direct investment**
- **Joint Commission International**
- **local population**
- **medical tourism special economic zones**
- **special economic zones**
- **United Nations Conference on Trade and Development**

REFERENCES

Adabi, K., Stern, C. S., Weichman, K. E., Garfein, E. S., Pothula, A., Draper, L., & Tepper, O. M., (2017). Population health implications of medical tourism. *Plastic and Reconstructive Surgery, 140*(1), 66–74.

Aggarwal, A., (2006). Special economic zones: Revisiting the policy debate. *Economic and Political Weekly, 41(*43, 44), 4533–4536.
Béland, D., & Zarzeczny, A., (2018). Medical tourism and national health care systems: An institutionalist research agenda. *Globalization and Health, 14*(1), 68.
Bernsein, A., (2012). Special economic zones: Lessons for South Africa from international evidence and local experience. *Edited Proceedings of a Round Table Convened by the Center for Development and Enterprise, The Center for Development and Enterprise*. Johannesburg.
Buhalis, D., (2000). Marketing the competitive destination of the future. *Tourism Management, 21*(1), 97–116.
Carayannis, E. G., Ferreira, F. A., Bento, P., Ferreira, J. J., Jalali, M. S., & Fernandes, B. M., (2018). Developing a socio-technical evaluation index for tourist destination competitiveness using cognitive mapping and MCDA. *Technological Forecasting and Social Change, 131*, 147–158.
Cohen, I. G., (2014). *Patients with Passports: Medical Tourism, Law and Ethics* (p. 18). New York: Oxford University Press.
Crane, B., Albrecht, C., Duffin, K. M., & Albrecht, C., (2018). China's special economic zones: An analysis of policy to reduce regional disparities. *Regional Studies, Regional Science, 5*(1), 98–107.
Crooks, V. A., Cohen, I. G., Adams, K., Whitmore, R., & Morgan, J., (2015). Inbound medical tourism to Barbados: A qualitative examination of local lawyers' prospective legal and regulatory concerns. *BMC Health Services Research, 15*(1), 291.
Crouch, G. I., & Ritchie, J. R. B., (1999). Tourism, competitiveness, and societal prosperity. *Journal of Business Research, 44*(3), 137–152.
D'Hauteserre, A. M., (2000). Lessons in managed destination competitiveness: The case of Foxwoods casino resort. *Tourism Management, 21*(1), 23–32.
Debata, B. R., Patnaik, B., Mahapatra, S. S., & Sree, K., (2015). Interrelations of service quality and service loyalty dimensions in medical tourism. *Benchmarking: An International Journal, 22*(1), 18–55.
Derrien, M. M., Cerveny, L. K., & Wolf, K. L., (2019). The human health dimensions of sustainable tourism. In: *A Research Agenda for Sustainable Tourism*. Edward Elgar Publishing.
Dwyer, L., & Kim, C., (2003). Destination competitiveness: Determinants and indicators. *Current Issues in Tourism, 6*(5), 369–414.
Dwyer, L., Mellor, R., Livaic, Z., Edwards, D., & Kim, C., (2004). Attributes of destination competitiveness: A factor analysis. *Tourism Analysis, 9*(1, 2), 91–101.
Ege, S. D., & Uslu, Y. D., (2018). An assessment of competitive factors in medical tourism. *International Journal of Research in Business and Social Science (2147–4478), 7*(2), 14–21.
Farole, T., & Akinici, G., (2011). *Special Economic Zones: Progress, Emerging Challenges, and Future Directions.* The International Bank for Reconstruction and Development, The World Bank, Washington DC.
Farole, T., (2008). *Special Economic Zones: Performance, Lessons Learned, and Implications for Zone Development.* The multi-donor investment climate advisory service managed (FIAS) by the International Finance Corporation (IFC) and supported by the Multilateral Investment Guarantee Agency (MIGA) and the World Bank (IBRD), The World Bank, Washington DC.
Farole, T., (2011). *Special Economic Zones in Africa Comparing Performance and Learning from Global Experience*. The International Bank for Reconstruction and Development, The World Bank, Washington DC.

Gan, L. L., & Frederick, J. R., (2011). Medical tourism facilitators: Patterns of service differentiation. *Journal of Vacation Marketing, 17*, 165–183.

Guiry, M., & Vequist, D. G., (2011). Traveling abroad for medical care: U.S. medical tourists' expectations and perceptions of service quality. *Health Marketing Quarterly, 28*(3), 253–269.

Haderlein, A., Rauch, C., Kirig, A., & Wenzel, E., (2007). *100 Top Trends: Die Wichtigsten "Driving Eorces" des Kommenden Wandels*. Kelkheim: Verlag Zukunftsinstitut.

Hodges, J. R., Turner, L., & Kimball, A. M., (2012). *Risk and Challenges in Medical Tourism: Understanding the Global Market for Health Service* (p. 68). Santa Barbara, California: Praeger.

Hsu, C. C., & Sandford, B. A., (2007). The Delphi technique: Making sense of consensus. *Practical Assessment, Research and Evaluation, 12*(10).

Ji, M., Li, M., & King, B., (2015). The impacts of China's new free-trade zones on Hong Kong tourism. *Journal of Destination Marketing and Management, 4*(4), 203–205.

Ji, M., Li, M., & King, B., (2016). Incremental Effects of the Shanghai Free-trade zone: An internet informed assessment of Hong Kong's tourism competitiveness. *Journal of China Tourism Research, 12*(1), 24–41.

Jing-dong, Y., & Eden, L., (1992). Export processing zones in Asia: A comparative study. *Asian Survey, 31*(11).

Jones, C. A., & Keith, L. G., (2006). Medical tourism and reproductive outsourcing: The dawning of a new paradigm for healthcare. *International Journal Fertility Women's Medicine, 51*, 251–255.

Kotler, P., Haider, D. H., & Rein, I., (1993). *Marketing Places: Attracting Investment, Industry, and Tourism to Cities, States, and Nations*. New York: Free Press.

Lee, C. F., & King, B., (2009). A determination of destination competitiveness for Taiwan's hot springs tourism sector using the Delphi technique. *Journal of Vacation Marketing, 15*(3), 243–257.

Lunt, N., Smith, R., Exworthy, M., Green, S. T., Horsfall, D., & Mannion, R., (2011). *Medical Tourism: Treatments, Markets and Health System Implications: A Scoping Review*. Retrieved from: http://www.oecd.org/els/health-systems/48723982.pdf (accessed on 14 July 2021).

MacReady, N., (2007). Developing countries court medical tourists. *Lancet, 369, 1849–1850.*

Mainil, T., Platenkamp, V., & Meulemans, H., (2011). Health and Medical Tourism: The discourse of medical tourism in the media. *Tourism Review, 66*(1, 2), 31–44.

Marrakech, (2009). *Towards Best Practice Guidelines for the Development of Economic Zones* (pp. 1–15). A contribution to the ministerial conference by working group 1: MENA-OECD investment program.

Miller, G., (2001). The development of indicators for sustainable tourism: Results of a Delphi survey of tourism researchers. *Tourism Management, 22*, 351–362.

Monga, C., (2011). *Cluster-Based Industrial Parks: A Practical Framework for Development Economics Vice Presidency Action*. The structural transformation unit, Africa region; and development economics vice presidency, The world bank.

Murray, M., (2013). *What are Special Economic Zones*. Retrieved from: https://kb.osu.edu/bitstream/handle/1811/54558/1/Special_Econ_Zones_NK_Reform_Adam_Kong_Senior_Thesis.pdf (accessed on 14 October 2021).

Naidu, A., (2009). Factors affecting patient satisfaction and healthcare quality. *International Journal of Health Care Quality Assurance, 22*(4), 366–381

Ramirez De, A. A., (2007). Patients without borders: The emergence of medical tourism. *International Journal of Health Services, 37*(1), 193–198.

Reisman, A. D., (2010). *Health Tourism: Social Welfare Through International Trade.* Cheltenham: Edward Elgar Publishing.
Ritchie, J. R. B., & Crouch, G. I., (2000). The competitive destination: A sustainability perspective. *Tourism Management, 21*(1), 1–7.
Ritchie, J. R. B., & Crouch, G. I., (2003). *The Competitive Destination: A Sustainable Tourism Perspective*. New York: CABI Publications.
RoyChoudhury, K., (2010). Review paper special economic zones in China. *SIES Journal of Management, 7*(1), 114–120.
Schalder, C., & Peters, M., (2012). Determinants of health tourism competitiveness: An alpine case study. *Tourism, 60*(3), 17, 307–323.
Shiji, O., (2016). Economic impact of tourism in India. *International Journal of Social Science, 5*(1), 35.
Smeral, E., (1996). Globalization and changes in the competitiveness of tourism destinations. In: Keller, P., (ed.), *Globalization and Tourism* (pp. 391–415). St. Gallen: AIEST.
Smith, R., Martinez, A. M., & Chanda, R., (2011). Medical tourism: A review of the literature and analysis of a role for bilateral trade. *Health Policy, 103*(2, 3), 276–282.
Soundarapandian, M., (2012). *Development of Special Economic Zone in India: Impact and Implication* (pp. 253–255). New Delhi: Concept Publishing Company.
Terzi, E., Kern, T., & Kohnen, T., (2008). Complications after refractive surgery abroad. *Ophthalmology, 105*, 474–479.
UNCTAD, (2019). *Special Economic Zones*. Retrieved from: https://unctad.org/en/PublicationChapters/WIR2019_CH4.pdf (accessed on 14 July 2021).
Unti, J. A., (2009). Medical and surgical tourism: The new world of health care globalization and what it means for the practicing surgeon. *Bulletin of the American College of Surgeons, 94*, 18–25.
Wang, H. Y., (2012). Value as a medical tourism driver. *Managing Service Quality, 22*(5), 27, 465–491.
Wang, J., (2013). The economic impact of special economic zones: Evidence from Chinese municipalities. *Journal of Development Economics, 101*, 133–147.
Wei, G., (2012). Special economic zone and technological learning: An analytical framework. *Economic and Finance Review, 2*(3).
Wongkit, M., & McKercher, B., (2016). Desired attributes of medical treatment and medical service providers: A case study of medical tourism in Thailand. *Journal of Travel and Tourism Marketing, 33*(1), 14–27.
York, D., (2008). Medical tourism: The trend toward outsourcing medical procedures to foreign countries. *J. Contin. Educ. Health Prof., 28*, 99–102. doi: 10.1002/chp.165.
Zeng, D. Z., (2012). China's special economic zones and industrial clusters: The engines for growth. *Journal of International Commerce, Economics and Policy, 3*(3).

CHAPTER 6

RENEWAL CHALLENGES: THE CASE OF THE REPUBLIC OF GEORGIA

LALI ODOSASHVILI and ALI A. POORANI

The Lerner College of Business and Economics, University of Delaware, Newark, DE–19711, USA, E-mails: lalio@udel.edu (L. Odosashvili), poorani@udel.edu (A. A. Poorani)

ABSTRACT

This chapter uses the Republic of Georgia to showcase how the country can renew and revive a lost reputation. To help to cultivate a strategic focus, and policy objectives to restore Georgia's health, wellness, and medical tourism (MT) as a competitive and viable destination: (i) the authors of this chapter contacted three top specialists in the Georgian travel industry to gain an insight regarding the challenges that Georgia's medical and health tourism is facing; (ii) the authors performed a comprehensive analysis of Georgia's natural healing assets, things that Georgia historically and culturally have been proud of—e.g., thermal waters, mud, climate; the natural, environmental landscape; medical services; and the country's carrying capacity; and (iii) additionally, the authors performed and presented five SWOT (strengths, weaknesses, opportunities, and threats) analyzes to validate if Georgia as a whole, and which specific regions within the country are best poised to revive and compete regionally and globally. Based on these analysis and input from the experts, we propose four main destinations, namely Tskaltubo, Borjomi, Ureki, plus the capital Tbilisi as viable destinations where the government and private sector efforts can be focused. Finally, we proposed a two-phase planning process with the hope that Georgia, one day, gain back her past reputation in health and wellness tourism.

6.1 INTRODUCTION

According to Georga (2018), "Georgia has much potential for health tourism development, and to a limited degree, medical tourism (MT). How well these targeted source countries [Saudi Arabia, Kuwait, Iraq, Azerbaijan, Armenia, Kazakhstan, Kyrgyzstan, Uzbekistan, Ukraine, and Russia] will respond to Georgian efforts in attracting foreign patients will depend on whether Georgia manages to adjust to and sustain international practices and service levels. There is much more to do, however to come up to the standards found in the best European and Asian countries."

The above script underscores Georgia's challenges to renew itself as a reputable health, wellness, and MT destination. Georgia, with its rich and diverse resources, has the potential for developing MT and can easily rival the world's leading health and wellness destinations. On a small territory, 69,700 km^2, one can find thermal and mineral waters, curative mud, mountain-climate, sea-climate, magnetic sand, and a perfect climate for people with different needs. Beyond the natural resource for healing, because of low prices and/or softer regulations, Georgia is already a popular destination for dentistry and teeth prosthesis, cosmetic medicine, excimer laser vision correction, treatment of ischemic heart disease (stenting, shunting), treatment with Bacteriophages, reproductive health services, among the neighboring countries (Verulava, 2016). Given these promises, is Georgia good enough as a top-tier destination to attract medical tourists? And what needs to be done to achieve that? In this chapter, we address and analyze Georgia's health and wellness assets, medical assets, and potential to become an attractive market for medical, health, and wellness destination (Figure 6.1).

6.2 BACKGROUND

It was said that MT is a new direction of the travel industry, however, people from all around the world have been searching for medical treatment from unknown times. If not international, but domestic, MT existed as long as medicine itself (Health Tourism, n.d.).

Documentation of the hot springs in the center of the capital of Georgia, Tbilisi, goes back to the 10th century when the Arab geographer, Abu Dulaf, noted in his diary that, "the water in Tbilisi boils without fire." But recent archeological excavations indicate to an even earlier settlement—Roman-style baths with plumbing that date to the first century (Asulo, 2012). The

Abanotubani baths are built along Persian traditions, but unlike the water in Persian baths (which are manually heated) the water in Tbilisi naturally comes from underground with a temperature ranging between 75° and 107° Fahrenheit (Figure 6.2).

FIGURE 6.1 Georgia: Country Map.

Not only in the capital, but in a number of other parts of the country unique places can be found. "There are up to 2000 mineral and thermal springs found throughout the country," and those resources were well developed and used over the years (Georgian National Tourism Administration, 2016). Development of resorts in the country began at the turn of the 19–20th centuries. During this period, tourism was developed in Borjomi and Abastumani, and later at the end of the 19th century on the Black Sea coast. In the Soviet Union times, Georgia was one of the most popular holiday destinations, and its resorts were recognized as one of the best medical sites "...due to its scenic beauty, its mild climate and a large number of health resorts." (Cappucci, 2015). Reliable data on tourism in SSSR are difficult to find. The most conclusive figures point to 4.5 million visitors annually in the late 1980s, of which about 2 million headed to sanatoria and health resorts. In the City of Tskaltubo, Georgia's abandoned paradise (Vardiashvili, 2019) the sanatoriums were packed full and even had a special bathhouse for Stalin.

Though still a number of them are functional, yet not developed enough to offer high-quality service (Kvachantiradze, 2016). So, how can Georgia gain back its acclamation? Or, can it? (Figure 6.3).

FIGURE 6.2 The Abanotubani Baths.

6.2.1 IMPORTANT NUMBERS AND DEVELOPMENT POSSIBILITIES

Tourism in Georgia in 2016 increased by 19% (UNWTO Tourism Highlights, 2017), which means more and more people explored the country and its different resources. In 2017, Georgia had 7.9 million tourists which represented 17.6% growth in 1 year. It is important to highlight, from that number, only 2.7% visited the country for health and medical care (The Ministry of Internal Affairs of Georgia, 2018). Yet, this is a significant number for Georgia, since the country is less known for MT than a touristic destination. Tourists traveling for medical reasons usually do not travel alone. Based on the MTA Patient Survey, in fact, 83% of patients travel with a companion and 90% of patients or their companions engage in tourism activities. Nearly 80% of the demand for medical travel is driven by cost savings (Godwin, 2012). All this information gives us reason, hope, and motivation that Georgia has potential for renewing its past reputation in health and wellness and perhaps develop a sound MT program. The country's affordability as compared to most European countries; its rich culture and history; its exceptional natural tourism assets; and it is ease of access present much optimism that these ambitions are achievable.

FIGURE 6.3 Stalin Sanatorium.

6.2.1.1 *MAIN CHALLENGES*

There are also a number of challenges that Georgia faces in her attempt to revive herself as a viable health, wellness, and MT destination. According to a recent study by Verulava, and Jorbenadze (2018):

> "Experts note that there is no comprehensive database of MT potential in Georgia. There are fewer advertisements about medicinal resorts. Moreover, most of the tourism agencies offer tourists lesser medical tours. Travel companies and medical sector should cooperate; unite efforts to develop MT in the country. Considering this, it is necessary to create a complete electronic database about MT potential in Georgia. It should include comprehensive information on clinical centers (the types of services and their prices) that will be available by the internet."

To close this void, in this chapter, we have identified the main health, wellness, and medical assets of the country, and explored regions that need serious strategic focus in order to be revived and renewed as competitive destinations. We did this by reviewing the history of major targets, conducting a SWOT analysis for each target and prioritize where the most efforts be expended. We also have exhibited the major challenges identified by experts in this study. Based on their recommendations and our analysis, we suggest that all stakeholders and the community leaders use the PPP MT toolkit, provided in Chapter 13, to further analyze each sector to guide strategic development and tactical planning, if the country is seriously considering becoming a competitive health, wellness, and MT destination.

6.3 MEDICAL TOURISM (MT) ASSETS AND CORE INFRASTRUCTURE ANALYSIS

One of the most important steps in determining the attractiveness of a destination for medical, health, and wellness tourism is identifying and analyzing its main assets and core infrastructure. According to a research report by Wellness Tourism Worldwide (2011), The natural resources and healing assets are the key factors in the Eastern European region. The top three assets in a rank-order are:

- Natural healing assets/evidence-based Medicine (e.g., thermal water, mud, climate, etc.);
- The nature/environmental/landscape; and
- Medical services.

Based on the study of literature, Georgia processes all three assets outlined above. Beyond Georgia's rich historic bathes, spa towns and coastal localities, and other therapeutic landscapes, the country now offers key medical specialties, advanced treatments and hospitals appropriate for MT. Therefore, we can segment Georgia's medical landscape into two distinct capacities: *Medical As*sets and *Health and Wellness Assets.*

6.3.1 MEDICAL ASSETS

Latest observations have shown that most all the clinics and hospitals are located in the capital-Tbilisi. According to Georgian National Tourism Administration, nowadays, Georgia is offering:

1. **Cardiology/Cardiac Surgery:** Cardiology is one of the most developed fields in Georgia. This is due to active collaboration with the leading centers and universities of Europe and the United States of America, which gives the possibility to timely implementation of up-to-date treatment methods and technologies.
2. **Ophthalmology:** More than 2500 surgical and laser treatments of different eye pathologies are performed in ophthalmology clinics of Georgia.
3. **Plastic Surgery:** These clinics in Georgia offer diverse procedures for both men and women.
4. **Dentistry:** This clinic in Georgia are equipped with modern technologies, offering customers a wide range of services. Professionals actively participate in local or international conferences in order to raise their qualifications and constantly get information about innovations and technologies implemented in dentistry. This field is one of the most developed in Georgia.
5. **Phage Therapy:** Recently, the bacteria, causing various infectious diseases, have worked out resistance against even the most powerful antibiotics. Due to the absence of an effective method of treatment a number of patients die annually. At the same time, almost all bacteria have their own natural killers. There are specific viruses, not dangerous for human beings, that are called bacteriophages. Bacteriophage is a Greek word, which means the eater of bacteria. The first phage, like this, was discovered by the famous Canadian scientist Felix D'Erele in 1917 (Summers, 2017).

According to Zhvania et al. (2017), "Bacteriophage therapy is an alternative therapy that has long been successfully used to treat *Staphylococcal* infections, in some parts of the world, including Georgia, and is drawing new interest worldwide in this era of antibiotic resistance."

Phages are able to multiply only in specific bacteria, which bear their genetic code. After penetrating into these bacteria, hundreds of new pages are created. They dissolve the "host" bacteria and start looking for the new bacteria. By means of this kind of chain reaction, the phage multiplies until all the bacteria are destroyed. After that, the phages leave the body as they are unable to multiply without bacteria. Phages are safe for the cells of human beings, animals, and plants. Institute of Microbiology in Tbilisi has re-established itself as a leader in bacteriophage research and its various applications in humans, as well as in animal, plant, and environment.

6.3.2 *THE NATURAL HEALING ASSETS AND THE NATURAL, ENVIRONMENTAL LANDSCAPE OF GEORGIA*

In Table 6.1, we have identified seven important natural health and wellness therapies and landscapes that offer therapeutic benefits found in Georgia. We have also listed their successive regions and popular resort towns within them. These seven natural therapeutic resources and landscapes are: balneological, sea climate, climate-balneological, mineral waters, mountain-balneological, mountain-climatic, and mud resorts.

TABLE 6.1 The Natural Healing Assets and the Natural, Environmental Landscape of Georgia

SL. No.	Type	Popular Resort Towns	Regions
1.	**Balneological:** Balneology is the study of the arts, sciences, and cultures of natural mineral waters for drinking, soaking baths, steams, and inhalations for health, wellness, and therapeutic benefits (Balneology Association of North America).	Tbilisi Grigoleti Nasakirali Shovi Tskaltubo	Tbilisi Guria Guria Racha-Lechkhumi Imereti
2.	**Sea Climate:** Thalassotherapy, or seawater therapy is a medical approach based on the systematic use of seawater, sea products, algae, and coastal climate. This therapy has provided relief for several ailments, and provides numerous sea-derived pharmaceuticals (Charlier and Chaineux, 2009)	Grigoleti Ureki	Guria Guria
3.	**Climate-Balneological:** Both climate and thermal springs purport to have health benefits, from easing aches and pains, to less obvious cures like skin conditions, asthma, arthritis, high blood pressure, and more (Kagaku, 2001).	Borjomi Nabeghlavi Nunisi Tsaghveri	Samtskhe-Javakheti Guria Imereti Samtskhe-Javakheti
4.	**Mineral Waters:** Hydrological science demonstrate the varied physiological effects of mineral waters (Weisz, 2001).	Borjomi Nabeghlavi Nasakirali Sairme Surami Tsaghveri	Samtskhe-Javakheti Guria Guria Imereti Shida-Kartli Samtskhe-Javakheti
5.	**Mountain-Balneological:** The quality of the surface-atmosphere in the mountain climate-balneological by means of climate and landscape to treat patients suffering from bronchial asthma (Frank, 2012).	Abastumani Sairme	Samtskhe-Javakheti Imereti

TABLE 6.1 *(Continued)*

SL. No.	Type	Popular Resort Towns	Regions
6.	**Mountain-Climatic:** Hypoxic environment such as high altitude is reported various health benefits, such as reduced cancer death and Leukemia mortality rates. The available data indicate that residency at higher altitudes is associated with lower mortality from cardio-vascular diseases, stroke, and certain types of cancer (Burtscher, 2014).	Bakuriani	Samtskhe-javakheti
		Bakhmaro	Guria
		Beshumi	Adjara
		Mzetamze	Samtskhe-Javakheti
		Surimi	Shida-Kartli
		Tsemi	Samtskhe-Javakheti
		Medical wellness resort biol	Tbilisi
7.	**Mud Resorts:** Mud is one of the most often used natural materials for preventive, healing, and cosmetic reasons. Its benefits include anti-inflammatory and anti-aging activity as well as moisturizing effects (Spillioti et al., 2016)	Akhtala	Kakheti

By studying the 11 regions outlined in Table 6.1, it emerges that three regions frequently appear to offer therapeutic resources. Those are the priority "1" destinations. We also add the capital Tbilisi due to the most offerings in medical and healthcare facilities for MT purposes.

These regions are:

- Samtskhe-Javakheti;
- Guria;
- Imereti; and
- Tbilisi-the capital.

While there are a number of different options for developing wellness tourism across the regions, logically, those three regions and Tbilisi seem potentially more suitable for further development of wellness tourism across the country and deserve priority as we further narrow down the assets of the regions.

6.3.3 NARROWING THE FIELD

To further narrow down, as shown in Table 6.1, each of the three regions have a number of resort towns that may be considered for renewal. However, we suggest prioritizing one resort town in each region that has the greatest potential and most assets for health and wellness tourism. To identify and select those resort towns, we used four important criteria:

1. **History:** How well-known was the resort in the past? It is much easier to remind people of something with a past glory, and something that they were familiar with, than to promote an unknown destination. Georgia can attract more tourists from neighboring countries by simply reminding them of the past experiences, or people they knew visited those resorts, at least once (post-Soviet countries). For the new markets, like Europe and America, the message will be that the country has experience in the field.
2. **Location/Accessibility:** How easy it is to reach from the international airport. As mentioned earlier, Georgia is blessed with many health and wellness destinations, but easy to reach locations have better opportunities to be developed faster than those far in the mountains. Accessibility is a major factor and motivator in choosing a destination and is well recognized by research.
3. **Diversity of Health and Wellness Assets:** Many resorts offer more than one component of wellness tourism. For this research, resort towns that offer diverse assets, such as climate, sea, sand, mineral waters, and/or thermal waters were selected as potentially better suited for development.
4. **Tourism Attractions:** Possibility for visitors to partake in other tourism activities. As it was already mentioned, most of the MT travelers do not travel alone. This means partners, family members, or other significant ones will need activities to be involved with during a treatment period. Opportunity to see the country's specific culture, history, and nature might play a role while choosing a destination for medical treatment.

Other criteria for branding a destination, such as quality of services will be considered later across the board.

6.3.3.1 *PRIORITY ONE DESTINATION TOWNS*

Based on the above criteria, history, accessibility, diversity of assets, and tourism attraction, we selected three resort towns plus the capital Tbilisi as priority one destinations for medical, health, and wellness tourism as follows:

- Tskaltubo in Imereti region;

- Borjomi in Samtskhe-Javakheti region;
- Ureki in Guria region; and
- The Capital Tbilisi.

Obviously, beyond the attributes mentioned above, history, location/accessibility, diversity of health and wellness assets, and tourism attractions, other key services, such as hospitable attitudes, customer orientation, staff competencies, quality of accommodations, and other services are essential for branding a destination as reported by WelDes (2014) in an article, Developing a Competitive Health and Well-being Destination.

In addition to the above assets analysis, to evaluate the readiness of Georgia for a medical, health, and wellness tourism, we conducted a Delphi analysis to identify specific challenges that the country is currently facing. Further section describes the procedures and findings.

6.4 IDENTIFYING CHALLENGES: RESEARCH AND METHODOLOGY

Now that we had analyzed Georgia's landscape and assessed among the many destinations which ones are best suited for medical, health, and wellness travel, the next step was to identify what challenges Georgia faces to renew and rebrand its medical, health, and wellness tourism. To achieve this goal, we designed a limited-scope Delphi technique. Two-leveled research was conducted to highlight the main problems and challenges of MT in Georgia.

1. **First Round:** An open-ended question, "What are the main problems and challenges for the medical and health and wellness tourism in Georgia?" were sent to three specialists of tourism field currently working in medical-recreational tourism:
   i. **Tamar Maisuradze:** Head of department of tourism product and small-scale infrastructure development;
   ii. **Tamar Kakhidze:** Chief specialist of department of tourism product and small-scale infrastructure development; and
   iii. **Maia Maruashvili:** Director of Ltd. Profkurort—a travel agency that represents the legal successor of the Republican Council of Health Resort's Management since 1948.
2. **Second Round:** For the second stage, respondents were asked to rank-order problems and challenges by their importance.

6.5 RESULTS

We report the results of the Delphi survey in the following subsections.

6.5.1 DELPHI SURVEY ROUND-ONE

During the first stage, all three respondents were asked, via e-mail, to answer the following question: "What are the main problems and challenges for the medical and health and wellness tourism in Georgia? Please list." After receiving their answers, they were categorized into 11 different problems/challenges listed below:

- Lack of connection between medical institutions and resort zones;
- The absence of a state strategy for the development of MT;
- Education (retraining medical personnel, resortology);
- Study of resorts and re-determination of medicinal factors;
- Rehabilitation of resort zones, improvement of infrastructure, modernization according to modern standards;
- Awareness of local and international markets (determining and informing target groups, marketing campaigns, positioning, gaining reliability);
- Chaotic development of resorts;
- The absence of technological methodology;
- Lack of multi-profile services;
- Strong rival in the region (Turkey);
- Lack of international certified clinics.

6.5.2 DELPHI SURVEY ROUND-TWO

In the round-two, we send the list of challenges back to respondents to rank-order them from most important to least important challenges. Results are shown in Table 6.2.

TABLE 6.2 Delphi Round Two Results

SL. No.	Problems/Challenges
1.	Education (retraining medical personnel, resortology)*
2.	Study of resorts and re-determination of medicinal factors.
3.	The absence of a state strategy for the development of medical tourism.
4.	Chaotic development of resorts.
5.	Rehabilitation of resort zones, improvement of infrastructure, modernization according to modern standards.
6.	Lack of connection between medical institutions and resort zones.
7.	The absence of technological methodology.
8.	Awareness of local and international markets (determining and informing target groups, marketing campaigns, positioning, gaining reliability).
9.	Lack of multi-profile services.
10.	Lack of international certified clinics.
11.	Strong rival in the region (Turkey).

**Resortology is a science that studies the effects of various physical properties on the body as a whole and on specific human diseases. Doctors in this area are called balneologists (Portnov, 2018).*

6.5.3 DELPHI SURVEY ANALYSIS

Beyond the educational and technological issues, analyzing the opinions of the experts reveal that Georgia's medical, health, and wellness tourism is fundamentally suffering from lack of planning, joint efforts between public and private sector, and chaotic development.

First four challenges primarily demand government resources. These include: (i) retraining medical personnel; (ii) re-determination of medicinal factors (scientific support for claims); (iii) the absence of a state strategy for the development of MT; and (iv) a chaotic development of resorts.

Other issues, such as lack of connection between medical institutions and resort zones, awareness on local and international markets, lack of multi-profile services and lack of international certified clinics mostly require private sectors engagement or better public, private sector and community coordination and support.

Obviously, this coordination and planning first requires a rigorous analysis of the destinations' strengths, weaknesses, opportunities, and threats (SWOT analysis) using available literature and opinions of experts. In the next section, we perform a brief SWOT analysis for the destinations selected for renewal to further assess their potential for renewal.

6.6 PLANNING FOR GEORGIA'S MEDICAL, HEALTH, AND WELLNESS TOURISM

Developing any destination for MT without planning, adequate investment, and waiting that they will develop on their own, is not the smartest choice. Public sector authorities should work with destinations and develop a clear objective and strategy. To achieve this, we suggest a comprehensive SWOT analysis to identify what attractions and assets are unique and authentic in Georgia? What opportunities are there that will make medical, health, and wellness tourists choose Georgia over other more developed countries? What are some challenges, weaknesses that destinations may face while entering this niche? and how should they be avoided or controlled? In the next segment, we will perform a SWOT analysis to clarify the answers for those questions. Let us first look at Georgia as a country.

6.6.1 SWOT ANALYSIS GEORGIA

6.6.1.1 STRENGTHS

Georgia is one of the safest countries to travel to in the world. With the crime index, it stands next to Belgaum and Nederland (Crime Index for Country, 2019). Georgia is the least expensive destination in Europe. People are friendly and welcoming, hospitality is a part of the Georgian tradition saying that, "guest is from the god." All these are essential factors that attract international tourists. In addition, as was described in previous sections, Georgia's recent developments in MT; existing spa centers and resorts; lenient laws; beautiful natural and environmental natural recourses; rich history and culture; delicious foods; and unique wine-making traditions are the strengths supporting Georgia's medical, health, and wellness tourism renewal.

6.6.1.2 WEAKNESSES

Georgia was a popular destination for Soviet Union countries in the 20th century. After the country gained independence, some resorts were abandoned or used as shelters for refugees, and some just stopped functioning. A number of the tourism destinations started to develop after the Rose revolution in 2003. Because of the laps in development, some of the resorts are far from meeting the standards required to attract tourists from developed countries. According to the Georgian national tourism administration, which is working on a brochure for MT, the country has only 15–17 clinics (excluding dentistry), ready to provide services for foreigners, most of them located in the capital-Tbilisi. And there are only 18 spa and wellness centers, from which 14 are located in cities chosen for this research. Meanwhile, there is only one travel agency working on MT and only one internationally accredited hospital by Joint Commission International (JCI) reported on April 28, 2017 (Health Tourism, 2019). It means that tourists visiting Georgia for medical reasons, mostly find places on their own which can be a barrier while choosing a destination. Other list of weaknesses identified by experts are listed in Table 6.1. We encourage readers to visit Chapter 14 about the important roles that medical travel facilitators play.

6.6.1.3 *OPPORTUNITIES*

"There were about 14 million medical tourists in 2017 and put the revenue and growth numbers in the middle of their ranges, at $50 billion and 20%. This multiplies out to about 60 million medical tourists in 2025, producing revenues of $260 billion, or about $4,333 per person" (DeMicco, 2018). Those are predictions and expectations which are close to reality when looking at how fast MT as a field is growing. If taken into account that travelers' interests are growing towards the small, unique countries like Georgia, it is more likely that with a good promotion and branding the country can regain its niche in a competitive market. There are different market segments, which needs to be studied and aimed, like baby boomers, Chinese growing wealthy class, Middle East countries and neighbor countries, etc.

Turning weaknesses to opportunities is a key factor that provides a road map where to start. Our experts in this study have identified the key challenges earlier in this chapter. A combined effort by public and private sectors can help by focusing on education, studying the resorts and services, as well as medicinal factors, rehabilitating sanatoriums, improving infrastructure, modernization according to recent standards and working on improving communication between a medical and tourism field representatives are some of the main challenges that can be turned to opportunities for Georgian health, wellness, and MT. Ultimately, development of medical and wellness tourism is expected to improve the living conditions of the local population, create jobs and improve the quality of hospitals/resorts.

6.6.1.4 THREATS

Georgian National Tourism Administration (2018) had lunched 2018 *Emotions are Georgia* champagne which was developed, "During 6 months of intense work, a complex Artificial Intelligence tool was detecting, analyzing, and sorting the most genuine human emotions through social media posts of more than 7,000,000 travelers, who shared their impressions about Georgia. These were then used to create the most emotional and accurate guidebook of the country." Though this is a great support for MT, in the 2025 tourism development plan, which describes Georgian tourism 10-year strategy, there is not even a word about MT. This may be a potential threat indicating to the lack of support and commitment from the government. There are also competition threats posed by neighboring countries in the region, such as Turkey, Iran, and others, which makes a serious and timely intervention by government and public sector highly essential.

6.7 SWOT ANALYSIS: FOUR SELECTED DESTINATIONS IN GEORGIA

Now that we have analyzed Georgia's overall feasibility as a destination for medical, health, and wellness tourism, and have prioritized which regions, and resort towns within the region deserve the most investment, we perform a more in-depth SWOT analysis for each of the four destinations: (i) Tskaltubo is a city next to Great Caucasus Mountains on a lowland of Imereti region; (ii) Borjomi is a mountain resort located in Lesser Caucasus Mountains; (iii) Ureki is located next to Black Sea; and (iv) the capital Tbilisi.

6.7.1 TSKALTUBO

Tskaltubo has already been widely popular as a spa and resort town. Documentation of Tskaltubo as a resort town, can be dated back to 1245–1250s. For the first time, in 1782, in the European edition of the "Berlin's Society of Natural Sciences'" collection, information about Tskaltubo's healing waters was published in the German language. Starting in 1809, materials on mineral sources of Tskaltubo was published in both Russian and Western European press (Tsiskarishvili, 2011). The resort Tskaltubo has been officially established as a *balneological resort* since 1920.

6.7.1.1 STRENGTHS

Based on the act that the resort was known during the Soviet Union era, can play a big role in promoting the destination. It is always easier to remind people about the destination than to offer a totally new perception. The bathhouse of Stalin can be a strong selling point for attracting visitors. Because of historical experience, locals are aware how to handle tourists and how to offer various services. Therefore, retraining them will be easier, and their willingness to accept tourists is higher than in other rural areas. Distance from the international airport in Kutaisi is only 22 kilometers-for the most part it is a highway, so the city is easy to access. There are already 8 active sanatoriums/hotels offering a variety of services, and others are under reconstruction. City government is actively working with investors to develop recreation zones. Latest developments include a central park, golf courses, and an artificially made lake. Around the city are various touristic attractions like Prometheus underground caves, Martvili, and Okatse canyons, and Sataplia Nature Reserve. According to World Atlas, Kutaisi, the 2nd largest city in Georgia, is only 7.8 miles away and, "... remains a cultural center with several cultural sites including Kutaisi State Historical Museum, Kutaisi Museum of Sport, and Kutaisi State Historical Archives." The city is served by different transportation networks, including the David the Builder Kutaisi International Airport, rail terminals, and roads. Kutaisi city celebrates a holiday known as Kutaisoba, which is marked by crowds gathering in the central park to celebrate together.

6.7.1.2 WEAKNESSES

The main weaknesses are the same as ones reported by the experts in this study. The quality of services is not at par with the levels of developed countries; there are hardly any clear standards or modern regulations; no world-known or trusted organizations have ever certified any of the sanatoriums; only 8 out of 22 sanatoriums/hotels are currently functional; and, the City's overall image is abandoned and quiet. Fortunately, given the successful history of the resort town and its experience in health and wellness tourism, it is possible to turn the odds to prolific opportunities.

6.7.1.3 OPPORTUNITIES

As soon as sanatoriums become active again, like during the Soviet Union era, it will attract more tourists, and investors will show more interest in development. As reported by Brayman (2018), "Developers are now eying the potential of the sanatorium experience for the international tourist. Tskaltubo is on their radar. The growth of wellness tourism worldwide, and Tskaltubo's proximity to national parks and medieval villages in west Georgia, offers multiple investment opportunities. The consulting company Kohl and Partner plans to construct nine new hotels, an amusement park, renovate medical centers, and develop a river transport system. Vichy, a French company with 13 spa resorts worldwide, is also rumored to be involved." Such stimulating activities will be interesting for younger generations which can encourage family tourism as well.

Having Stalin's bathhouse is one of the most invaluable opportunities for promoting the city. For Gori, 75 miles from Tskaltubo, the city where Stalin's house museum is located, tourism is the main source of income. In 2017 director of museum, Liana Okromelidze, in the interview with *Business Media* mentioned that museum is hosting 500–600 tourists per day, most of them from Israel, Russia, and Iran (Gogberashvili, 2017).

6.7.1.4 THREATS

Renovating sanatoriums require sizable investment, the same applies to developing new hotels and resorts. This might lead to increased prices for services and affordability challenges for locals. In addition, prices can reach to a point when the city will no longer be competitive with competing countries. This means investors entering this field may face longer payback periods to recoup their investment, which is 2–3 years typical for Georgians' character.

6.7.2 BORJOMI

Borjomi is a unique historic resort in South-Central Georgia famous for its mineral water springs, warm climate, and forests. All of this had made Borjomi favorite summer destination for the 18th-century Russian and Georgian aristocracy and gave it the nickname "the Pearl of the Caucasus." (Kvanchilashvili, 2018). Tourism industry development in Borjomi started by building the first hotels and resorts in the mid-1800s (Akhalaia, 2019). Later, after Georgia broke out from Soviet Union, the post-Soviet economic crisis jeopardized resort development. Nowadays, Borjomi has an international reputation for its geothermal spring mineral waters. Borjomi's mineral waters were first mentioned as early as the 15th century, but gained wider notoriety when the daughter of Evgeny Golovin, Viceroy of the Russian Czar in the Caucasus, was cured there. The Yekaterinsky spring and the Yevgeniyevsky springs are named after them, respectively (Adventure, 2018). Borjomi mineral water has become a well-known brand and a calling card for Georgia. The sulfuric water's healing properties are proven to be directly beneficial to the digestive system and metabolism. Today, Borjomi mineral water is sold in over 30 countries (Giorgadze, 2017). Near to the central park of Borjomi, there are number of spa centers, resorts, and mineral spring pools. The place is perfect for families. Central Park has lots of adventure opportunities for kids. The National Park offers hiking, horse riding, biking, snowshoes, cultural, and educational tours. The routes function throughout the year (Figure 6.4) (Agency of Protected Areas, 2018).

FIGURE 6.4 Crowne Plaza Borjomi: SPA Resort.

6.7.2.1 STRENGTHS

Because of the mineral waters, and well-known Borjomi bottle waters, Borjomi is the most famous resort in Georgia right now. It has a history of being a resort and is most developed among the others. Distance from two international airports is short, 135 kilometers from Kutaisi and 170 kilometers from Tbilisi airport and easy to access from both points. Borjomi is also 68 kilometers from the border with Turkey and 120 kilometers from the border with Armenia. Unlike Tskaltubo, Borjomi can offer activities to every age group. Tourists can enjoy national park's various difficulty level routes; Adventures Central Park, horse riding, kayaking, and rafting on mountains' fast river; skiing at resort Bakuriani during winter months; Abastumani Observatory; and many others in and around the city itself. The resort offers spa and wellness therapies, climate therapy, mineral waters, and hot spring.

6.7.2.2 WEAKNESSES

While Borjomi is the nearest resort to different borders and it is easy to access, still there is no straight flight or fast train to the city. Travelers can use two types of public transports, the train, which has not been renovated since the Soviet Union era and takes a long time to cover the short distance; and minibuses, which are not the most comfortable way to travel, especially for visitors who are traveling for medical reasons.

6.7.2.3 OPPORTUNITIES

As Borjomi is the most popular destination nowadays, it has the best opportunity for fast renewal. If all market players like destination management organization (DMO), government, private companies, and local community work together, the city has an ideal location for international as well as domestic MT development.

6.7.2.4 THREATS

After a war in 2008 in a forest nearby Borjomi, in Borjom-Kharagauli national park, there was a huge fire which destroyed a significant part of the forest. After that, in 2017, there was one more forest fire which took almost a week to be extinguished. Georgia used help from Turkey and Azerbaijan to control the fire because the country does not have enough proper equipment to stop fires. Most mountains in Georgia are covered with forests. This can be both an opportunity for visitors, and a threat to the country where summers are getting hotter every year because of global warming.

6.7.3 UREKI

Forest-dimming has started in the west of the country since 1928. The region was known for the cultivation of citrus trees and tea. In addition, for the first time, the benefits of magnetic sand were noticed by quarry workers who were using sand in nearby factories. Soon therapeutic properties of sand drew the attention of the community, medical professionals, and research centers. Artificially created magnetic fields are widely used in medicine. Some people use magnet therapy for treating pain, such as foot, back, or joint pain (University of Michigan Health). In-resort Ureki, you can find naturally created low-intensity magnetic field. Above that, the resort has a subtropical climate, sea, and Sun from May to October.

6.7.3.1 STRENGTHS

Having unique magnetic sand is the biggest strength of the resort. As in most sandy beaches water is not deep, so Ureki is the most popular sea destination for families. It is close to Batumi (60 km) and Kutaisi (117 km) international airports. The city is easy to access by highway, or by a fast, comfortable train. Near to the city is Kolkheti National Park. "National Park is created in order to protect and maintain Kolkheti wetland ecosystems of International Importance" (Agency of Protected Areas, 2018). In the territory of the park, different activities can be found like river and lake boat tours, sports fishing, bird watching, etc.

6.7.3.2 WEAKNESSES

The region Guria, where Ureki is located in the least developed region for international tourism. Given the potential demand for therapeutic properties of sand, there are a few small, privately owned, and old facilities for getting treatment; and, there are no modern supporting facilities and expertise for sand therapy. Hotels nearby hardy offer anything except accommodations.

6.7.3.3 OPPORTUNITIES

Ureki is near the two ports of Batumi and Poti. Both ports can facilitate ferries and small cruise ships bringing more tourists to the resort. Hotels can create packages including updated brochures. The city potentially can become a prime destination for follow-up medical procedures if packaged with Batumi's medical facilities-Batumi is the second big city in Georgia poised for MT. Batumi hosted the 12th international exhibition of tourism and hotel equipment and 2nd international exhibition of MT in 2019. The basic objective of the exhibit was to establish medical innovations and develop tourism fields (Expo Med/Expo Batumi, Travel, 2019).

Ureki is a promising destination for travelers for recreational and health and wellness tourism in Georgia.

6.7.3.4 THREATS

During any planning process, competition in the market should be considered. Threats of similar alternative destinations, such as Bulgaria, which is already well known for its golden sands, are eminent. Without supporting facilities for sand therapy, and a serious planning as well as a campaign for promoting Ureki, the potentials might be lost. In addition, due to the therapeutic effects of sand-therapy, a number of mature sand-therapy products are introduced to the market, such as indoor sand-therapy, breaking the geographical boundaries moving geothermal bath practically to anywhere, including at the spa, resorts, nursing homes, hospitals, convalescent hospitals, beauty salons, and other leisure health clubs. In addition, there are Heat sand-therapy that can control the sands' temperature and keep it in the preferred range (Tian, Liu, and Yan, 2013).

6.7.4 TBILISI

Tbilisi is the capital of Georgia, established in the 5th century (Kutateladze, 1900). "With chaotic yet charming cobblestone streets, dome-shaped bathhouses steaming with sulfuric waters, and crumbling Soviet factories repurposed as hipster hotels, Tbilisi is a study in contrasts… From its medieval fortress walls to its buzzing new luxury boutiques, this East-meets-West city offers something for everyone" (Figure 6.5) (Kamin, 2018).

FIGURE 6.5 Tbilisi Capital of Georgia.

6.7.4.1 STRENGTHS

Tbilisi is the most visited destination in Georgia, according to Georgian National Tourism Administration's annual reports. In 2018, 53.3% of tourists visited Tbilisi (National Statistics Office of Georgia, 2019). The Capital is also the host of different museums, cultural events, festivals for different times of the year, and attractions that will capture the interest of travelers in any age, culture, or interest categories.

Tbilisi is accessible by Tbilisi International Airport. It is close to the Russian border, 167 kilometers, which follows a scenic highway through mountains. It is also easy to access from Armenia-the closest border is within 76 kilometers and a close distance from Azerbaijan, where it is only 62 kilometers to the border.

As we mentioned earlier, Tbilisi has the most hospitals and it is famous for its hot spring waters. Also, Institute of Microbiology, a leader in bacteriophage research, is located in the capital.

6.7.4.2 WEAKNESSES

While Tbilisi is the best option from an accessibility perspective and possess numerous resources, it is also the most populated city in Georgia. Rush hours, lots of traffic, weak regulations for car engines, all of the above have resulted in high levels of air pollution. In May of 2018, Georgian Journal publish an article discussing World Health Organizations' research and informed population about nitrogen dioxide and the harm it can cause for the citizens. This is the biggest challenge and weakness of the city as it is still continuing to grow.

6.7.4.3 OPPORTUNITIES

Tbilisi is surrounded by mountains. Therefore, it is possible to have spa resorts, wellness, and recreational centers, away from the center city, yet a 30-minute drive from touristic attractions and downtown. There are opportunities for medical institutions in the private sector to acquire internationally recognized accreditations, such as JCI, to improve reliability; invest in modern medical equipment; work with medical staff to improve language and service skills; as well as train hospitality and other tourism-related segments to deliver high standards of service and customer experience.

6.7.4.4 THREATS

Tbilisi is the capital city. This means any kind of political or economic crisis will hit here first, which can be a concern. Also, as mentioned earlier, if the government does not act in a timely fashion, air pollution will continue to be a critical threat. As we reported earlier, tourism in Georgia is on the rise, and most visitors come to see Tbilisi. Therefore, this can lead the capital to overtourism and a threat to the local population. The term, overtourism refers to "…the impact of tourism on a destination, or parts thereof, that excessively influences perceived quality of life of citizens and/or quality of visitors experiences in a negative way" (World Tourism Organization, 2018). Therefore, managing the carrying capacity of Tbilisi requires deliberate planning by all stakeholders and the community.

6.8 ANALYSIS

On the upside, a comprehensive review of SWOT analysis of Georgia and four major cities underscores that Georgia overall can revive and revitalize its lost reputation in health, wellness, and to some degree in MT. The country's natural therapeutic resources and landscapes reinforces establishing health tourism using Balneological, Sea Climate, Climate-Balneological, Mineral Waters, Mountain-Balneological, Mountain-Climatic, and Mud Resorts as the regions' signature treatments. These terms will need to be simplified and branded so that the country can easily reach customers, initiate clear value propositions, and ways to sell these valuable health and wellness assets. Additionally, with some substantial investment and planning, Tbilisi can complement the health and wellness tourism and brand the capital as a MT destination, particularly capitalizing on its research and accomplishments in *Phage Therapy*.

On the downside, feedback from experts underscores that there is a fragmented approach to developing Georgia as a health, wellness, and MT. The lack of public and private sector coordination is apparent. The private sector is not well-developed and tourism laws are the same since 1996. To renew health, wellness, and MT requires public sector playing a significant role in fostering development in this field and encouraging private sector for action. Our Delphi survey respondents proposed some key factors that need immediate attention and are necessary from the get-go, such as retraining medical personnel (Resortology), clear strategy and commitment for revamping MT, and reformulating current chaotic development of resorts. All these require governments intervention and public private partnerships. Government efforts for initiating clear strategies and funding research in this field will entice investors for further development.

Coordination of tourism activities and cooperation from the public and private sectors are essential for a seamless medical travel experience. Our analysis further showed that even though there are a number of remarkable medical facilities, travel agencies, and great health and wellness resorts, they are not connected to each other, and more often than not, one does not even know about another's existence. Rehabilitation of resort zones, improvement of infrastructure, modernization, establishing connection between medical institutions and resort zones, creating awareness and branding in local and international markets (PR, marketing campaigns, positioning, etc.), offering multi-profile services, and international certified clinics are challenges which primarily demand not only private sector's engagement, but also must be facilitated by the government support.

Overall, Georgia has an incredible opportunity to enter the MT market and find its unique niche, but only it can be achieved by engaging all stakeholders in deliberate planning.

6.9 PLANNING FOR BRANDING

An honest, credible planning and sustainable brand identity and positioning will help Georgia and the four designated destinations to achieve their health, wellness, and MT aspirations. For more details, please refer to Chapter 13, A toolkit for Branding.

6.10 CONCLUSIONS AND RECOMMENDATIONS

In this chapter, first we addressed Georgia's natural healing assets (e.g., thermal waters, mud, climate, etc.), the natural, environmental landscape, and medical services of Georgia followed by a limited Delphi study exploring the challenges that the country is facing in their ambitions towards becoming a medical, health, and wellness destination. Then we performed a SWOT analysis of Georgia as a whole, to further discover its strengths, weaknesses, opportunities, and threats (SWOT) for becoming a MT destination. The analysis revealed enormous potential and opportunities for this aim, but also there were a number of weaknesses and treats that demand attending, if the country is seriously exploring to become a competing MT destination in the region.

Second, through a comprehensive health and wellness assets identification and analysis, we concluded that three regions with three distinct resort towns plus the capital, Tbilisi, were best situated for initial efforts in investment and development. Prioritizing these four, namely Tskaltubo, Borjomi, Ureki, plus the capital Tbilisi will expedite branding Georgia and the four destinations as potential targets for health, wellness, and MT.

Third, through the SWOT analysis for each destination, we learned more details about each and their assets and capacity within the country. The result showed plenty opportunities to aim at popularizing these destinations and their potential for becoming serious competitors in the area. However, each town faces its own weaknesses and threats as well. The weaknesses and threats we identified were congruent with the challenges and problems propositioned by the experts in the Delphi survey.

Fourth, we proposed a comprehensive two-phase planning steps to involve all stakeholders; identify touchpoints in each destination; develop their brand image; brand positioning strategy; establish priorities; communicate the brand; and develop criteria for measuring success.

Finally, we recommend the Georgian government designate these destinations as medical tourism special economic zones (MTSEZs) to attract domestic and foreign investment. For detail on establishing MTSEZs, please refer to Chapter 6.

KEYWORDS

- **Delphi survey**
- **destination management organization**
- **Joint Commission International**
- **medical tourism**
- **medical tourism special economic zones**
- **strengths, weaknesses, opportunities, and threats**

REFERENCES

Akhalaia, M., (n.d.). History. In. *Blog about Borjomi*.

Asulo, S., (2012). *Taking the Waters: The Sulfur Bathes of Tbilisi*. Retrieved from: https://georgiaphiles.wordpress.com/2012/10/04/taking-the-waters-the-sulfur-baths-of-tbilisi/ (accessed on 14 July 2021).

Borjomi, Georgia Overview, (n.d.). *In Advantour*. Retrieved from: www.advantour.com/georgia/borjomi.htm (accessed on 14 July 2021).

Borjomi-Kharagauli National Park, (n.d.). *In Agency of Protected Areas*. Retrieved from: www.apa.gov.ge/en/protected-areas/cattestone/bordjom-xaragaulis-erovnuli-parkis-administracia (accessed on 14 July 2021).

Brayman, L., (2018). *Finding Peace in GEORGIA'S Last Soviet Spa*. Retrieved from: https://www.curiositymag.com/2018/08/24/georgia-soviet-spa-bathhouse-6-tskaltubo/ (accessed on 14 July 2021).

Burtscher, M., (2014). Effects of living at higher altitudes on mortality: A narrative review. *Aging and Disease, 5*(4), 274–280.

Cappucci, M., Pavliashvili, N., & Zarilli, L., (2015). New trends in mountain and heritage tourism: The case of upper Svaneti in the context of Georgian tourism sector. *Geo Journal of Tourism & Geosites, 15*, 65–78.

Charlier, R., & Chaineux, M. P., (*2009*). The healing sea: A sustainable coastal ocean resource: Thalassotherapy. *Journal of Coastal Research, 25*(4), 838–856.

Crime Index for Country, (2019). In NUMBEO. Retrieved from: www.numbeo.com/crime/rankings_by_country.jsp (accessed on 14 July 2021).

DeMicco, F. J., Cetron, M. J., & Davies, O., (2018). *Strategic Management for Hospitality & Travel: Today and Tomorrow* (Vol. 453, p. 163). N.P.: Kendall Hunt Publishing.

Expo Med/Expo Batumi, Travel, (2019). Retrieved from: https://www.showsbee.com/fairs/ExpoMed-Batumi.html (accessed on 14 October 2021).

Facts & Statistics, (2014). *In Medical Tourism.* Retrieved from: www.medicaltourism.com/Forms/facts-statistics.aspx (accessed on 14 July 2021).

Frank, A., (2012). The air cure town: Commodifying Mountain air in alpine central Europe. *Central European History, 45*(2), 185–207.

Georga, M., (2018). Georgia's developing medical tourism ambition. *International Medical Travel Journal.* Retrieved from: https://www.imtj.com/articles/georgias-developing-medical-tourism-ambition/ (accessed on 14 July 2021).

Georgian National Tourism Administration, (2016). *Georgia Your Spa & Wellness Destination.* Tbilisi, Georgia.

Georgian National Tourism Administration, (2018). *Georgian National Tourism Administration Launch AI to Help Create Accurate Guidebook Based on Tourist Emotions.* Retrieved from: https://www.prnewswire.com/news-releases/georgian-national-tourism-administration-launch-ai-to-help-create-accurate-guidebook-based-on-tourist-emotions-893117247.html (accessed on 14 July 2021).

Georgian National Tourism Administration, (2018). *Georgian Tourism in Figures* (2017th edn., Vol. 48, pp. 11–12).

Giorgadze, L., (2017). *Dgeidan Borjomi Aluminis Qilitac Gamova [From Now on Borjomi Will Produce Water in Aluminum Cans Too].* In Samkhretis Karibche *[South Gate].* Retrieved from: www.sknews.ge/index.php?newsid=13419 (accessed on 14 July 2021).

Godwin, N., (2012). *There's Untapped Opportunity for Travel Agents in Medical Tourism.* Travel Weekly. Retrieved from: https://www.travelweekly.com/Travel-News/Hotel-News/Theres-opportunity-for-travel-agents-in-medical-tourism (accessed on 14 July 2021).

Gogberashvili, E., (2017). *Stalinis Saxl-muzeums Dgeshi 600-mde Ucxoeli Vizitori Stumrobs [About 600 Foreign Tourists Visit Stalin's Museum Daily].* In Business Media Georgia. Retrieved from: www.bm.ge/ka/article/stalinis-saxl-muzeums-dgeshi-600-mde-ucxoeli-vizitori-stumrobs/12405 (accessed on 14 July 2021).

Health Tourism, (n.d.). *The History of Medical Tourism.* Retrieved from: https://www.health-tourism.com/medical-tourism/history/ (accessed on 14 July 2021).

Kagaku, T., (2001). Medical balneology and climatology in Europe. *Japanese Journal of Physical Fitness and Sports Medicine, 50*(1), 7–18.

Kamin, D., (2018). *36 Hours in Tbilisi.* New York Times. Retrieved from: https://www.nytimes.com/2018/06/07/travel/what-to-do-in-tbilisi.html (accessed on 14 July 2021).

Kutateladze, A., (1900). *An Overview of Georgian History* (Vol. 1, p. 38). Tbilisi: St'amba ts'ignebis gamomtsemel kartvelta amkhanagobisa. From: www.dspace.nplg.gov.ge/bitstream/1234/191351/1/Konspeqti_Saqartvelos_Itoriidan.pdf (accessed on 14 July 2021).

Kvanchilashvili, E., (2018). *Interview with Andrey Vlasov, General Manager of Crowne Plaza Borjomi.* Forbes Georgia. Retrieved from: www.Forbes.ge/news/4535/Where-Hospitality-Meets-The-Nature (accessed on 14 July 2021).

Medical Tourism to Georgia, (n.d.). *In Health Tourism.* Retrieved from: https://www.health-tourism.com/medical-tourism-georgia/ (accessed on 14 July 2021).

Portnov, A., (2018). *Health Resorts*. Retrieved from: https://iliveok.com/health/health-resorts_105218i16051.html (accessed on 14 July 2021).

Rogers, K., (2016). *Medical Tourism*. In Encyclopedia Britannica. Retrieved from: www.britannica.com/science/medical-tourism (accessed on 14 July 2021).

Spilioti, E., Vargiami, M., Letsiou, S., Gardikis, K., Sygouni, V., Koutsoukos, P., Chinou, I., et al., (2016). *Biological Properties of Mud Extracts Derived from Various Spa Resorts*. Retrieved from: https://www.researchgate.net/publication/305480946_Biological_properties_of_mud_extracts_derived_from_various_spa_resorts (accessed on 14 July 2021).

Summers, W. C., (2017). Félix Hubert d'Herelle (1873–1949): History of a scientific mind. *Bacteriophage, 6*(4), e1270090. doi: 10.1080/21597081.2016.1270090.

Tbilisi Among Dangerous Cities in Europe Where Air Pollution is High, (2018). Retrieved from: www.georgianjournal.ge/society/34468-Tbilisi-among-dangerous-cities-in-Europe-where-air-pollution-is-high.html (accessed on 14 July 2021).

Tian, X., Liu, J., & Yan, M., (2013). SWOT analysis and countermeasures research on the development of Turpan sand-therapy tourism resources. *Advanced Material Research*. doi: 10/4028/www.scientific.net/AMR.734-737.3295.

Tsiskarishvili, O., (2011). *Tskaltubos Unikaluri Mineraluri Tsklebi [Unique Mineral Waters of Tskaltubo]*. In Ekofact.

University of Michigan Health, Retrieved from: https://www.uofmhealth.org/health-library/aa141392spec (accessed on 22 September 2021).

Verulava, T., & Jorbenadze, R., (2018). *Medical Tourism in Georgia: Current Barriers and Recommendations*. Retrieved from: https://www.researchgate.net/publication/324528413 (accessed on 14 July 2021).

Verulava, T., (2016). Medical tourism and Georgia: Specifics of problems and recommendations. *Medical Tourism: Political, Social-Economical, Educational and Medical Problems*.

Weisz, G., (2001). Spas, mineral waters, and hydrological science in twentieth-century France. *Journal of History and Science Society*. Retrieved from: https://www.journals.uchicago.edu/doi/abs/10.1086/385278 (accessed on 14 July 2021).

WelDes, (2014). *Developing a Competitive Health and Well-Being Destination*. Retrieved from: https://globalwellnesssummit.com/wp-content/uploads/Industry-Research/Global/2015-developing-a-competitive-health-and-wellbeing-destination.pdf (accessed on 14 July 2021).

Wellness Tourism Worldwide, (2011). *4WR: Wellness for Whom, Where and What? Wellness Tourism 2020*. Retrieved from: https://globalwellnesssummit.com/wp-content/uploads/Industry-Research/Global/2011-wellness-tourism-worldwide-wellness-for-whom.pdf (accessed on 14 July 2021).

World Tourism Organization, (2017). *Tourism Highlights* (2017th ed., Vol. 16, p. 7).

World Tourism Organization, (2018). *'Over Tourism'? - Understanding and Managing Urban Tourism G Growth Beyond Perceptions: Executive Summary* (p. 4).

Zhvania, P., Hoyle, N. S., Nadareishivili, L., Nizharadze, D., & Kutateladze, M., (2017). Phage therapy in a 16-year-old boy with Netherton syndrome. *Frontier in Medicine*. Retrieved from: https://www.frontiersin.org/articles/10.3389/fmed.2017.00094/full (accessed on 14 July 2021).

CHAPTER 7

MEDICAL TOURISM DESTINATION SUCCESS FACTORS: THE CASE OF THREE CHINESE CITIES

ALI A. POORANI and NAN JAING

Alfred Lerner College of Business and Economics, Department of Hospitality Business Management, University of Delaware, 14 W., Main Street, Newark, DE–19716, USA, E-mail: poorani@udel.edu (A. A. Poorani)

ABSTRACT

In this chapter we use three-step city branding model to analyze China's medical travel history and the country's ambitions to become one of the leading medical travel destinations. To achieve this goal, we analyze the propensity of three major destinations, namely Beijing, Shanghai, and Sanya dispersed across mainland China, using a cluster mode. These three destinations are China's promising entry points each with unique characteristics and underway to deliver the country's ambitious MT goals. Based on our analysis, we offer several specific suggestions for each geographical location. We predict China by bringing various players, public and private, in each of these gateway cities can deliver first-class MT clusters and meet its domestic and international medical travel goals.

7.1 INTRODUCTION

Medical tourism (MT) is evolving. In Part 2 of this book, Chapters 3–5 demonstrated that healthcare receivers are becoming more and more diversified. They could be international or domestic patients; they could be approved by an insurance company; or, they could be employees of a company traveling as a part of their covered benefits. The patients' expectations are also changing; demanding higher quality services and a patient-centric approach to treatments. Chapters 16–19 address *Patient-Centric Approach: Managing Medical Traveler Expectations.* In this case study, we explore the changing landscape of China's MT.

In 2013, the authors studied China's ambitions and plans for becoming one of the top MT destinations in Asia (Jiang and Poorani, 2017). We began with conducting a *Three-Step City Branding* to identify if China could engineer a MT destination and, if so, what specific destinations should be cultivated first to achieve a clear and successful outcome. In this case analysis, we explore what the study concluded *then*, and where China is *now*. The following section explains the procedures we used to pinpoint China's readiness to becoming a renown MT destination.

7.2 THREE-STEP CITY BRANDING

The three-step model incorporates the main components of effective marketing strategy, namely a solid strategic planning; clear value positioning and differentiation; as well as detailed action plans with timelines and accountabilities.

7.2.1 STRATEGIC PLANNING

The guiding principles to follow in this model were to conduct market analysis and assess needs. Performance of competitors and benchmarking destinations were also taken into consideration (Metin, 2004). It also included a Strategic SWOT analysis, a comprehensive analysis of the market development, competitor performance, destination attributes, and consumer behavior.

7.2.2 VALUE PROPOSITIONS

Tourists travel abroad for medical treatment, not only for affordability and availability, but also more for value propositions. "Customers don't really buy what you do, but buy why you do it (Simon, 2013)." It is not simply building, branding, and waiting; it is critical to know what is your special value proposition, the unique selling point of each destination, the participants, and the community. Only if you know why you build it, will your customers know what they come for.

7.2.3 ACTION PLANS

The action plans included the quality of experiences including pre-arrival, encounter, and post-treatment activities of patients and their families. Quality issues were at the core of our analysis of China's MT ambitions, whether it is the quality of medical treatment or quality of tourism arrangements. Quality management aims for sustaining an ongoing improvement of the standard of facilities, products, and services provided (Kozak and Baloglu, 2011).

Using the three-step model, our analysis first probed into China as a whole (country) followed by three pioneer Chinese cities, Beijing, Shanghai, and Sanya in the Hainan Provence. These three destinations were identified using a detailed SWOT (strengths, weaknesses, opportunities, and threats) analysis as well as Health Care Cluster Model for potential entry points into the MT market.

7.3 CHINA MEDICAL TOURISM (MT) THEN AND NOW

Then, in 2013, global business travel spending was expected to expand in the next several years, with China leading the charge. According to statistics by Travel China Guide, China was projected to experience double-digit annual growth, reaching tourism revenues CYN 423 billion by 2017, a 15.9% increase over 2016. Since then, tourism revenues have continued to grow to reach CYN 513.9 billion in 2019, 8.2% growth over 2018. The ongoing growth of tourism in China was a sign of the country's aptitude for inbound and outbound tourists.

Can China engineer a MT destination with this promising outlook? To answer this question, we present a brief overview of China's overall SWOT:

1. **Strength:** China's MT assets, such as Traditional Chinese Medicine (TCM) and Chinese herbal medicine (CHM), in the forms of naturopathy (natural therapy), acupuncture, massage, footbath, and herbal beauty, have a solid foundation in Mainland. A stable political

and economic environment, and friendly local people are additional advantages as an international tourism destination. Medical tourists visiting China enjoy relatively low medical and lodging costs and easy access to relevant information. Besides the domestic economic growth, which provides powerful support for business development and international investment, China has all-scale accommodations, convenient public transportation, world-class medical teams, unique oriental culture, rich history, and tons of attractions. In 2013, there were 15 JCI (Joint Commission International) Certified hospitals in China, now, the number of JCI certified hospitals in Chia has reached 88, which is an indication of commitment to high quality hospitals.

2. **Weaknesses:** Since China is a relatively new player in the MT industry, there is still a lack of a brand identity that is optimal for medical/wellness tourism travelers, differentiating it from competitions (Poorani and Poorani, 2013). According to MT index (MTA, n.d.), China ranks 23 among the 41 top MT destinations. The main question is, can the country establish itself as a top-ranking MT destination?
3. **Opportunities:** Per our 2013 study, many Mainland Chinese chose Taiwan and South Korea for health care and cosmetic surgeries. Additionally, there were a growing number of patients who shopped around for medical care within China. Yet, they were forced to seek care elsewhere due to a shortage of quality facilities to support this demand. Coupled with an advancing retirement age, and increased patient dissatisfaction with public hospitals; China was faced with an urgent need to develop MT economic free zones within its own borders, developing advanced services, efficient infrastructure, and a strategy that emphasizes TCM and modern Western Medicine techniques.
4. **Threats:** Situated near Thailand, India, Malaysia, and Singapore, Mainland China faces a stark competition, primarily due to the fact that these destinations enjoy government funding and other auxiliary support. Though China cannot compete with its neighbors on cost, its booming economy, soaring travel, both domestic and international, and established infrastructure, MT in China can be viewed as one of the most underpenetrated segments in China. Our 2013 market study of China led us to the following conclusions:

    i. China had the potential to become a top MT destination in Asia;
    ii. Beijing, Shanghai, and Sanya in the Hainan Provence were identified as the top three entry points to lead MT in China and model destinations in the country.

Beijing and Shanghai had already started building medical teams, launching associations, and developing related systems. Sanya, by contrast, was a hot destination for domestic travel, yet needed more ideas and actions on branding itself as a MT destination. In the following section, we explain the metamorphosis of MT in the three destinations.

7.4 METAMORPHOSIS OF THE THREE REPRESENTATIVE DESTINATIONS IN CHINA

Beijing, Shanghai, and Sanya are dispersed across the Mainland, in Northeastern China, Southeastern China, and South China, respectively, and all have achieved worldwide recognition today. Figure 7.1 shows the geographical position of the three destinations.

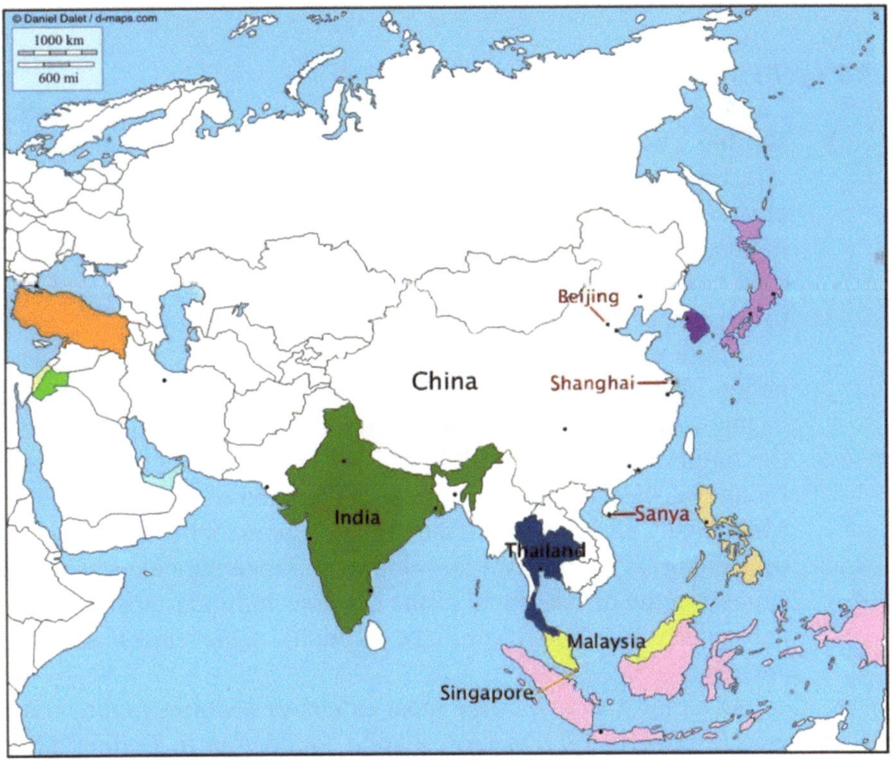

FIGURE 7.1 Location map of three Chinese medical tourism destinations.

7.5 CLUSTER MODEL

According to the MT Association's definition, "A Healthcare Cluster is generally an independent organization of hospitals, clinics, medical professionals and the government in a specific city, state, or region. A healthcare cluster is funded by all the participants in the healthcare cluster and represents the interests of all the members; the cluster may also be supported by government funding." (MTA, 2012) For instance, Swiss Health is the representative body for 27 leading hospitals, clinics, and resorts that provides marketing support, facilitates travel plans, and makes recommendations for international MT seekers in their exploration of Swiss MT. A more comprehensive discussion of building cluster models is explained in Chapter 13. All the three destinations' successes in building reputable medical destinations used some degree of public and private participation to reach their current prominence. What is the current status of medical tourism in these three destinations?

7.5.1 BEIJING

- **Beijing Then:** Based on our 2013 analysis, there was a need for an effective facilitator for Beijing to establish a well-defined healthcare cluster and assist in branding itself as an international medical destination. Contemporary training was needed for hospital and hotel staff to learn how to treat Medical Tourists, with specific strategies to target specific market needs.

  While we were analyzing Beijing's readiness as a top MT destination in 2013, the local government and associations had already began working together to establish a collaborative goal and plan for the future. In December 2012, Beijing Health Culture and Creative Industry Association and State Administration of TCM, and Beijing Municipal Commission of Tourism Development held a symposium on Beijing TCM Tourism Development. Governors expressed their support in the development of the Chinese cultural tourism industry, particularly TCM, by rewarding exceptional achievements and travel agencies.

  As a city with one of the most extensive histories in the modern world, TCM and CHM have a solid foundation in Beijing. Given Beijing's significance in the international market, TCM, and CHM are recognized around the world for their wide applications, high efficiency, minimal side effects, and low cost. Besides, TCM can be used as a tool to fight obesity and carry out anti-aging treatments for seniors.

In terms of unique tourism resources, Beijing was the cultural center of China, with an internationally recognized artistic community encompassing music, drama, dance, craft, sculpture, museums, libraries, and heritage architecture. All of these were early signs of Beijing's readiness to launch a formidable task.

- **Beijing Now:** Today, Beijing has established itself as a premier MT destination in China. Among the many hospitals, five reputable and JCI accredited hospitals, such as Jian Gong Hospital, Beijing United Family Hospital, Beijing Yanhua Hospital, Puhua International Hospital, ReLife International Medical Center now are offering a variety of modern and traditional treatments, in various languages including, Chinese, English, Russian, Arabic, Japanese, Korean in order to accommodate the needs of inbound and outbound medical tourists. In addition, as of 2018, there were 16 medical tourism facilitators (MTFs) in Beijing (Yuan et al., 2018).

  Just recently, in July 2019, the 14th China International MT exhibit attracted more than 1,200 world-renowned hospitals and medical institutions from 38 countries and regions to Beijing, receiving strong support from the state. Events like this is hoped to put Beijing on the frontline of MT.

7.5.2 SHANGHAI

Shanghai is the most developed city in Mainland China, typified by a fascinating mix of East and West. As an international city, Shanghai has the charm and power to introduce the latest technology, promote world corporations and draw talented people to work there. "Creative people, in turn, don't just cluster where the jobs are. They cluster in places that are centers of creativity and also where they can afford and like to live (Florida, 2002)." Shanghai is such a place, full of chances and ideas, and gathering place of universities and global enterprises. Give Shanghai a chance, then she will wow the world.

Shanghai is China's largest metropolis city with a number of well-recognized tourism assets, such as business, media, fashion, and technology. Among other tourism assets are a wide array of accommodations, multiple transportation options, and the support of government agencies who are promoting it as a leading health tourism destination.

- **Shanghai Then:** In addition to competition coming from neighboring countries and other regions from within China, such as Beijing and Hainan, in our 2013 analysis, Shanghai needed a clear positioning and an effective information-sharing cluster in partnership with local government, MT association of China, and other important stakeholders. It also needed additional modern hospitals with multilingual staff, multicultural approach to help better serve internal guests. Next step was to develop more cooperating tourist centers overseas, carry out more foreign customers-targeted campaigns to broadcast the city programs, and meanwhile keep eyes on post-service standards to maintain customer relationship.
- **Shanghai Now:** The municipal government has proposed to enhance the urban landscape, improve tourist facilities, make the Huangpu River tour a flagship tourism projects, including developing MT and customer-friendly services to attract millions of inbound travelers. For example, Shanghai International Medical Center (SIMC), opened in 2014 and managed by Parkway Health, uses JCI standards and modern services, bestows a global vision for healthcare. This comprehensive hospital located in the Shanghai International Medical Zone of Pudong New District; is 15 km away from the Lujiazui Finance and Trade Zone, and 17 km from the Pudong International Airport (http://www.1uchina.com/en.php?/company/id-40/index.html).

  SIMC public and private project is a Cleveland Clinic Connected medical center providing value-based, patient-centric healthcare, that will enhance patient experience and improve recovery times while attracting and retaining the best medical professionals from across China and the rest of the world (Figure 7.2) (https://worldarchitecture.org/article-links/ecncm/foster-partners-releases-design-for-a-greenery-hospital-in-shanghai.html).

In additions, according to Gresham Smith, "In an effort to position Shanghai as a world-class medical destination, the Chinese government launched a design competition challenging architects across the globe to create a blueprint for a new international medical city that would not only attract medical tourists, but also entice expatriates and Chinese nationals alike."

The first phase of the project, which represents an investment of 10 billion yuan ($1.46 billion), is expected to be completed by 2020. Shanghai International Medical City, a medical campus consisting of 77 acres and 5 million square feet of healthcare spaces to accommodate 3,000 beds, five hospitals and specialty clinics is proposed by Gresham Smith, which will include a park-like setting, sustainable design, the most energy-efficient, reusing rainwater and green roofs, as well as the most recent healthcare esthetic designs. The proposed facility will allow the site's private and public medical groups to share diagnostic equipment, eliminating the expense of each hospital and clinic providing its own (https://www.greshamsmith.com/project/shanghai-international-medical-city/).

FIGURE 7.2 Shanghai's New Hong Qiao International Medical Center (SIMC) campus. *Source:* https://newsroom.clevelandclinic.org/2019/09/20/combining-nature-and-technology-luye-medical-and-cleveland-clinic-join-to-build-a-future-hospital-in-shanghai/

Both Shanghai projects offer ample green spaces that can contribute to short recovery times, better clinical outcomes for patients, and a comfortable workplace for medical staff as well as visitors. Once these plans are completed, Shanghai can certainly claim a major destination for MT.

7.5.3 SANYA IN HAINAN

The third destination analysis included Sanya in Hainan Provence. Sanya is a tropical coastal city and the second largest on the Hainan Island. Hainan was the first International Tourism Island in China and had been given the nickname "Hawaii of the Orient." Sanya was blessed with strong government support, an easy Visa policy for group tourists, and two tax-free shopping areas. The government of Hainan intended to expand the province's MT industry by introducing tax incentives for international investment. Sanya was also at the forefront of China's hotel boom, with luxury resort and hotel facilities everywhere, providing a good foundation for the hospitality sector in the MT cluster.

On the downside, based on some of the observations made in 2013 study, the hospitals were relatively low tech with a few renown surgeons; there were no JCI accredited hospitals; there were language barriers in the healthcare system to accommodate various international visitors; and, Sanya was not recognized as an established and unique brand for MT.

On the upside, given the city's healthy living philosophy, strong lodging presence and attractive beaches, the opportunity for building a MT cluster seemed promising to make Sanya an internationally recognized MT brand.

- **Sanya and Hainan Now:** Sanya's healthcare sector has grown rapidly in recent years. In 2017, Yi Shan Li He Rehabilitation Hospital was opened at the Leaguer MOHO Plaza. As a value propositioning, the hospital offers TCM. Leaguer MOHO Plaza (see Figure 7.3) is a multipurpose complex that offers accommodation, shopping, dining, entertainment, and healthcare all at the same place. The rehabilitation hospital provides convenience and all other auxiliary services demanded by both domestic and international patients seeking TCM treatment.

FIGURE 7.3 Leaguer MOHO Plaza.

The plans for the major development of the island's modern MT destination began in 2013, when an experimental area for international MT was created by the government in a nearby village in the island's east coast known as, Boao Lecheng International MT pilot zone (see Figure 7.4). Some of the benefits of MT economic free zones are flexible policies, new regulatory models and management systems for investment, financing, fiscal, and taxation, financial innovation, entry, and exit, to stimulate investment. This MT free economic zone would provide the latest medical services, medical equipment with standards compatible to those in Europe, Japan, and the USA. From all the evidences gathered, not only Sanya, but the whole Hainan Island is destined to be the premier MT destination in China.

According to China Daily, (https://www.chinadaily.com.cn/a/201901/08/WS5c33dfa6a31068606745f5a9.html) the National Development and Reform Commission's focus will primarily be on medical treatment, medical cosmetology, rehabilitation, and health promotion; and, introducing Hainan to internationally renowned institutions, insurance providers and medical tourism intermediaries (MTIs) (facilitators) to build a high-end foundation for the international MT market. Main specialties will include cancer treatment and rehabilitation, fertility therapy, cardiovascular, and cerebrovascular disease treatment and rehabilitation, environmental healing, and sports recuperation, TCM and spiritual retreat, according to China Daily.

FIGURE 7.4 Aerial photo taken on March 20, 2019, shows the Lecheng international medical tourism pilot zone in Boao town of Qionghai city, south China's Hainan province.
Source: http://www.xinhuanet.com/english/2020-04/12/c_138969752.htm. © 2000-2022 XINHUANET.com

China Daily also reported, "27 medical projects have been completed or are under construction in the pilot zone, and another 38 projects have passed medical technology appraisals." By 2030, the zone is expected to be a world-class destination for international MT and innovation platform for medical science and technology.

7.6 CONCLUSION

Beijing, Shanghai, and Sanya were three promising entry points that have started MT in China. To provide quality services, China needed to bring together various players in each touchpoint, develop detailed analyzes, and solicit not only commitment but also enthusiasm from all involved to create detailed action plans and new milestones. Our recent analysis supports that China is underway to deliver its MT goals. With their distinctive characteristics, all three have created advanced infrastructure, international insights, and other basic tourism assets and accolades that international MT demands. They are model cities for other Chinese municipalities.

As a gateway to China, these three destinations, could communicate clear value propositions, and commit to providing excellent services at all touchpoints. The unique geographical locations of these three destinations will allow China opportunities to create its first-class MT clusters to meet its domestic needs and go beyond by drawing patients across borders.

Finally, in our assessment, China could not achieve prominence in such scales without incorporating private and public resources, creating successful international and national partnerships with an eye on quality and detailed planning.

KEYWORDS

- **Chinese herbal medicine**
- **Joint Commission International**
- **medical science and technology**
- **Shanghai International Medical Center**
- **strengths, weaknesses, opportunities, and threats**
- **traditional Chinese medicine**

REFERENCES

Florida, R., (2002). *The Rise of the Creative Class, Basic Books*. New York, NY.

Jiang, N., & Poorani, A., (2017). Medical tourism today and in the future for China. In: DeMicco, F. J., (ed.), *Medical Tourism and Wellness: Hospitality Bridging Healthcare (H2H)*. Apple Academic Press.

Medical Tourism Magazine. An Overview of the Benefit Market in China. https://www.magazine.medicaltourism.com/article/an-overview-of-the-benefit-market-in-china (accessed on 14 October 2021).

Poorani, A., & Ryan, P., (2013). *The Role of Hotel Industry in Medical & Wellness Tourism.* Medical tourism magazine. Retrieved from: https://www.magazine.medicaltourism.com/article/future-wellness-tourism-industry-preparing-hospitality-students (accessed on 14 October 2021).

Yuan, J., et al., (2018). *Investigation and Analysis of the Content of China Medical Travel Facilitator Websites*. Retrieved from: http://www.yiyaodaobao.com.cn/EN/abstract/abstract1096.shtml (accessed on 14 July 2021).

CHAPTER 8

A GEO MAPPING PROJECT FOR MEDICAL TRAVEL AND WELLNESS: OPPORTUNITY TO ENHANCE TOURISM REVENUE IN COLORADO

FREDERICK J. DEMICCO,[1] SOPHIA LINN,[2] and AKIKO YOSHIMI[3]

[1]*Visiting Professor, the University of South Florida-Sarasota-Manatee; and Visiting Scholar, the Department of Human Dimensions of Natural Resources at Colorado State University, Fort Collins, Colorado; and Aramark Chair Emeritus, the University of Delaware, Delaware, USA, E-mail: FDeMicco@udel.edu*

[2]*Assistant Director, Geospatial Centroid at Colorado State University, Colorado, USA*

[3]*Colorado State University, Colorado, USA*

ABSTRACT

Medical tourism (MT) is one of the hottest niche markets in the hospitality and tourism industry. Through 2025, MT is expected to grow by 20% per year, reaching 60 million people and a market value in the neighborhood of $260 billion (DeMicco, Cetron, and Davies, 2018).

The researchers gathered data about health and wellness facilities throughout the state. These data, including facility type, addresses, and types of services, were then geocoded so that they could be mapped and shared online, thereby making them more discoverable to potential health seekers.

There is an opportunity to increase tourism revenue in Colorado and in other states in the U.S. from Medical Tourists and health and wellness seekers. In fact, Colorado has an extensive network of high quality medical, health, and wellness providers throughout the state. Therefore, Colorado can be a destination of choice for medical tourists and medical travelers via interstate, intra-state, and those also traveling from international destinations. This study attempted to identify some of the potential medical, health, and wellness destinations around the state. The goal is to expand the use of these facilities in Colorado, thereby enhancing overall Colorado tourism.

8.1 INTRODUCTION TO THE CONCEPT OF MEDICAL TOURISM (MT) AND MARKET PROJECTION

Medical, health, and wellness tourism is one of the hottest niche markets in the hospitality and tourism industry. The Organization for Economic Cooperation and Development (OECD) estimates that nearly 50 million patients travel abroad for medical services each year.

Revenue estimates for MT are variable. In 2016, one analyst placed the market for medical tourism (MT) at only $5 billion a year. We believe this is considerably low. In 2016, VISA, and Oxford Economics carried out a major study of MT. The first version of their report put annual revenues at $349 billion worldwide. A hasty revision brought the figure down to around $50 billion per year. In mid-2017, the latest version available online suggests that MT brings in global revenues of $100 billion annually. This appears to be the highest credible number.

This report also projects revenue growth of 25% per year, another number we believe marks the high end of credibility. Starting at $100 billion and expanding at that pace, the MT market could be worth about $600 billion in 2025. This seems very optimistic, but not totally impossible.

Tentatively, then, we will cast our vote with Patients Beyond Borders, an online clearinghouse of information about MT. They estimate that 12 to 14 million people travel internationally for care each year. Only one in 10 go for medical treatment, and 70% travel abroad for cosmetic surgery or other elective procedures. The remainder seek a variety of "wellness" services like spa visits, supervised exercise, and dieting. The company estimates that medical tourists would spend an average of $3,800 to $6,000 per visit, including direct costs for medical services, accommodations, and local and international transportation. This puts revenues between $40 billion and $60 billion per year. The firm's best guess is that the market is now growing by 15 to 25% annually.

For a quick analysis, let us accept that there were about 14 million medical tourists in 2016 and put the revenue and growth numbers in the middle of their ranges, at $50 billion and 20%. This multiplies out to about 60 million medical tourists in 2025, producing revenues of $260 billion, or about $4,333 per person. We find these numbers much easier to accept. Until better data becomes available, we will make this our official forecast: Through 2025, expect MT to grow by 20% per year, reaching 60 million people and a market value of about $260 billion (DeMicco, Cetron, and Davies, 2018).

8.2 WHY GET HEALTHCARE ELSEWHERE?

The United States (U.S.) has some of the best medical care in the world. A benefit of MT is comfort. A private room in a clinic or hospital with the amenities of a four-star hotel is a lot easier to stay in than the bare two- and four-bed cells typical of American hospitals.

Ordinary sight-seeing also is a factor. Why spend time in a hospital at home when you can combine your procedure with a good vacation at lower cost? Almost 90% of patients or their companions take advantage of the standard attractions while traveling for medical care.

There are other compelling reasons to go abroad, especially for critical care. In some regions, state-of-the-art medical facilities are hard to come by, if they exist at all. For this reason, patients throughout the Middle East regularly travel to Jordan or Dubai, Asia, and even the U.S. for complicated surgery; close to 1 million Indonesians make a yearly trip to Malaysia or Singapore for routine matters as well as more challenging treatments; and 100,000 Russians annually find care in Israel, Turkey, and the U.S.

In other countries, the public health system is overburdened, and needed care can be delayed, particularly for difficult, uncommon, or expensive procedures. In Britain, regulations require the National Health Service to provide surgery no more than 18 weeks after diagnosis. In 2016, more than 193,000 patients waited longer for their medical procedure. In Canada, the waiting list for neurosurgery in 2016 averaged 47 weeks after referral by a general practitioner. Esoteric procedures can take even longer. Waiting times of three months or more are common for stem-cell transplants needed for more than 80 diseases, including several blood cancers. In one well-publicized case from 2016, an 18-year-old girl with acute myeloid leukemia died while waiting for a transplant 8 months after learning that several matching donors were available. Until May 2017, cancer patients who had relapsed after chemotherapy were barred from stem-cell transplants that might have saved their lives.

Stem cells bring up another reason to leave home for care. Some procedures that are either banned or not yet approved may be available in other countries for patients who want and can afford them. Canadians with multiple sclerosis, for example, must leave the country for treatment of abnormal blood drainage from the brain and spinal cord, which a few specialists believe is a factor in the disease. And in the U.S., stem cell transplants are accepted only for patients with certain cancers and a few other disorders. Despite this, nearly 600 clinics across the country offer them for purposes ranging from repair of spinal cord injuries to breast augmentation. However, insurance rarely covers these procedures, and costs of $25,000 or more can make it worth searching for less expensive providers in other countries.

8.3 MEDICAL TOURISM (MT) MAPPING FOR THE STATE OF COLORADO

Colorado has an extensive network of medical, health, and wellness providers. Thus, in addition to its many other tourism opportunities, Colorado can be a destination of choice for medical tourists and medical travelers inter-state, intra-state, and from international destinations. This mapping study identified some of the potential medical, health, and wellness destinations around the state of Colorado with the goal of expanding the use of these medical, health, and wellness facilities in Colorado by making their locations discoverable online. This in turn could enhance access to the facilities, thereby increasing Colorado tourism revenue.

For this study, we partnered with a research librarian at Colorado State University as well as with staff from the Geospatial Centroid, a GIS and mapping service center also in the library. The research and data gathering were conducted by during an 8-week period during fall 2018.

The intention of the project was to gather information, in a semi-structured format, about health and wellness facilities within the state, and then to map these facilities by geocoding their addresses so they could be viewed in a digital map. The list of facilities could potentially include the following; medical centers, hospitals, specialized clinics, spas, medical marijuana (MMJ) centers, etc. The research design examined geographic portions of the state. These included the northeast, the southeast, northwest, and the southwest rural areas. In addition, we gathered information for the Metro Denver region; the Colorado Springs metro area; and also focused on the Fort Collins, Loveland, and Greeley areas (Figure 8.1).

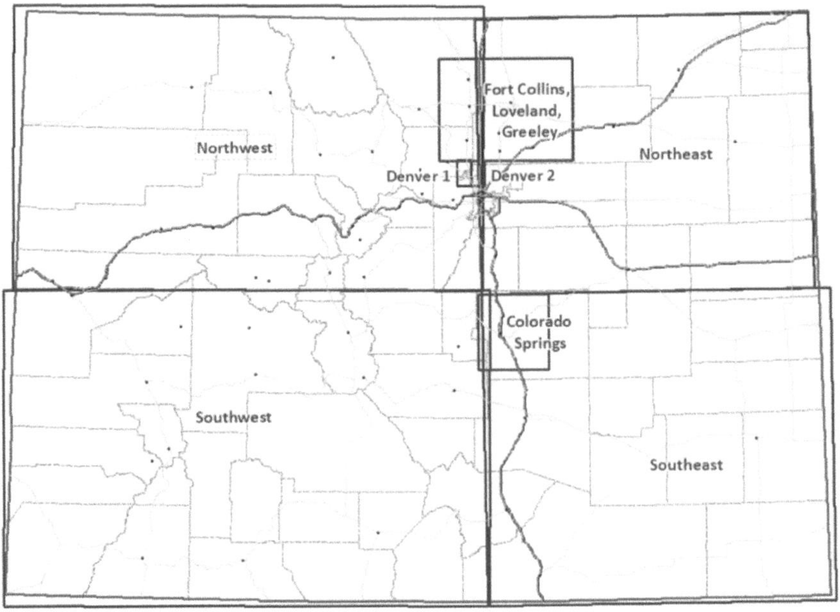

FIGURE 8.1 Approximate division of groups for gathering information about Colorado medical tourism facilities.

8.3.1 INSTRUCTIONS TO THE STUDENT TEAMS

The first step in this process was to brainstorm what type of data, or pertinent details, would be useful for someone who was seeking information about MT opportunities in Colorado. A long list of possibilities was generated, but ultimately the following essential elements were selected:

- Facility name;
- Address*;
- City*;
- Zip*;
- URL or website link;
- Point of service code (POS);
- Walk-in or appointment only;
- Specialty;
- Age of patients accepted;
- Other comments.

Note: * = required for geocoding process.

For each one of these essential elements, possible values were discussed to maintain some consistency in data entry. For instance, while facility Name would likely be a unique text entry, the POS was a coded value between 1 and 6 that represented a certain type of facility:

1 = Inpatient hospital;
2 = Outpatient hospital;
3 = Independent clinic;
4 = Rehabilitation facility;
5 = Indian health service;
6 = Other.

Similarly, it was decided to group the age of patients' options to five categories:

- Children;
- Adult;
- Elderly;
- All; and
- Unknown.

Once this general structure was determined, the investigators created a spreadsheet template. The researchers gathered information for their region using the various resources provided either online or through the library and filled out their spreadsheet to the best of their ability (Figure 8.2).

Mapping of Medical Tourism Colorado, Fall 2018.
MTM Course graduate student project with Dr. Frederick DeMicco, Visiting Scholar to CSU
with Sophia Linn, Geospatial Centroid at CSU

Each student team (n = 8), will fill in a DATA INPUT spreadsheet (see tab below) for their assigned region of Colorado
Fill in as much as you can; not everything will be available for all facilities.
Add any comments or suggestions to the last column in the DATA INPUT spreadsheet.

Save your version of this spreadsheet as follows:
CO_MedicalTourism_GroupX (where X is your group number)

Questions?

Sophia.Linn@colostate.edu

A	B	C	D	E	F	G	H	I	J	K
Facility Name	Address	City	Zipcode	URL	Point of Service Code	**(Values for POS)** 1 = Inpatient Hospital 2 = Outpatient Hospital 3 = Independent Clinic 4 = Rehabilitation Facility 5 = Indian Health Service 6 = Other	Walk in or Appointment only	Age of Patients accepted	**(Values for Patient Ages)** Children Adult Elderly All Unknown	Other Comments

FIGURE 8.2 Spreadsheet template for gathering details about health service facilities in Colorado.

Once we collected the data, the CSU Geospatial Centroid compiled these data into one master spreadsheet, and made edits both for consistency and for usability in the geocoder. A total of 228 facilities were recorded. The master spreadsheet was loaded into the GIS software program, ArcGIS 10.6, and then the addresses were geocoded using the Esri World Geocoder service. This service takes text-based addresses and converts them into points on a map (Figure 8.3).

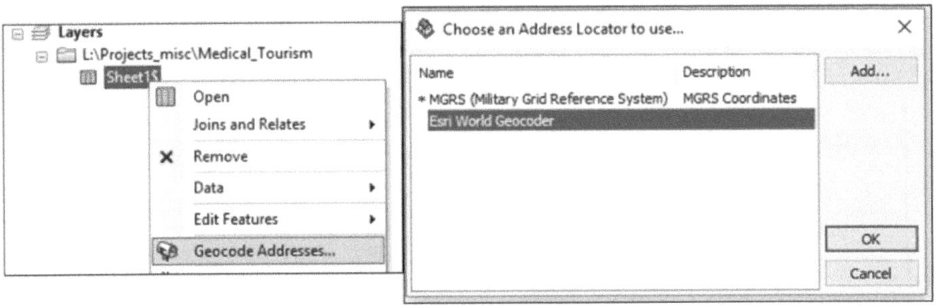

FIGURE 8.3 Geocoding in ArcGIS 10.6.

Using the Canvas course management system, in Files, student teams will see the GIS Excel spreadsheet for this MT GIS. A spreadsheet was created using Microsoft Excel where the data would be gathered by the 8 student teams for entry into the GIS software. We used ArcGIS 10.6 and ArcGIS Online for the geocoding and online mapping (Figure 8.4).

Once the locations are geocoded, the points—and the attribute data tied to them can then be used in other mapping applications. As a proof of concept, we created a preliminary interactive map using ArcGIS Online using the Web App Builder. Note that the information shown in the "pop-up" is the very data that was originally put into the spreadsheet templates by the students (Figure 8.5).

This interactive map provides an initial example of a platform where MT facilities could be gathered and visualized for the purpose of providing potential medical tourists with a way to discover the many diverse services provided in Colorado. By combining a resource like this, together with links and overlays of other tourist facilities in Colorado, tourists may well be drawn to Colorado for their MT experience.

Here are some medical and health databases that we used for our searches for this MT GIS project. Listed below are a few options from the State of Colorado for searching for medical and health sites around Colorado information related to the GIS project:

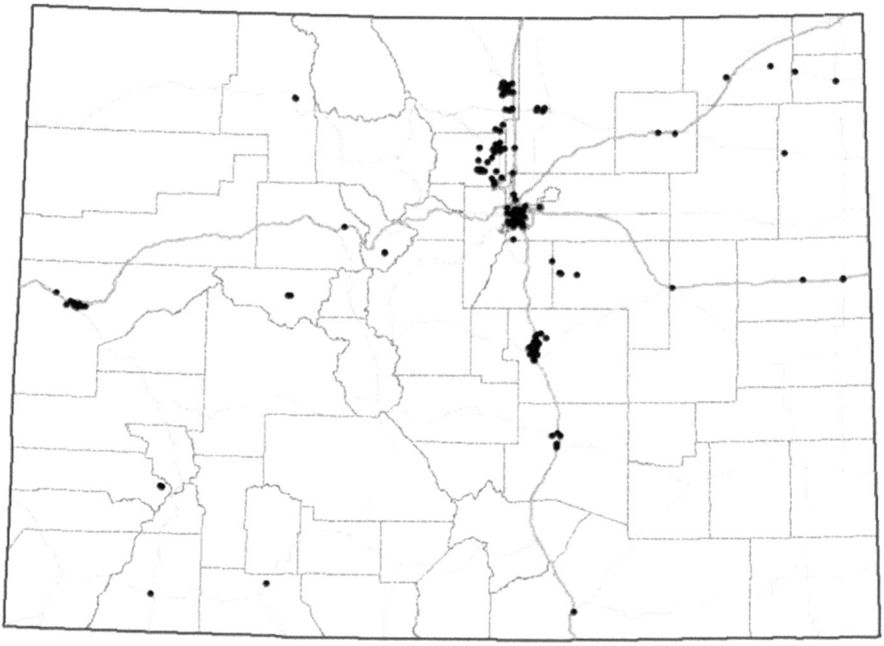

FIGURE 8.4 Results of geocoding. Initial map of medical tourism facilities in Colorado.

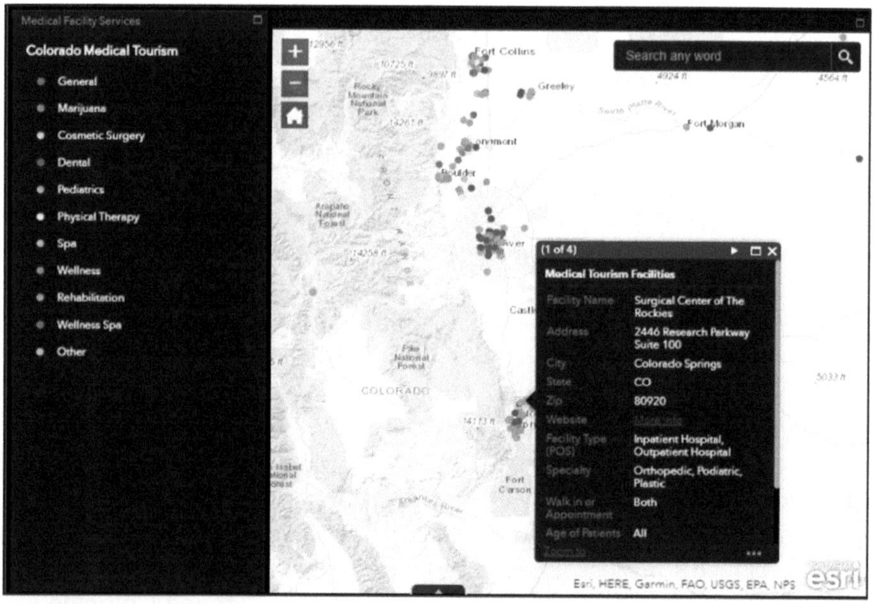

FIGURE 8.5 Preliminary web mapping application for medical tourism facilities in Colorado.

- https://www.colorado.gov/pacific/cdphe/find-and-compare-facilities (links to an external site).
- https://www.colorado.gov/pacific/cdphe/facilities-and-providers-type (links to an external site).
- https://www.colorado.gov/pacific/cdphe/medicalmarijuana (links to an external site).
- https://apps.colorado.gov/dora/licensing/Lookup/LicenseLookup.aspx (links to an external site).

Note: Courtesy of Jocelyn Boice, MSLIS in the CSU Library as we build our GIS map of MT in Colorado.

8.4 CONCLUSIONS

This study mapped and described some of the potential medical, health, and wellness destinations around the state of Colorado. As discussed above, an important first step in this process was to brainstorm what type of data, or pertinent details, would be useful for someone who was seeking information about MT opportunities in Colorado. A long list of possibilities was generated, but ultimately a total of nine essential elements were selected. Other elements for the database could include; the elevation of a healthcare facility (for example, high altitude could be problematic and detrimental in certain types of ear surgery), lodging availability for medical tourists and travelers near a health facility, etc.

The goal of this project was to expand the use of these medical, health, and wellness facilities in Colorado to other states such as Florida, a mecca for MT and wellness travel, thereby enhancing overall C tourism. Interactive maps are very useful tools that can be used in many other tourism, hospitality, and travel efforts to bring awareness to a geographical area. In this case, the state of Colorado vis-à-vis medical, health, and wellness tourism. The overarching goal is to enhance tourism awareness of an area and thereby increase the recognition of a geographical area.

The authors wish to thank our graduate students in the CSU Masters in Tourism Management Program for their help in gathering the data: Yiran Jiang; Sean Grabosky, Brent Heffron, Su Jeon, Roichelle Marble, Mela Montoya, Enkhchimeg Nergui, Tatum Neubert, Katelyn Novak, Sam Pallazza, Kevin Perko, Sarah Withers, Jinhe Xu, and Ivy Yang.

KEYWORDS

- **medical mapping**
- **medical tourism**
- **online mapping**
- **Organisation for Economic Co-operation and Development**
- **point of service code**
- **travel health**

REFERENCES

ArcGIS 10.6 Desktop Software, (2018). Esri, Redlands, CA.

ArcGIS Online, (2018). Esri, Redlands, CA.

DeMicco, F. J., Cetron, M., & Davies, O., (2018). Medical tourism trends. In: *Strategic Management: Today and Tomorrow*. Electronic Publication from Kendall-Hunt Publishers, Iowa, USA.

CHAPTER 9

TRADE-OFF ANALYSIS OF HEALTH AND WELLNESS TOURISM DESTINATION ATTRIBUTES: AN OUTBOUND U.S. CONSUMERS' PERSPECTIVE

RADESH PALAKURTHI[1] and FREDERICK J. DEMICCO[2]

[1]*Professor and Dean, I.H.G. Chair of Excellence, The Kemmons Wilson School of Hospitality and Resort Management University of Memphis, 3700 Central Ave, Suite 140, Memphis, Tennessee–38152, USA, Tel.: (901)-678–3430, Fax: (901)-678–0034, E-mail: rplkrthi@memphis.edu*

[2]*Executive Director and Professor of HRM, Northern Arizona University, USA*

ABSTRACT

The purpose of this research was to analyze the trade-offs that potential U.S. consumers make while choosing an overseas health and wellness tourism destination-seeking of such services overseas is defined as outbound health and wellness tourism. The research utilized conjoint analysis methodologies to quantify the relative preferences of the significant decision-making factors in outbound health and wellness tourism. This research considered 0 factors identified previously as being significant for outbound health and wellness tourism decision-making: Quality of Medical Care; Relative Cost Savings; Comparative Wait Time; Geographic Distance; Economic Conditions; Tourism Attractiveness; Political Situation; Shared Culture and History; English Language Prevalence; and Enabling Entities. This research was limited to the perspectives of outbound U.S. consumers of health and wellness tourism.

9.1 INTRODUCTION

According to the latest figures, wellness tourism was a $535 billion industry in 2015 and has been growing at a 14% clip, a rate that is twice as fast as the global tourism expenditures (Global Wellness Institute, 2018). During the same year, across the globe, people made about 700 million wellness tourism trips, a growth of about 15% from the previous 2 years (Global Wellness Institute, 2018). In spite of such a rapid growth, there is limited understanding of the complex interplay that occurs among the key driving factors for engaging in wellness tourism, especially while seeking healthcare overseas.

This research was designed to answer the question of what trade-offs potential health and wellness tourism consumers make while choosing an international destination for a medical procedure, or for a health treatment or care that they need. This study limited the scope to U.S. health and wellness tourism consumers but expanded the understanding by linking consumers' demographics, beliefs, and preferences to the underlying drivers for such travel. With the rapid growth in the health and wellness tourism industry-especially outbound U.S. consumers-understanding the influence of such factors on the core motivators for travel will help governments and medical tourism (MT) providers make more coordinated marketing efforts and also help to develop products that are more in sync with the consumers' needs.

Traveling for health and wellness is an important trend in the tourism industry (Pearce, Filep, and Ross, 2010). Sociological trends suggest that not only are people more concerned for their own health and wellbeing but are also willing to travel great distances to purchase wholesome products and holistic services that enrich their lives, enhance their psychological sense, and improve their physical being (Smith and Puczko, 2013; Henson, Guy, and Dotson, 2015). The concept of the contemporary tourist seeking more in-depth and meaningful experiences at a destination has been well documented (Crouch, 2004; Saraniemi and Kylänen, 2011). However, the health and wellness industry are constantly evolving as newer products and services are brought to market and many more destinations compete for a share of the growing revenues. The mix of players and products has created a confusion in the minds of the consumers as they try to align the health and wellness tourism destination's attributes with their travel motivations (Smith and Forgione, 2014). Providing appropriate messaging and reliable information may ameliorate the situation-an outcome assured through adequate research.

Health tourism can be defined as the act of traveling to receive general or specialized medical treatment at domestic or international institutions in conjunction with tourism programs. When such an engagement involves travel outside the country, it is termed as outbound health tourism. Traveling overseas for medical care can be broadly classified into three categories: *critical care* such as organ transplant or open-heart surgery; *acute care* that is serious but not life-threatening such as hip replacement or gastric bypass surgery; and, *elective care* that is fully optional such as cosmetic surgery, dental procedures, or plastic surgery (Henson, Guy, and Dotson, 2015). On the other hand, *wellness tourism* involves people in good health traveling domestically or abroad to improve their physical and psychological health by engaging in a variety of services and activities at specialized institutions. The motives for such travel can range from stress relief to attaining spiritual enlightenment.

In the context of health and MT, several streams of research have covered issues such as, the profile of patrons (Gan and Frederick, 2011; Guy, Henson, and Dotson, 2014; Lunt and Carrera, 2010); the travel motivational dynamics (Alsharif, Labonte, and Lu, 2010; Dehdashti Shahrokh, and Nakhaei, 2016); the decision-making factors (Singh, 2012; Crooks et al., 2011); and, the important information sources (Heung, Kucukusta, and Song, 2010; Crooks, Kingsbury, Snyder, and Johnston, 2010). However, a clearer understanding of the decision-making process for choosing a health and wellness tourism destination is needed (Henson, Guy, and Dotson, 2015). Smith and Forgione (2007) suggested that health and wellness tourists first choose a destination and then choose a facility or the medical professionals they need. The model was only applied for domestic travel, leaving a gap in our understanding in the international realm. A conceptual model proposed by Lee, Heesup, and Lockyer (2012), postulated that the behavior intent of the health and wellness tourist depended on their attitude, subjective norms, and the perceived control they have. Heung, Kucukusta, and Song (2010) used a demand and supply model to conceptualize the process of decision-making in health and wellness tourism. While all the previous research were conceptually sound, none took fully into account the normal interplay that happens among the decision variables during the choice process. The closest work applying this approach was undertaken by Levary (2011) who used hierarchical analysis methods to evaluate multiple tangible and intangible decision-making criteria.

9.2 RESEARCH METHODOLOGY

The research methodology used in this research is based on established methods in consumer behavior. A conjoint analysis study was conducted to assess the consumer's subtle preferences for the key factors for choosing an overseas health and wellness tourism destination. By using a hypothetical health and wellness destination profile as the basis for such an understanding, a real-life simulation was created where consumers considered multiple attributes at once while making a choice. Analyzing the choices made by the respondents across many such hypothetical health and wellness tourism destination profiles, the elusive trade-offs that the consumer were making were highlighted. Comparing such trade-offs across different market segments throws light on the differences in the decision-making models across the different market segments. Applying such acute understanding of the market segments will have far-reaching practical implications for the health and wellness tourism industry, mainly by helping develop strategies for enhancing not only the reach but also the overall destination experience of the health and wellness tourism consumer through better product design.

The primary research for this project was undertaken using an online survey of U.S. consumers that had either engaged in or were willing to engage in outbound wellness tourism in the future. In addition to answering questions about their motivations, beliefs, and preference for engaging in outbound health and wellness tourism, the respondents were also provided with a series of conjoint profiles of potential health and wellness tourism destination options in order to rate their willingness to patronize such destinations. Evaluating their willingness to patronize the destinations based on the profiles described allowed this research to identify the health and wellness tourism profile that is the most preferred across all the respondents. Such analysis was conducted at many levels-by the respondents' demographics, motivations, beliefs, and preferences to give a more nuanced understand of the attributes that truly motivate the consumers to engage in health and wellness tourism. However, the results only at the aggregate level are presented in this chapter.

Previous research had identified the motivations for engaging in health and wellness tourism (Singh, 2012; Adams, Snyder, Crooks, and Johnston, 2015). While some researchers focused on the conceptual decision-making models in wellness tourism (Smith and Forgione, 2014), none have attempted to quantify the relative importance of the key attributes involved in such decision-making, especially from the prospect of the trade-offs that are made while making the final choice. This research closes the gap in understanding the decision process for outbound health and wellness tourism.

9.3 RESEARCH OBJECTIVES

There is a lack of understanding of the complex decision-making processes used by consumers while choosing a health and wellness tourism destination. While previous research had identified some key factors for such decisions, the interplay between them has not been studied. Following the previous, the stated goals of this research project were:

- To determine the relative importance of the 10 key destination choice factors identified in health and wellness tourism previously;
- To quantify the subtle trade-off among the 10 key factors that consumers make while making the final health and wellness tourism destination choice;
- To investigate the differences in the decision-making choice models of the consumers based on their preferences for health and wellness travel.

An extensive review of literature has identified the key variables (attributes) such consumers take into account while making a health and wellness destination choice. The key factors identified through research include: Quality of Medical Care; Relative Cost Savings; Comparative Wait Time; Geographic Distance; Economic Conditions; Tourism Attractiveness; Political Situation; Shared Culture and History; English Language Prevalence; and Enabling Entities (Dehdashti and Nakhaei, 2016; Iordache, Ciochină and Roxana, 2013). Table 9.1 identifies the key variables and provides a description of the options (levels) within each variable.

This research applied conjoint methodology to determine the relative importance of the key variables and the incumbent trade-offs that consumers make while choosing a health and wellness tourism destination. Conjoint analysis is an established technique applied in research for evaluating the value proposition or the utilities of products with multiple critical factors (Kohli and Sukumar, 1990). Conjoint analysis has been used for a wide range of applications, including product design, price testing, and service development plans. The underlying benefit is the evaluation of the key attributes and their levels to tease out relationships and variable interplay that would otherwise be opaque. For example, one of the early and seminal studies that evaluated the attributes for designing a new hotel brand for a popular hotel

TABLE 9.1 Key Health and Wellness Tourism Decision-Making Attributes and Their Levels

SL. No.	Attributes	Number of Levels	Attribute Level Type/Title	Attribute Level Description
1.	Quality of medical care at destination	3	Excellent	Internationally accredited medical facility, U.S. Medical Board-certified physicians, Extensive medical facilities and use of technology, and excellent aftercare
			Very Good	Some U.S. Medical Board-certified physicians, very good medical facilities, use of technology and aftercare
			Good	Good medical facilities, use of technology with some aftercare
2.	Relative cost savings compared to similar medical procedure or care in the U.S.	6	15%	15% cost savings to patient compared to similar medical procedure/care in the U.S.
			30%	30% cost savings to patient compared to similar medical procedure/care in the U.S.
			45%	45% cost savings to patient compared to similar medical procedure/care in the U.S.
			60%	60% cost savings to patient compared to similar medical procedure/care in the U.S.
			75%	75% cost savings to patient compared to similar medical procedure/care in the U.S.
			90%	90% cost savings to patient compared to similar medical procedure/care in the U.S.
3.	Comparative wait time for similar medical procedure or care in the U.S.	3	1–90 days	1–90 days average time patient has to wait for similar medical procedure/care in the U.S.
			91–180 days	91–180 days average time patient has to wait for similar medical procedure/care in the U.S.
			180+ days	180+ days average time patient has to wait for similar medical procedure/care in the U.S.

TABLE 9.1 *(Continued)*

SL. No.	Attributes	Number of Levels	Attribute Level Type/Title	Attribute Level Description
4.	Geographic distance of medical provider	4	Short-haul flight	2–4 hours of flight time from home
			Medium-haul flight	4–8 hours of flight time from home
			Long-haul flight	8–12 hours of flight time from home
			Ultra-long-haul flight	12+ hours of flight time from home
5.	Economic condition of medical treatment destination	3	Upper grade-established economy	An economy characterized by a high level of industrialization, solid infrastructure, high level of income and economic growth.
			Medium grade-speculative economy	An economy characterized by moderate infrastructure, income growth but some signs of economic distress.
			Lower grade-highly speculative and risky economy	An economy characterized by underdeveloped infrastructure, low levels of income and economic growth.
6.	Tourism attractiveness of destination	3	Very attractive	Abundant natural resources such as beaches, and mountains; fully developed tourism infrastructure such as hotels and transportation; great variety of food, leisure, and recreational activities; and rich culture and heritage.
			Attractive	Many natural resources such as beaches, and mountains; good tourism infrastructure such as hotels and transportation; adequate variety of food, leisure, and recreational activities; and rich culture and heritage.
			Somewhat attractive	Adequate natural resources such as beaches, and mountains; underdeveloped tourism infrastructure such as hotels and transportation; low variety of food, leisure, and recreational activities; and rich culture and heritage.

TABLE 9.1 *(Continued)*

SL. No.	Attributes	Number of Levels	Attribute Level Type/Title	Attribute Level Description
7.	Political situation at medical treatment destination	3	Very safe and secure	Low corruption and good rule of law
			Reasonably safe a secure	Noticeable corruption and reasonable application of the rule of law
			Relatively unsafe and insecure	High corruption and unreliable application of the rule of law
8.	Shared culture and history with U.S. at medical destination	3	High share	Similar culture, norms, and beliefs as in the U.S. Most locals can relate to U.S. history and current affairs.
			Moderate share	Somewhat similar culture, norms, and beliefs as in the U.S. Reasonable number of locals can relate to U.S. history and current affairs.
			Minimum share	Different culture, norms, and beliefs compared to the U.S. Few locals can relate to U.S. history or current affairs.
9.	English language prevalence at medical destination	3	Native English speakers	Almost everyone in the country is fluent in English.
			Non-native English speakers	A comfortably large number of locals speak English. Day-to-day communications are manageable.
			Non-English speakers	Only a small percentage of the locals speaks English. Local language interpreter will be needed.
10.	Enabling entities	3	Employer-sponsored	Sponsored (not required) by employer that offers cash and other incentives to use the foreign medical facility instead of the ones in the U.S.
			Intermediaries facilitated	Arranged by several intermediary companies and brokers that arrange travel, find medical facilities for treatment, etc.
			Insurance company recommended	Strongly recommended (not required) by medical insurance company that offers incentives to seek medical treatment and care at facilities abroad.
	Number of profiles possible:	**157,464**	(157,464 is the product of all the number of levels of each attribute)	

chain applied conjoint methodology (Wind, Green, and Shifflet, 1989). In that study, the researchers evaluated 50 hotel factors with 167 levels in all ranging from building shape to lounge atmosphere. Through an empirical evaluation, the researchers developed the now very successful *Courtyard by Marriott* hotel brand. Similarly, researchers have identified the most preferred performing arts tourism products as perceived by tourists using conjoint techniques (Kim, Chung, Petrick, and Park, 2018; Ross, Norman, and Dorsch, 2003).

In this study, the researcher utilized an online survey methodology to evaluate the decision-making preferences of health and wellness tourists from the U.S. A paid online panel of Qualtrics' U.S. residents was used as the population for this research. The potential respondents were screened for their age (over 18 years) and their experience with or their willingness to use overseas health and wellness facilities personally. A total of 357 usable survey respondents were collected in order to have adequate date for the complex analysis needed for a conjoint study. Following the screening questions, the survey instrument had three additional sections-health and wellness status questions, conjoint destination profile evaluation section, and a socio-demographic questions section.

In this conjoint section, the respondents were presented with series of profiles (cards) of health and wellness destinations (two at a time) and were asked to state if they would choose either or none of the destinations as a potential location for health or wellness care. In this study, there were 10 key attributes with a total of 34 different levels across the attributes. Using just the 10 attributes, and multiplying all the levels together, a total of 157,464 different profiles of health and wellness destinations were obtained. A profile/card is a listing of one combination of levels for each of the 10 key attributes. Table 9.2 shows an example of a destination profile. Since it is impossible for all the respondents to state their preference for all the profiles determined by the research design, using an orthogonal design, the number of health and wellness profiles that each respondent had to evaluate was reduced to 22 each using a fractional factorial statistical design. A total of 7854 different destination profiles were evaluated by the respondents through this study. This study also considered the type of health or wellness issue for the hypothetical travel to the destination and a "Reason for Travel" question was added with four levels: Critical care, Acute Care, Elective Care, and Wellness Care. Assessing the destinations choices by reason for travel could give more insights into the type of destinations consumers would choose based on the gravity of their health condition.

TABLE 9.2 Example of Health and Wellness Destination Profile for Respondent Rating

Attribute	Level	Description
Quality of medical care at destination	Very good	Some U.S. Medical Board-certified physicians, very good medical facilities, use of technology and aftercare
Relative cost savings compared to similar medical procedure or care in the U.S.	75%	75% cost savings to patient compared to similar medical procedure/care in the U.S.
Comparative wait time for similar medical procedure or care in the U.S.	91–180 days	91–180 days average time patient has to wait for similar medical procedure/care in the U.S.
Geographic distance of medical provider	Long-haul flight	8–12 hours of flight time from home
Economic condition of medical treatment destination	Medium grade-speculative economy	An economy characterized by moderate infrastructure, income growth but some signs of economic distress.
Tourism attractiveness of destination	Very attractive	Abundant natural resources such as beaches, and mountains; fully developed tourism infrastructure such as hotels and transportation; great variety of food, leisure, and recreational activities; and rich culture and heritage.
Political situation at medical treatment destination	Reasonably safe a secure	Noticeable corruption and reasonable application of the rule of law
Shared culture and history with U.S. at medical destination	Minimum share	Different culture, norms, and beliefs compared to the U.S. Few locals can related to U.S. history or current affairs.
English language prevalence at medical destination	Non-native English speakers	A comfortably large number of locals speak English. Day-to-day communications are manageable.
Enabling entities	Employer-sponsored	Sponsored (not required) by employer that offers cash and other incentives to use the foreign medical facility instead of the ones in the U.S.

GIVEN THE PROFILE OF THE REASON FOR YOUR TRAVEL AND THE HEALTH AND WELLNESS DESTINATION PROFILE DESCRIPTION, WOULD YOU CHOOSE THIS DESTINATION? YES NO

Analysis of data was conducted using SPSS software. The data was analyzed at both the univariate and multivariate level. At the aggregate level, the respondent's self-explicated utility for each of the levels of the attributes was determined. Such analysis gave the relative importance of each of the attributes. Conjoint regression analysis was used to evaluate the marginal change in utility that the consumers had as they switched between the different attribute level choices. Such an analysis quantified the relative change in utility of the consumers as they switched from one level of an attribute to another. An analysis of the impact of the socio-demographics and the other respondent characteristics on the respondent self-explicated utilities was also conducted in the final stage.

9.3.1 SIGNIFICANCE OF RESEARCH

The rapidly evolving changes faced by the stakeholders in the health and wellness tourism industries are two-fold: the increasing technological savviness and information gathering ability of the consumers is enabling them to directly seek information and contact the providers in the industry no matter how far they are. On the other hand, the increasing healthcare costs domestically in the U.S. is putting pressure on corporation and insurance companies to consider outbound resources for health and wellness. With the two issues converging, it becomes critical for all stakeholders to understand more clearly what the decision-making processes are for the consumer. In addition, smart tourism demands that we develop products and services that are truly needed by the consumer (Gretzel, Sigala, Xiang, and Koo, 2015; Sultana, Dale, and Syed, 2012). This research is an attempt to help the health and wellness tourism practitioners gain a better understanding of their consumers in the U.S.

Consequently, this research has implications at both the academic and practical leaves. It uniquely applies the conjoint methodology to advance the understanding of the health and wellness destination choice decision-making at both the individual and aggregate levels. The methodology applied in this research provides an additional tool for academic researchers in the discipline. At the practical level, this research provides useful information for operators in three ways. First, it quantifies the relative importance of the key health and wellness destination attributes. Second, it identifies the most preferred destination profile- and the least preferred-for effective destination product design. Additionally, the relative preferences in terms of the consumers' total utility for all the destinations that lie between the best and worst rated profiles can also be calculated for comparative purposes with existing destinations in the market. Third, this research allows the operators to evaluate the preferences within socio-demographic sub-markets or markets that are closer to the existing target group. Such an understanding will allow them to amend their offerings to better meet the needs of their market segments.

9.4 FINDINGS

The results for this study are presented in three subsections. In the first, the profile of the respondents is described along with their preferences for travel overseas for health and wellness reasons. The profile of the respondents was quite similar to the average overseas medical tourist as identified by previous studies. In the second subsections, the relative importance of the 10 key profile factors for choosing an overseas health and wellness destination are assessed. The relative cost savings was found to be the most important driver for choosing an overseas health and wellness destination. The final subsection presents the core results of the relative preferences of the factor levels within each of the key destination profile attributes. The evaluation of the results helped determine the destination profile that is most preferred (highest total utility) across all the respondents.

9.4.1 RESPONDENT PROFILE

The socio-demographic profile of the respondents in this study is shown in Table 9.3. The demographic profile was compared with the attributes that were available from previous research. There was only a very slight difference in the proportion of responses by gender in this study-overseas medical tourists tend to be slightly more males (about 55% vs. 45%). About 75% of the respondents were below the age of 55 years compared to the general medical tourist population of between 75% and 80%. Similarly, about 75% of the respondents also had up to a bachelor's degree compared to between 75% and 78% of medical tourists. About half (49.86%) of the respondents were married, with another 15% divorced and the remaining being single (about 35%). The sample of respondents is slightly different than the profile of an average medical tourist where, about 50% are single, 35% married and the rest divorced. Overall, the respondents were quite similar to the profile of an average medical tourist.

TABLE 9.3 Respondent Profile

Gender	Male	Female	Other		
Percentage of Total:	50.14%	49.58%	0.28%		
Age Distribution	18–34 Years	35–54 Years	55+ Years		
Percentage of Total:	34.82%	41.23%	23.96%		
Race	**White**	**Black or African American**	**Others**		
Percentage of Total:	76.04%	11.42%	12.53%		
Employment Status	**Employed Full Time**	**Employed Part-Time**	**Unemployed or Other**		
Percentage of Total:	56.82%	14.21%	28.97%		
Education	**High School or Less**	**2-Year Degree or Some College**	**4-Year Degree**	**Professional Degree or PhD**	
Percentage of Total:	13.65%	31.76%	32.31%	22.28%	
Annual Household Income	**Less Than $30K**	**$30K–$60K**	**$60K–$90K**	**More Than $90K**	
Percentage of Total:	19.50%	27.86%	22.84%	29.81%	
Marital Status	**Married**	**Widowed**	**Divorced**	**Separated**	**Never Married**
Percentage of Total:	49.86%	3.90%	14.76%	2.23%	29.25%

Table 9.4 presents the results of the respondents' experience and their preference related to health and wellness travel overseas. Interestingly, more of the respondents reported traveling overseas solely for the purpose of health (medical care) compared to wellness care (29.25% vs. 23.25%). On an average, they also reported that a personal cost savings of about 60% would make them to seriously consider going overseas for health and wellness care. Overall, only about 2/3rd (about 67%) of the respondents reported being in good or excellent overall health status; about 12% reporting that they were in poor health. Although fewer number of the respondents reported actually traveling overseas for wellness care than healthcare, there seems to be a higher level of interest in traveling overseas for wellness care (30.73%) compared to health care (between 18.44% and 28.49%). Health care travel included treatment for all three types of care: Elective, Acute, and Critical.

TABLE 9.4 Travel Overseas for Health and Wellness

Have you ever traveled overseas for the sole purpose of seeking medical treatment or care?

Medical Care Travel	**Yes**	**No**
Percentage of Total:	29.25%	70.75%

Have you ever traveled overseas for the sole purpose of engaging in any wellness-related activities such as spa treatments, meditation programs, stress management seminars, etc.?

Wellness Travel	**Yes**	**No**
Percentage of Total:	23.25%	76.75%

What percentage of cost savings personally to you would make you to SERIOUSLY CONSIDER medical treatment or care overseas (i.e., engage in medical tourism)?

Medical Costs	**Mean**
Personal Savings (%)	57.46%

How would you generally categorize the status of your overall health?

Overall Health	**Excellent**	**Good**	**Average**	**Poor**
Percentage of Total:	23.46%	43.73%	20.89%	11.92%

How likely are you to seek care at an overseas medical tourism facility for the following types of care?

Type of Care	**Wellness Care**	**Elective Care**	**Acute Care**	**Critical Care**
Extremely Likely (%)	30.73%	28.49%	18.44%	22.63%

The respondents also reported the types of healthcare treatments they would most definitely seek overseas as illustrated in Figure 9.1. The top three treatments and procedures that they are most likely to consider include, oncology/cancer treatment (14.08%), cosmetic/plastics/esthetic surgeries and procedures (12.08%), and orthopedic/spine surgeries and treatments (12.08%). The respondents reported that that they were least likely to travel overseas for vision care treatment and surgeries (5.03%).

Figure 9.2 presents similar results for the respondents' likelihood to travel overseas for various wellness-related treatments and care. The top three wellness-related treatments for which the respondents are definitely likely to travel overseas are: spa and beauty programs (15.52%), body, and mind programs (15.04%), and for spiritual connections (14.02%). The respondents are least likely to travel overseas to pursue a health-related wellness program such a for an integrated medical/diagnostic health check-up or for weight-loss programs (9.15%).

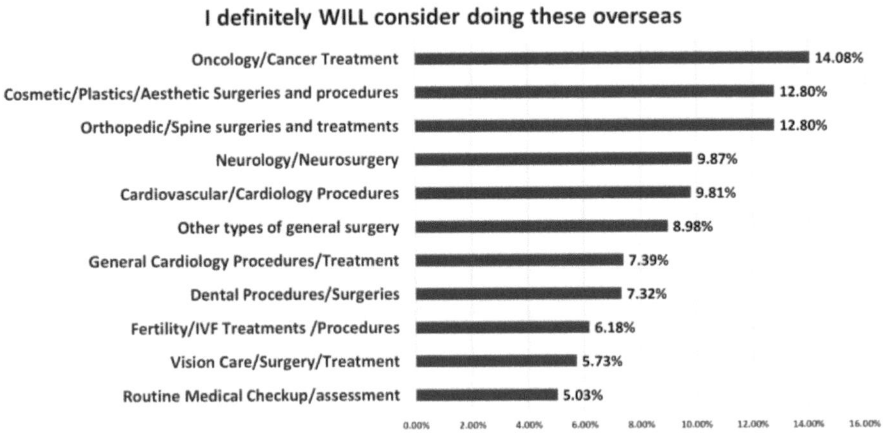

FIGURE 9.1 How likely are you to seek care at an overseas medical tourism facility for the following types of care?

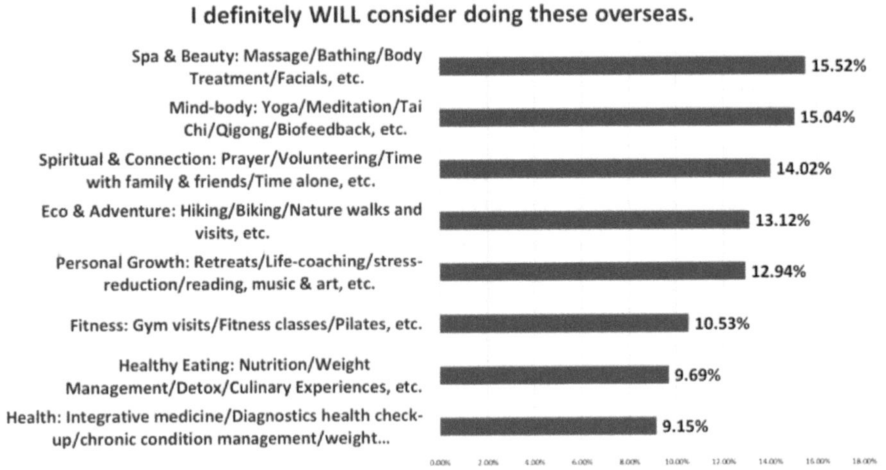

FIGURE 9.2 How likely are you to seek care at an overseas wellness tourism facility for the following types of care?

9.4.2 *RELATIVE IMPORTANCE OF KEY FACTORS*

One of the core outcomes of conjoint analysis is the determination of the relative importance of the key attributes in the study. The 10 key attributes considered in this study include: Quality of Medical Care; Relative

Cost Savings; Comparative Wait Time; Geographic Distance; Economic Conditions; Tourism Attractiveness; Political Situation; Shared Culture and History; English Language Prevalence; and Enabling Entities. Figure 9.3 shows the relative importance of the key factors in order of importance.

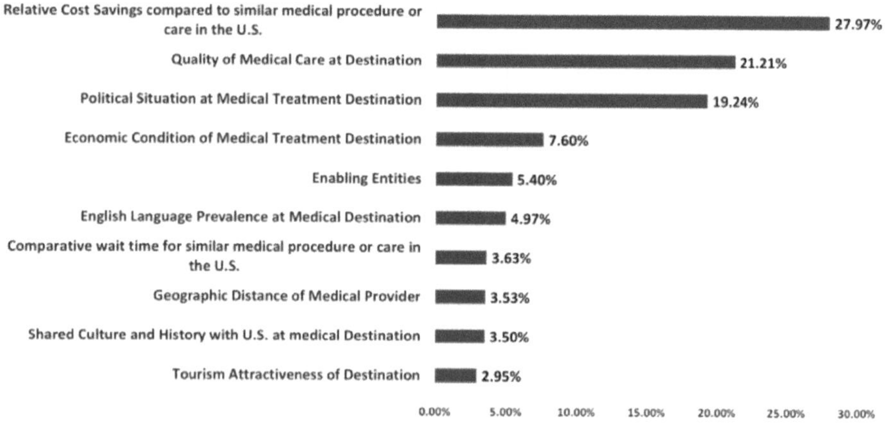

FIGURE 9.3 Relative key factor importance.

Figure 9.3 shows that the most critical factor for choosing an overseas health and wellness destination is the relative personal cost savings that can be gained by the person considering the treatment abroad. In all, this factor by itself accounts for 27.97% of the decision-weight for choosing a final destination. Similarly, the quality of medical care at the destination and the political situation at the destination account for the next two highest weights, 21.21% and 19.24%, respectively. The top three factors by themselves make up about 2/3rd (68.42%) of the decision-weight for choosing an overseas health and wellness destination. The key to gaining prominence in marketing a destination seems to be dependent on these three factors.

Interestingly, the respondents reported that the least important factor for their decision in choosing a health and wellness destination was the tourism attractiveness of the destination (2.95%). Similarly, the factors that demonstrated relatively less importance for health and wellness destination choice include: shared culture and history with U.S. (3.50%), geographic distance (3.53%), comparative wait time (3.63%), English language prevalence (4.97%), and enabling entities (5.40%).

9.4.3 ATTRIBUTE LEVEL TRADE-OFF RESULTS

While the previous section presented the results at the aggregate level, this section highlights the change in marginal utility as consumers potentially change their choices between the different options within each of the 10-key health and wellness destination attributes. Figure 9.4 shows charts (A through J) that highlight the change in utility as consumers shift from one attribute level to another. The change in utility is discussed separately for each of the attributes in this section.

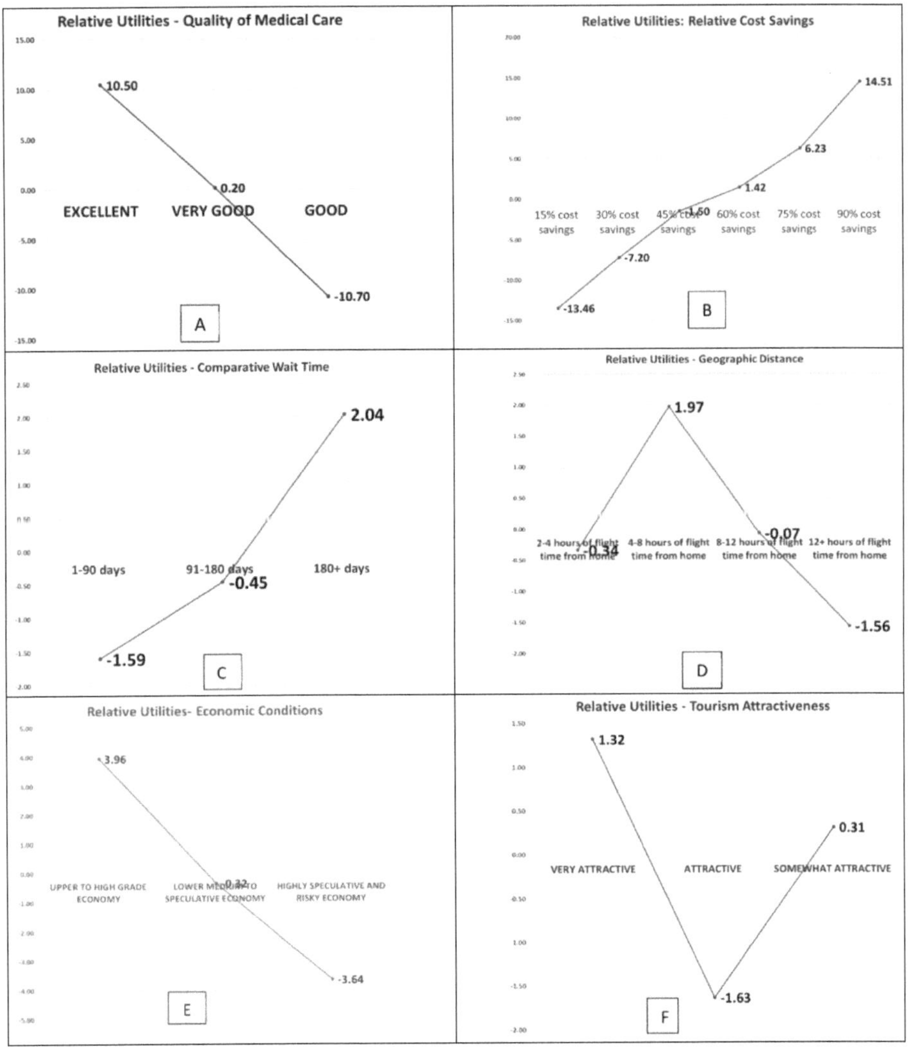

FIGURE 9.4 *(Continued)*

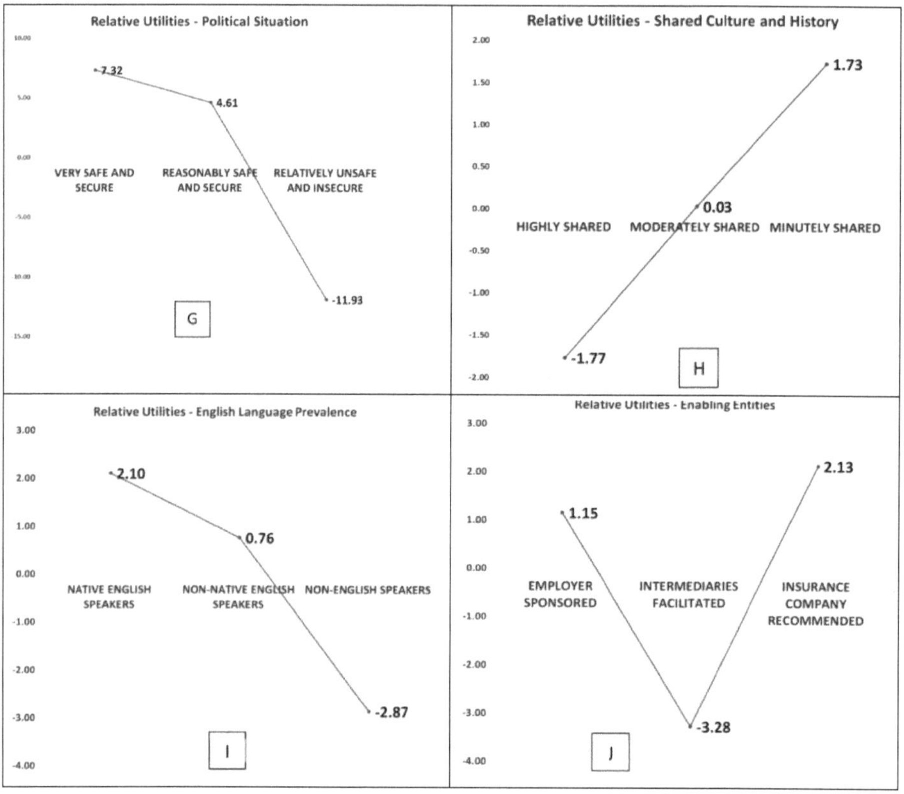

FIGURE 9.4 Relative utilities for attribute levels.

9.4.4 QUALITY OF MEDICAL CARE

Figure 9.4(A) shows the expected change in utility with change in the level of the quality of medical care. The quality of medical care was described at three levels-Excellent (Internationally accredited medical facility, U.S. Medical Board-certified physicians, Extensive medical facilities and us of technology, and excellent aftercare), Very Good (Some U.S. Medical Board-certified physicians, very good medical facilities, use of technology and aftercare), and Good (Good medical facilities, use of technology with some aftercare).

The results show that consumers had a utility of –10.70 when the quality of care was Good and increased to 0.20 when it was Very Good. However, the utility was a strongly positive, 10.50 units when the quality of medical care was Excellent. The consumers do not seem to gain any meaningful utility (strong preference) unless the quality of medical care at the overseas destination is Excellent.

9.4.5 *RELATIVE COST SAVINGS*

The relative cost savings compared to a similar procedure in the U.S. was compared in terms of the savings in personal costs to the potential consumer. Figure 9.4(B) shows the cost savings and the related utility derived by the consumers. The cost savings were graded from 15% to 90% in 15% increments.

As expected, the trend in increasing utility with increasing savings for the consumer is noticed. However, a negative utility is noticed when consumers derive only up to 45% savings compared to the U.S. for a similar procedure or care. Positive utility is only noticed beyond a 45% savings compared to the U.S. The utility gradually increasing as savings escalate up to 90% (utility of 14.51). From the chart, it is clear that a minimum of 45% savings have to be offered in order for the foreign destination to be considered a viable option. However, this level of utility may be outweighed by other more attractive (positive utility) attribute levels in the destination profile. In other words, a trade-off can occur where lower savings (therefore lower utility) may be more than compensated by other attractive options in the other attributes.

9.4.6 *COMPARATIVE WAIT TIME*

The time that a patient has to wait in order to receive the same treatment or care compared to the foreign health and wellness destination was considered as one of the 10 key attributes. The three levels for the attribute were identified as, 1–90 days (1–90 days average time patient has to wait for similar medical procedure/care in the U.S.), 91–180 days (91–180 days average time patient has to wait for similar medical procedure/care in the U.S.), and 180+ days (180+ days average time patient has to wait for similar medical procedure/care in the U.S.). Figure 9.4(C) shows the utilities for the three levels.

The utility derived is lower for a shorter duration of time the patients have to wait for a similar procedure in the U.S., indicating their preference to have the procedure or care in the U.S. if the wait time is not extraordinarily long. In fact, the utility for wait times for the two levels of up to 180 days of wait time were negative, (–1.59 and –0.45) indicating that the consumers are willing to seriously consider having the treatment or care overseas only if they had to wait for more than 180 days for a similar treatment in the U.S.

The utility for the wait time increased to 2.04 if the patient had to wait for the treatment or care for more than 180 days and it was immediately available at the overseas health and wellness facility.

9.4.7 GEOGRAPHIC DISTANCE OF DESTINATION FROM HOME

Research shows that one of the key variables considered for evaluating health and wellness options overseas is how far the destination is from the home of the consumers. For this study, the distance was measure in terms of how long the flight time was to get to the destination from home. The attribute was defined at four levels: 2–4 hours, 4–8 hours, 8–12 hours and 12+ hours. Figure 9.4(D) shows the utilities associated with the four levels for geographic distance.

The results show that the highest utility (1.97) derived by the consumers was when the distance of the health and wellness destination was between 4 and 8 hours away from home. When the destination was less or more than that distance, the respondents showed negative utility. The results show that the consumers are willing to travel reasonably far to receive the healthcare or wellness treatment and any long-haul travel will require compensation using other attributes such as an increase in cost savings or an improvement in the quality of medical care.

9.4.8 ECONOMIC CONDITIONS AT DESTINATION

Previous research has shown that the level of economic development had an impact on the decision for choosing an overseas health and wellness destination. For the purpose of this research, the economic conditions at the destination were defined using three levels of development: Upper Grade-Established Economy (An economy characterized by a high level of industrialization, solid infrastructure, high level of income and economic growth), Medium Grade-Speculative Economy (An economy characterized by moderate infrastructure, income growth but some signs of economic distress), and Lower Grade-Highly Speculative and Risky Economy (An economy characterized by underdeveloped infrastructure, low levels of income and economic growth). Figure 9.4(E) shows the utilities derived by the consumers based on the different levels of economic development at the overseas health and wellness destination.

The results show that the utility derived by the consumers is strongly correlated with the economic condition at the health and wellness destination; higher levels of economic development led to higher levels of utility. However, as expected, a Highly Speculative and Risky Economy led to negative utility of –3.64 while a Medium Grade-Speculative Economy led to just 0.32 units of utility. Only when the economic condition was Upper Grade-Established Economy, the utility of the consumers was strongly positive at 3.96.

9.4.9 TOURISM ATTRACTIVENESS OF THE HEALTH AND WELLNESS DESTINATION

Invariably, it is common for the overseas health and wellness facilities to package the tourism attractiveness of the destination as a value-added benefit for patronizing the facility. Such options are often made available to both the patients and the family members by including them in the price of the healthcare or wellness treatment. The level of attractiveness of the destination was defined at three levels in this research: Very Attractive (Abundant natural resources such as beaches, and mountains; fully developed tourism infrastructure such as hotels and transportation; great variety of food, leisure, and recreational activities; and rich culture and heritage), Attractive (Many natural resources such as beaches, and mountains; good tourism infrastructure such as hotels and transportation; adequate variety of food, leisure, and recreational activities; and rich culture and heritage), and Somewhat Attractive (Adequate natural resources such as beaches, and mountains; underdeveloped tourism infrastructure such as hotels and transportation; low variety of food, leisure, and recreational activities; and rich culture and heritage). Figure 9.4(F) shows the utilities derived by each of the three levels of tourism attractiveness.

Interestingly, this research reveals that the utility derived by the consumers is highest when the tourism attractiveness of the destination is Very Attractive (1.32), however, the utilities were positive when the destination was Slightly Attractive (0.31) while they were negative when the destination was Somewhat Attractive (–1.63). It is clear that the consumers most preferred a destination that was Very Attractive as a tourist destination. It should be noted that in spite of the emphasis on tourism in medicine and wellness, tourism attractiveness played the least important role among the 10 key attributes for choosing a health and wellness tourism destination as shown in Figure 9.3.

9.4.10 POLITICAL SITUATION AT THE DESTINATION

The geopolitical situation and unrest in many parts of the world have made many overseas travelers wary. For the purpose of this research, the political situation at the destination was defined at three levels: Very Safe and Secure (Low corruption and good rule of law), Reasonably Safe a Secure (Noticeable corruption and reasonable application of the rule of law), and Relatively Unsafe and Unsecure (High corruption and unreliable application of the rule of law). Figure 9.4(G) shows the utilities associated with the three levels of political situation at the destination.

The utility derived by the consumers increased with increasing level of safety and security at the destination. However, when the respondents perceived the destination as being Relatively Unsafe and Unsecure, they perceived negative utility (–11.93)-a relatively high level of non-preference. However, the consumers were more forgiving if the political situation was Reasonably Safe a Secure since they perceived a utility of 4.61. The utility was highest at 7.32 when the political situation was Very Safe and Secure.

9.4.11 SHARED CULTURE AND HISTORY WITH THE U.S.

The level of comfort that consumers feel in a foreign land because of a shared culture has been established as an important attribute in choosing an overseas destination. The intensity of shared culture and history was defined at three levels in this research: Highly Shared (Similar culture, norms, and beliefs as in the U.S. Most locals can relate to U.S. history and current affairs), Moderately Shared (Somewhat similar culture, norms, and beliefs as in the U.S. Reasonable number of locals can relate to U.S. history and current affairs), and Minutely Shared (Different culture, norms, and belief compared to the U.S. Few locals can related to U.S. history or current affairs). Figure 9.4(H) shows the utilities associated with the three levels of shared culture and history.

Interestingly, Figure 9.4(H) shows that the utility derived by the consumers is highest when the shared culture and history is the least (1.73) indicating that the consumers are viewing the opportunity to go overseas for healthcare and wellness as also an opportunity to learn more about foreign cultures that are very different than that of the U.S. However, the consumers perceived only 0.03 units of utility when the shared culture and history was Moderately Shared with the U.S.; perceived utility actually was negative (–1.77) when shared culture and history was Highly Shared with the U.S.

The results suggest that the consumers are also looking for a cultural experience while they are overseas at a healthcare and wellness destination.

9.4.12 ENGLISH LANGUAGE PREVALENCE AT DESTINATION

The level of interaction in English that the potential consumers can have at the overseas healthcare and wellness destination is an important factor in their decision to choose a destination. The level of prevalence of English at the destination was defined at three levels for the purpose of this research: Native English Speakers (Almost everyone in the country is fluent in English), Non-native English Speakers (A comfortably large number of locals speak English. Day-to-day communications are manageable), and Non-English Speakers (Only a small percentage of the locals speaks English. Local language interpreter will be needed). Figure 9.4(I) shows the utilities derived with the three levels of English language prevalence at the destination.

The utility derived by the consumers increased with increasing levels of English language prevalence at the destination. When the providers at the health and wellness destination are native English Speakers, the perceived utility is highest (2.1) compared to when they are Non-Native English Speakers (0.76) or Non-English Speakers (–2.87). Comparing the results of this attribute with the shared culture and history attribute, it is interesting that the consumers would prefer destinations that are not similar in culture or history to the U.S. yet, they prefer that the providers to be able to speak English fluently.

9.4.13 ENABLING ENTITIES

Enabling entities are the sources that empower the consumer to make the decision to go overseas for seeking healthcare or a wellness treatment. For the purpose of this research, three sources were considered: Employer-Sponsored (Sponsored (not required) by employer that offers cash and other incentives to use the foreign medical facility instead of the ones in the U.S.), Intermediaries Facilitated (Arranged by several intermediary companies and brokers that arrange travel, find medical facilities for treatment, etc.), and Insurance Company Recommended (Strongly recommended (not required) by medical insurance company that offers incentives to seek medical treatment and care at facilities abroad). Figure 9.4(J) shows the consumer utilities derived when each of the three sources is utilized.

The results show that the consumers perceive most utility when their decision to seek health and wellness care overseas is Insurance Company recommended (2.13). However, they also perceive a slightly lower positive utility when the decision is Employer Sponsored (1.15). They do not seem to prefer Intermediaries Facilitated decisions since they perceived a strong negative utility of –3.28. Clearly, the consumers seem to prefer the involvement of their insurance company or their employer while choosing an overseas destination for health or wellness care.

9.4.14 *TRADE-OFF ANALYSIS*

From the above results, it is possible to derive the profile of the health and wellness destination that is most preferred by the consumers based on the total utility derived. Figure 9.5 shows the profile of the destination that is most preferred. The total consumer utility derived for this profile is 47.58 which is the highest that can be achieved for any combination of destination profiles considered in this study. However, since the destination profile can vary compared to the most preferred, the key question is, what is the total level of consumer utility perceived when the destination profile is different than the best suggested by this research? In other words, how do the consumers make their trade-offs among the attribute levels when the destination profile is less than optimal? In order to evaluate this condition, an example of the trade-off analysis is provided in this section by comparing two hypothetical destination choices.

Optimal Package

Attribute	Attribute Option
Quality of Medical Care at Destination	EXCELLENT: Internationally accredited medical facility, U.S. Medical Board certified physicians, Extensive medical facilities and us of technology, and excellent aftercare
Relative Cost Savings compared to similar medical procedure or care in the U.S.	90% cost savings to patient compared to similar medical procedure/care in the U.S.
Comparative wait time for similar medical procedure or care in the U.S.	180+ days average time patient has to wait for similar medical procedure/care in the U.S.
Geographic Distance of Medical Provider	4-8 hours of flight time from home
Economic Condition of Medical Treatment Destination	UPPER TO HIGH GRADE ECONOMY: An economy characterized by high level of industrialization, solid infrastructure, high level of income and economic growth.
Tourism Attractiveness of Destination	VERY ATTRACTIVE: Abundant natural resources such as beaches, and mountains; fully developed tourism infrastructure such as hotels and transportation; great variety of food, leisure and recreational activities; and rich culture and heritage.
Political Situation at Medical Treatment Destination	VERY SAFE AND SECURE: Low corruption and good rule of law
Shared Culture and History with U.S. at medical Destination	MINUTELY SHARED: Different culture, norms and belief compared to the U.S. Few locals can related to U.S. history or current affairs.
English Language Prevalence at Medical Destination	NATIVE ENGLISH SPEAKERS: Almost everyone in the country is fluent in English.
Enabling Entities	INSURANCE COMPANY RECOMMENDED: Strongly recommended (not required) by medical insurance company that offers incentives to seek medical treatment and care at facilities abroad.

FIGURE 9.5 Profile of the most attractive health and wellness destination.

Figure 9.6 shows the profiles of two hypothetical health and wellness destinations and their respective total utilities perceived. By comparing the two destination profiles, it is clear that both the destinations are very different and have total utilities that are comparatively much less than the optimal for the market (47.58) as discussed above. Destination 1 has a total utility of 12.18 while Destination 2 is 42% greater at 17.26. While we know that between the two destinations, the consumers clearly prefer Destination 2 since it has a higher total utility, it is interesting to see the trade-off they are making to arrive at that choice.

ATTRIBUTE	Destination 1 Profile		Destination 2 Profile	
	Attribute Level	Level Utility	Attribute Level	Level Utility
Quality of Medical Care at Destination	Very Good	0.20	Excellent	10.50
Relative Cost Savings compared to similar medical procedure or care in the U.S.	75% Savings	6.23	45% savings	-1.50
Comparative wait time for similar medical procedure or care in the U.S.	91-180 Days	-0.45	91-180 Days	-0.45
Geographic Distance of Medical Provider	4-8 Hours Flight from Home	1.97	12+ Hours Flight from Home	-1.56
Economic Condition of Medical Treatment Destination	Highly Speculative and Risky	-3.64	Lower Medium to Speculative	0.32
Tourism Attractiveness of Destination	Very Attractive	1.32	Attractive	-1.63
Political Situation at Medical Treatment Destination	Reasonably safe and Secure	4.61	Very Safe and Secure	7.32
Shared Culture and History with U.S. at medical Destination	Moderately Shared	0.03	Moderately Shared	0.03
English Language Prevalence at Medical Destination	Non-Native English Speakers	0.76	Native English Speakers	2.10
Enabling Entities	Employer Sponsored	1.15	Insurance Company Recommended	2.13
	TOTAL DESTINATION UTILITY:	12.18	TOTAL DESTINATION UTILITY:	17.26

FIGURE 9.6 Profiles of two hypothetical health and wellness destination and the respective total utilities.

While Destination 1 is relatively closer geographically, has very attractive tourism, and offers a 75% savings in costs compared to the U.S., the rational consumer would still choose Destination 2 since they weight its excellent quality medical care, very safe political environment, and the native English-speaking ability of the service providers very highly. The superior cost savings at Destination 1 (75%) compared to Destination 2 (45%) is more than outweighed by the other factors such as excellent quality medical care and English language ability of the hosts at Destination 2. However, if

Destination 1 was able to offer a 90% savings in costs instead of the 75% (leaving all other options the same for both the destinations), then Destination 1 would have a total utility of 20.46 making its utility 16% higher than that of Destination 2 and therefore the more preferred option. Hence, it is possible for the destinations to vary their profiles in order to compete with the competition. Such strategies may not necessarily be based on cost savings since total utility can be enhanced using any of the 10 key attributes.

9.5 DISCUSSION

Choosing an overseas facility for health and wellness involves a complex interplay of various factors related to the destination. This research identified 10 such key factors and quantified their relative importance and the unique utilities derived from each of the levels (options) within the factors. The results provide a methodology to assess the relative preference of a rational consumer given a specific health and wellness destination profile. From such an analysis a key set of takeaways emerge that are cardinal to this research:

- As with any other consumer product or service, health, and wellness consumers seeking care or treatment overseas are invariably making trade-off choice among the attribute options in order to maximize their total utility from their own perspective. It is therefore possible for an overseas health and wellness destination to create a profile that could place it very competitive light in the market by modifying the levels of the key attribute options. The attractiveness of the destination could be managed better by positioning the destination profile competitively.
- The key attributes that make up two-thirds of the decision weight for choosing an overseas health and wellness destination are: cost savings compared to the U.S., quality of medical care, and the political situation at the destination. Since the facility may have limited to no influence in the local political situation, the destinations could focus more on blending the costs with the quality of medical care provided in order to arrive at a profitable model. Although the other key factors are important, the marginal improvement in utility derived from such efforts behooves the facilities to focus more on cost savings and quality of medical care instead of the rest of the attributes.
- Although we link medical care and wellness with tourism, this research reveals that the tourism attractiveness is the least important

attribute among the 10 key attributes considered for health and wellness destination choice. This result may mean that overseas health and wellness providers are better off focusing on the core attributes such as quality of medical care and costs and offer tourism-related services as a value-added service. Focusing primarily on tourism services as a means to attract potential consumers will have limited effect.

- The consumers prefer to have their insurance company or their employer enable their decision for choosing an overseas health and wellness destination. The subscription to such overseas services will grow as more and more employers and insurance companies make it easier for consumers to access such services.

This chapter presented the results at the aggregate level by discounting the effects of many other factors such as the consumers' socio-demographic characteristics, their health status, or their reason for travel. The impact of such factors will be discussed in a subsequent chapter.

KEYWORDS

- **health tourism**
- **medical tourism**
- **research methodology**
- **sociological trends**
- **tourism industry**
- **wellness tourism**

REFERENCES

Adams, K., Snyder, J., Crooks, V., & Johnston, R., (2015). Tourism discourse and medical tourists' motivations to travel. *Tourism Review, 70*(2), 85–96.

Alsharif, M. J., Labonte, R., & Lu, Z., (2010). Patients beyond borders: A study of medical tourists in four countries. *Global Soc. Policy, 10*(3), 315–335.

Crooks, V. A., Kingsbury, P., Snyder, J., & Johnston, R., (2010). What is known about the patient's experience of medical tourism? A scoping review. *BMC Health Serv. Res., 10*, 266–277.

Crooks, V. A., Turner, L., Snyder, J., Johnston, R., & Kingsbury, P., (2011). Promoting medical tourism to India: Messages, images, and the marketing of international patient travel. *Soc. Sci. Med., 72*, 726–732.

Crouch, D., (2004). *Tourist Practices and Performance.* Chapter 7 on: A companion to tourism. London: Blackwell Publishing LTD.

Dehdashti, S. Z., & Nakhaei, H., (2016). An entropy (Shannon) based approach for determining importance weights of influencing factors in selecting medical tourism destinations. *International Journal of Travel Medicine and Global Health, 4*(4), 115–121.

Gan, L. L., & Frederick, J. R., (2011). Medical tourists: Who goes and what motivates them. *Health Mark Q, 30*(2), 177–184.

Global Wellness Institute, (2018). *Statistics and Facts.* Retrieved from: https://www.globalwellnessinstitute.org/press-room/statistics-and-facts/ (accessed on 14 July 2021).

Gretzel, U., Sigala, M., Xiang, Z., & Koo, C., (2015). Smart Tourism: Foundations and developments. *Electron. Mark, 25*, 179–188.

Guy, B. S., Henson, J. N., & Dotson, M. J., (2014). Characteristics of consumers likely and unlikely to participate in medical tourism. *Int. J. Healthc. Manage, 1*(1), 12–24.

Henson, J., Guy, B., & Dotson, M., (2015). Should I stay or should I go?: Motivators, decision factors, and information sources influencing those predisposed to medical tourism. *International Journal of Healthcare Management, 8*(1), 4–14.

Heung, V. C. S., Kucukusta, D., & Song, H. A., (2010). Conceptual model of medical tourism: Implications for future research. *J. Travel Tour, 27*, 236–251.

Iordache, C., Ciochină, I., & Roxana, P., (2013). Medical tourism-between the content and socio-economic development goals. Development strategies. *Romanian Journal of Marketing, 1*(1), 31–42.

Kim, S., Chung, J. Y., Petrick, J., & Park, J. W., (2018). Determination of preferred performing arts tourism products using conjoint analysis. *Journal of Vacation Marketing, 24*(1), 44.

Kohli, A., & Sukumar, R., (1990). Heuristics for product-line design using conjoint analysis. *Management Science,* (12), 1464.

Lee, M., Heesup, H., & Lockyer, T., (2012). Medical tourism-attracting Japanese tourists for medical tourism experience. *J. Travel Tour, 29*(1), 69–86.

Levary, R. R., (2011). Multiple criteria approach to ranking medical tourism destinations. *Thunderbird Int. Bus. Rev., 53*(4), 529–537.

Lunt, N., & Carrera, P., (2010). Medical tourism: Assessing the evidence on treatment abroad. *Maturitas, 66*, 27–32.

Medina-Mun, D., & Medina-Mun, R., (2013). Critical issues in health and wellness tourism: An exploratory study of visitors to wellness centers on gran Canaria. *Current Issues in Tourism, 16*(5), 415–435.

Pearce, P., Filep, S., & Ross, G., (2010). *Tourists, Tourism, and the Good Life.* Milton Park, U.K.: Taylor and Francis Group.

Ross, S. D., Norman, W. C., & Dorsch, M. J., (2003). The use of conjoint analysis in the development of a new recreation facility. *Managing Leisure, 8*(4), 227–244.

Saraniemi, S., & Kylänen, M., (2011). Problematizing the concept of tourism destination: An analysis of different theoretical approaches. *Journal of Travel Research, 50*(2), 133–143.

Singh, N., (2012). Exploring the factors influencing the travel motivations of U.S. medical tourists. *Current Issues in Tourism, 16*(5), 436–454.

Smith, M., & Puczko, L., (2013). *Health, Tourism and Hospitality: Spas, Wellness and Medical Travel.* London: Routledge Publications.

Smith, M., & Puczko, L., (2014). *Health, Tourism and Hospitality: Spas, Wellness and Medical Travel.* London: Routledge Publications.

Smith, P., & Forgione, D., (2007). Global outsourcing of healthcare: A medical tourism decision model. *J. Inf. Technol. Case Appl. Res.*, *9*(3), 19–30.

Smith, P., & Forgione, D., (2014). Global outsourcing of healthcare: A medical tourism decision model. *Journal of Information Technology Case and Application Research, 9*(3), 19–30.

Sultana, L. A., Dale, M., & Syed, M. U. R., (2012). Crowdsourcing travel experience: A case study of user participation on the Tourism Australia Facebook page. In: *Proceedings of the Mediterranean Conference on Information Systems.* Guimaraes, Portugal.

Wind, J., Green, P. E., & Shifflet, D., (1989). Courtyard by Marriott: Designing a hotel facility with consumer- based marketing models. *Interfaces,* 1925–1947.

CHAPTER 10

MEDICAL TOURISM/TRAVEL IN INDIA: A COST COMPARISON OF PROCEDURES WITH THE UNITED STATES

FREDERICK J. DEMICCO[1] and JACKIE GUZMAN[2]

[1]*The Franke College of Business, School of Hotel and Restaurant Management, Northern Arizona University, USA*

[2]*Michigan State University, USA*

ABSTRACT

Home to over 1.3 billion people, India is known as an advanced and sought-after MT and medical travel destination around the world. India has the chance to grow its MT/travel industry more in the future through marketing and branding efforts and linking its medical centers with its hotels and the hospitality initiatives (H2H) India is renowned for (DeMicco, 2017).

10.1 INTRODUCTION

Medical tourism (MT) is traveling to another country to receive medical care or treatment. There are many different reasons why one might do this, such as to save money or to seek a treatment that is not available in one's home country. Immigrants sometimes return to their country of birth for a medical procedure because they feel more comfortable there or because they have family there that will help them during the recovery process. MT is a growing industry, and in 2017 alone, between 14 and 16 million people traveled outside their home country to receive medical treatment. It is estimated that the global MT market is worth between $45.5 and $72 billion (https://www.

cnn.com/2019/02/13/health/india-medical-tourism-industry-intl/index.html; DeMicco, 2017). Long known for its holistic approach to health and well-being, the popular Ayurvedic of India is world renown (Figure 10.1). Today for more serious medical procedures, India is a sought-after MT and medical travel destination.

FIGURE 10.1 Kairali: The Ayurvedic healing village.

Source: https://www.justkerala.in/kairali-ayurvedic-healing-village.html. © 2017 Culture of Kerala

10.2 DESTINATION: INDIA

Globally, India is the country of choice for many patients who choose to partake in medical tourism (MT) and medical travel. In 2015, the MT industry in India was valued at $3 billion and continued to grow (https://www.cnn.com/2019/02/13/health/india-medical-tourism-industry-intl/index.html). For Americans, getting medical procedures done in India often gives a savings of anywhere between 65% and 90% compared to prices for the same procedures in the U.S (https://www.cnn.com/2019/02/13/health/india-medical-tourism-industry-intl/index.html). Traveling abroad to get a medical procedure done can be risky, but India is a popular choice because the country is known for its high-quality doctors, many of whom have studied abroad (https://www.cnn.com/2019/02/13/health/india-medical-tourism-industry-intl/index.html). Not only that, but India is known for traditional homeopathic practices, which are being used to attract tourists as well (https://www.cnn.com/2019/02/13/health/india-medical-tourism-industry-intl/index.html). Some of these practices include Ayurveda, yoga, Unani, and Siddha.

There are approximately 3 million hotel rooms in India (https://hotelivate.com/travel-tourism/the-ultimate-indian-travel-hospitality-report-2019/). The national capital region of Delhi has the highest supply of rooms with over 14,000 in 2018. There are approximately 2,000 hotels in India and many surrounding medical facilities serving medical tourists and medical travelers for Hospitality Bridging Healthcare (H2H) (DeMicco, 2017).

Indian hospitals have taken a leaf out of hospitality brands' playbook by carefully branding themselves (Chekitan, 2012). A handful of hospital brands dominate the private medical services market in India. Brands such as Apollo and Fortis have positioned themselves as the leading providers of services to medical tourists. See Figure 10.2: The Apollo Hospitals and Figure 10.3: Luxurious Fortis Memorial Hospital in Gurgaon, just outside of New Delhi, India with the nearby Taj Hotel, Marriott, Radisson, and Park Plaza Hotels for serving medical tourists/travelers.

FIGURE 10.2 An Apollo hospital, India's largest hospital chain.

Source: https://www.laingbuissonnews.com/healthcare-markets-international-content/india-suneeta-reddy-sells-3-6-block-in-apollo-hospitals/. © 2022 Apollo Hospitals Enterprise Ltd.

FIGURE 10.3 The luxurious Fortis Memorial Hospital in Gurgaon, just outside of New Delhi, India, with the nearby Taj Hotel, Marriott, Radisson, and Park Plaza hotels for medical tourists/travelers.

Source: https://www.businessdestinations.com/move/travel-management/indias-luxury-hospitals-spur-health-tourism/. © Business Destinations 2022

10.3 PRICE COMPARISON: INDIA VS. THE US

The U.S. spends more on healthcare than any other wealthy country in the world, yet life expectancy in the United States (U.S.) is behind those of comparable countries, and many Americans find it difficult to pay these high price tags (https://www.hsph.harvard.edu/news/hsph-in-the-news/the-most-expensive-health-care-system-in-the-world/). Below we describe some of the most expensive medical procedures in the U.S. and compare the prices to the same procedures in India.

First, we look at transplants. Kidney and liver transplants are both very costly but necessary and life-saving procedures. Nearly 40,000 organ transplants occurred in the U.S. in 2019 alone, but as of March 2020 over 112,000 candidates are on the National transplant waiting list (https://www.organdonor.gov/statistics-stories/statistics.html). A kidney transplant costs about $46,760 in the U.S., (https://www.cbsnews.com/pictures/most-expensive-medical-procedures-without-insurance/) while ranging from $5,000-$22,000 in India (http://www.indiahealthcaretourism.com/average_cost_of_treatment.php). Meanwhile, a liver transplant costs about $101,240 in the U.S., (https://www.cbsnews.com/pictures/most-expensive-medical-procedures-without-insurance/) while ranging from $40,000-$56,000 in India (http://www.indiahealthcaretourism.com/average_cost_of_treatment.php).

A coronary artery bypass graft (CABG) is another very expensive procedure in the U.S. CABG is also known as heart bypass or bypass surgery, and it is used to treat coronary artery disease. Those with coronary artery disease have plaque buildup on the arteries near their heart, so CABG is a procedure where an artery is taken from one part of the body to replace these blocked arteries (https://www.hopkinsmedicine.org/health/treatment-tests-and-therapies/coronary-artery-bypass-graft-surgery). In the U.S., a CABG costs about $77,177, (https://www.cbsnews.com/pictures/most-expensive-medical-procedures-without-insurance/) while in India it only runs about $3,000–$9,500 (http://www.indiahealthcaretourism.com/average_cost_of_treatment.php).

Spinal fusion is what it sounds like: connecting two or more vertebrae in the spine, permanently eliminating motion between them. Patients have this done to improve stability, correct a deformity (such as with scoliosis), or to reduce pain (such as with a herniated disk) (https://www.mayoclinic.org/tests-procedures/spinal-fusion/about/pac-20384523#:~:text=Spinal%20fusion%20is%20surgery%20to,healing%20process%20of%20broken%20bones). In the U.S., a spinal fusion procedure costs about $67,826, (https://www.cbsnews.com/pictures/most-expensive-medical-procedures-without-insurance/) while in India it costs about $6,000–$13,000 (http://www.indiahealthcaretourism.com/average_cost_of_treatment.php).

Those suffering from arrhythmia (an abnormal heartbeat) may need an implantable pacemaker or defibrillator. A pacemaker helps keep the heart rate consistent, while a defibrillator monitors the heartbeat and kicks in when it detects a dangerous rhythm (https://medlineplus.gov/pacemakersandimplantabledefibrillators.html). In the U.S., having a pacemaker or defibrillator inserted costs about $58,281 (https://www.cbsnews.com/pictures/most-expensive-medical-procedures-without-insurance/) while in India, the procedure costs about $1,800–$4,500 (http://www.indiahealthcaretourism.com/average_cost_of_treatment.php).

Finally, we look at hip replacement surgery. Oftentimes people opt for a hip replacement when they are in so much pain that nothing other than surgery will be able to help them. Various conditions can damage the hip, such as osteoarthritis, rheumatoid arthritis, and osteonecrosis (https://www.

mayoclinic.org/tests-procedures/hip-replacement/about/pac-20385042). In the U.S., hip replacement surgery can cost about $25,000, while in India, one can have it done for $7,000 (Sengupta, 2011).

Procedure	Cost in the U.S. ($)	Cost in India (in U.S. Dollars)
Liver transplant	$101,240*	$40,000–$56,000**
Coronary artery bypass graft	$77,177*	$3,000–$9,500**
Spinal fusion	$67,826*	$6,000–$13,000**
Insertion of pacemaker or defibrillator	$58,281*	$1,800–$4,500**
Kidney transplant	$46,760*	$5,000–$22,000**
Hip replacement	$25,000***	$7,000***

**https://www.cbsnews.com/pictures/most-expensive-medical-procedures-without-insurance/.*
***http://www.indiahealthcaretourism.com/average_cost_of_treatment.php.*
****Sengupta (2011).*

10.4 CONCLUSION

MT and Wellness ties strongly into Hospitality Bridging Healthcare (H2H), a significant growth area for countries and careers (DeMicco, 2017). Home to over 1.3 billion people, India is known as an advanced and sought-after MT and medical travel destination around the world. India has the chance to grow its MT/travel industry more in the future, through marketing and branding efforts and linking its medical centers with its hotels and the hospitality initiatives (H2H) India is renowned for (DeMicco, 2017).

KEYWORDS

- **coronary artery bypass graft**
- **global medical tourism**
- **growing industry**
- **hospitality bridging healthcare**
- **medical tourism**
- **medical treatment**
- **United States**

REFERENCES

Chekitan, S. D., (2012). *Hospitality Branding* (p. 192). Ithaca, London: Cornell University Press.

DeMicco, F. J., (2017). *Medical Tourism and Wellness: Hospitality Bridging Healthcare (H2H)*. Apple Academic Press. https://www.amazon.com/Medical-Tourism-Wellness-Hospitality-Healthcare/dp/177188505X (accessed on 14 July 2021).

http://www.indiahealthcaretourism.com/average_cost_of_treatment.php (accessed on 14 July 2021).

https://hotelivate.com/travel-tourism/the-ultimate-indian-travel-hospitality-report-2019/ (accessed on 14 July 2021).

https://medlineplus.gov/pacemakersandimplantabledefibrillators.html (accessed on 14 July 2021).

https://www.businessdestinations.com/move/travel-management/indias-luxury-hospitals-spur-health-tourism/ (accessed on 14 July 2021).

https://www.cbsnews.com/pictures/most-expensive-medical-procedures-without-insurance/ (accessed on 14 July 2021).

https://www.cnn.com/2019/02/13/health/india-medical-tourism-industry-intl/index.html (accessed on 14 July 2021).

https://www.hopkinsmedicine.org/health/treatment-tests-and-therapies/coronary-artery-bypass-graft-surgery (accessed on 14 July 2021).

https://www.hsph.harvard.edu/news/hsph-in-the-news/the-most-expensive-health-care-system-in-the-world/ (accessed on 14 July 2021).

https://www.laingbuissonnews.com/healthcare-markets-international-content/india-suneeta-reddy-sells-3-6-block-in-apollo-hospitals/ (accessed on 14 July 2021).

https://www.mayoclinic.org/tests-procedures/hip-replacement/about/pac-20385042 (accessed on 14 July 2021).

https://www.mayoclinic.org/tests-procedures/spinal-fusion/about/pac-20384523#:~:text=Spinal%20fusion%20is%20surgery%20to,healing%20process%20of%20broken%20bones (accessed on 14 July 2021).

https://www.organdonor.gov/statistics-stories/statistics.html (accessed on 14 July 2021).

Sengupta, A., (2011). Medical tourism: Reverse subsidy for the elite. *Signs, 36*(2), 312–319. doi: 10.1086/655910.

PART IV

Marketing and Branding

CHAPTER 11

THE THINGS I LEARNED ABOUT BUILDING A BRAND IN THE MEDICAL TOURISM SECTOR

ILAN GEVA

President at Ilan Geva and Friends, and Adjunct Professor at the Graham School, University of Chicago and DePaul University, Chicago, Illinois, USA; 2252 Washington Dr., Northbrook, IL–60062, USA, E-mail: ilan@ilanandfriends.com

ABSTRACT

Medical tourism (MT) is actually not a new phenomenon, or a new business category. As early as 2000 years ago, people were traveling to seek medical treatments and solutions in other geographic regions, and there were many known destinations that provided the services.

What changed in the last 10 years is obviously the technology, and with it the realization that for certain destinations, MT could be lucrative.

Forecasters have predicted a boom in the business, and up until recently, some extremely exaggerated reports valued the MT market to be worth 439 billion dollars! Nonsense (http://global-mtm.com/medical-tourism-industry-valued-at-439-billion-poised-for-25-annual-growth/).

Unfortunately for all of us in the business, there is no credible data to support any claim, and they're are very few academic institutions involved in teaching the basics required for managing MT efforts.

Many of us, who are active in the business, bring our previous experiences from other disciplines. Some come from health insurance, others from tourism, some from pharma and others from nursing, and yes there are few doctors involved.

The main bodies of knowledge come at this point from keen observation, frequent travel, and visits to as many hospitals as possible.

I am part of that group that is fortunate enough to be included in many conferences and site inspection of destinations, hospitals, and clinics that promote MT. My impressions are included in these reviews.

11.1 INTRODUCTION

Since 2008, I have been invited to speak at many medical tourism (MT) conferences around the world. Among the countries I have visited were England, Spain, Poland, Croatia, Romania, Czech Republic, Kosovo, Panama, Mexico, Malaysia, Japan, China, India, Israel, Turkey, USA, The Philippines, Korea, Portugal, and more…

I had the privilege of meeting and providing consulting services to ministers, heads of MT boards and councils, CEOs of hospitals and clinics, and conducting workshops about the topic.

This hands-on interaction with the MT community and stakeholders gives me a unique observation point, and when combined with my lifelong career as a brand builder and marketer, allows me to develop my own opinions without relying on academic papers (which are scarce to begin with).

Brand equity is the key to success in today's competitive MT market. Ilan Geva, a global brand communications expert, looks at how destinations can best market themselves for medical travel success.

11.2 MEDICAL TOURISM (MT)

Two magical words firmly etched onto our business radar for the last 7 years thanks to a Deloitte report, MT, Consumers in search of value, 2008 (https://www.academia.edu/9144718/Medical_Tourism_Consumers_in_Search_of_Value_Produced_by_the_Deloitte_Center_for_Health_Solutions) which provided a high-impact picture of the potential opportunities afforded by travel for medical care. The reaction to this report was far-reaching, with many hoping to make quick money from the MT industry, while at the same time reaching exotic destinations and assisting the rapid development of healthcare in poorer nations. The report cited a potential $100-billion market, with an average annual growth rate of 25%, and a lot of people were taken in by these figures, still basing their business decisions on figures that are an illusion. So, what is the reality when it comes to supply and demand in this global industry, and how are countries marketing themselves to those who are willing to travel for medical care? At the moment, around 90 nations have declared themselves 'MT destinations.' They are essentially competing for the same global patient, offering very similar medical solutions and services, and claiming to charge lower prices than the competition. The simple market factors of supply and demand already show a clear tilt towards supply, while demand is barely measurable.

Marketing is an essential element to get right, then, if countries are to stand out in a fairly crowded global marketplace. If we examine the offering of other products on the global open market, we find that consumer products with similar multinational distribution and an equivalent price point (cars, for example) are professionally marketed. But medical services are known for being marketed in a slightly different way because medical conditions are not planned or budgeted for in advance. MT marketing, in particular, is deficient in government support, budget allocation, legal, and ethical implementation.

That said, where governments do support MT in their country, and where their marketing efforts are more professional and cohesive, these countries and their hospitals are more often found on lists of the 'best hospitals in the world.' A look at the MTQUA's list, *Medical Quality Travel Alliance, World's best hospitals for Medical Tourism* for example, https://mtqua.org/top-10-worlds-best-hospitals-for-medical-tourists-list/ provides an example of what this particular organization believes are the best hospitals, according to its criteria. The list contains no U.S. hospitals, but it does contain hospitals in countries including Malaysia, Turkey, Thailand, Singapore, and India-whose governments all support MT at various levels.

11.3 GOVERNMENT SUPPORT

On 14 August 2013, Bloomberg BusinessWeek mentioned the topic of government involvement in MT. It stated: "India, which offers the world's biggest savings for U.S. medical tourists, is losing clients to Singapore and Thailand as visa rules and greater [public] awareness of drug-resistant germs that spread from the South Asian nation scare away patients. Government neglect means India may fail to tap the $40-billion market that's expanding 25% a year, said Josef Woodman, founder of the guidebook *Patients Beyond Borders*" (Joseph, 2008). In the same article, Prathap Reddy, a cardiologist who built a $2 billion hospital chain over 30 years, said: "[The government has] done everything to ruin our prospects of becoming a tourism center. I once said India should become the global healthcare destination now, I am swallowing those words. It could grow 10-fold in the next 5 years, if only the government would facilitate it, the way others have."

Here, you have a controversial government action which causes delays in the processing of medical visas-contradicting every marketing logic.

At the same time, some governments have worked on visa issues and have enacted laws to make access to the country's medical care easier for those from abroad by relaxing certain visa rules. Thailand, for example, relaxed its immigration laws with the specific aim of promoting MT. New regulations in the country allow nationals from Bahrain, Kuwait, Oman, Qatar, Saudi Arabia, and the UAE to stay in Thailand for 90 days without a visa. Such medical tourists are allowed to be accompanied by up to three individuals who are also Gulf state nationals. Nationals from Argentina, South Korea, Peru, Brazil, and Chile are also eligible for the 90-day visa exemption.

It is claimed that 60,000 Omanis visited Thailand for medical care in 2012, and with these relaxed rules in place, numbers for this year are expected to be nearer 75,000. In a further bid to expand MT, some of the country's hospitals have also established offices in Muscat so they can better target Omanis seeking care. In addition, Oman Air has added more weekly flights on the Muscat-Bangkok route.

Thailand also promotes MT through its appearance at various roadshows, and its government officials are traveling the world to spread the word. I have been to one of their roadshows in the U.S., where several high-ranking officials were present, but they were from the Tourism Authority, spoke limited English and presented long and statistics-loaded presentations.

Medical SEA, a Malaysia-based online publication dedicated to MT in Asia, published an article in August 2013, predicting that MT in Thailand will grow by 24.4% between 2013 and 2015. The country houses 28 hospitals with JCI accreditation, and this information is used in the marketing materials of the Tourism Authority of Thailand. With current political unrest, all predictions are off.

Similar activities are taking place in countries like Malaysia and Turkey. Turkey boasts 50 JCI-accredited hospitals, some of which look and feel like five-star hotels. During official visits with the Health Ministry in Ankara, I have received approximately 10 kilograms of printed materials, mostly in Turkish. There is no doubt that the Turkish government is extremely involved in promoting MT, and sees it as a major financial category, but coordinating its efforts is a bit of a challenge, especially since private hospitals compete with government hospitals, who then compete with university hospitals, and so on.

Furthermore, Turkey was planning special Free Zones for medical practices, whereby foreign doctors would be allowed to practice in Turkey without speaking Turkish, and without paying taxes. This plan did not come about, as the minister advocating such MT-related activities was moved into a new post … but more on that later. Again, political unrest, and monetary instability in turkey can place all such activity in jeopardy.

In Malaysia, where the government created a separate entity to coordinate and promote MT-the Malaysia Healthcare Travel Council (MHTC) Its efforts are proving extremely successful, and Malaysia is enjoying a steady and growing stream of patients from neighboring countries and beyond. The MHTC even has its own website; which reads: "We are proud to be among one of the only countries within the region where MT is promoted by the government. This, thus, provides medical tourists the assurance on quality care, regulations, safety standards and the governing laws within this industry."

The previous MHTC's CEO was a tireless executive, who traveled the world, lectured, promoted, and got involved with any source and resource that can help Malaysia be at its best. And the facts speak for themselves. Malaysia provides good healthcare services not only to its citizens but also to many patients in the region. Thus, an important factor to observe in the model Malaysia has developed is that foreign patients do not interfere, push aside, or postpone regular services to locals.

The unfortunate events leading to Malaysia Airlines demise, and the way the government dealt with that crisis did not help the Malaysian brand reputation.

In another country, Israel, a public scandal erupted, where senior doctors demanded payments "under the table" in addition to the quoted hospital fees. Obviously, any marketing and brand equity assets can be wiped instantly following such events.

Israeli MT efforts are also facing hard realities due to the fact that the location is considered unsafe. In addition, previous governments made attempts to block the MT industry by imposing 15% taxes on MT activities. The fall of the Russian currency had a serious impact on the fortunes of the MT market, since 99% of the patients came from Russia, Ukraine, and other Russian speaking countries.

Altogether, Israel is a negative example of a market with great potential, good hospitals and excellent doctors, and bad brand management of the MT business.

11.4 WASTED EFFORTS

I personally believe that the choices medical tourists make are generally driven by country/destination first, and doctor/hospital second. So, it is important for countries to cohesively and effectively market themselves as MT destinations. Price is not necessarily the main motivator for patients to travel in search of a medical solution.

Many countries invest much of their marketing efforts in participating in MT conferences in different parts of the world, of which there are around two per month. Unfortunately, few of these events are bringing patients to any of the participating countries, simply because those conferences are not marketed to attracting patients. They attract, primarily, healthcare officials, suppliers, marketers, and consultants, hospital administrators, clinics, tourism boards and, to a lesser degree, insurance companies, equipment manufacturers, investors, wellness providers and the like.

By spending huge amounts on sponsorships, booth rental, build, and design, staff, printed materials, gifts, and mementos, MT promoters are literally wasting significant amounts of money without any return on their investment. Recently, I participated in a conference where Poland had a big, fancy booth, with about 15 people on the floor, and nothing to sell. Poland does not have any known superior medical practices that present potential for being a competitor in this category. The only reason for being there is the fact that Poland has created a National Marketing Unit within the Economic Promotion Department, promoting polish MT by the polish information and Foreign Investment Agency.

They have a budget, or a grant, and they have to spend it-simple. The only way Poland is hoping to market itself is by bringing "buyers" on familiarization trips to some of the main hospitals in the country.

During just a couple of months in 2013, we have seen the cancelation of two MT conferences-one in Ivory Coast, the other in California- and the signs of conference fatigue are evident. The best solution for countries wishing to make an impression, then, is by building 'medical brand equity'- and the way to do this is to build a good brand.

11.5 GETTING IT RIGHT

With about 90 countries vying for business and claiming the ability to provide medical travel services, there is a degree of confusion as to the 'best place to go' for medical travelers. Let us look, then, at where different countries' strengths lie with regards to their brand.

I would say, without a doubt, the U.S. has the strongest brand equity in terms of 'centers of medical excellence.' Brands like Mayo Clinic, Cleveland Clinic, and Johns Hopkins have been busy building their brand equity for years, much before the term 'medical tourism' was fashionable. A single hospital in Thailand built a good reputation some years ago, investing in infrastructure and marketing, and put the country on the MT map; but does this mean that Thailand is now a 'center of excellence'? A similar example exists in the Philippines, where one hospital is an absolute shining example of excellence in medical service, but that does not rub on the entire country. In many cases, few renowned hospitals and their international marketing efforts have made their country one of the world's top MT destinations. Malaysia is investing in its medical services sector in a unique and significant way, making the entire healthcare marketplace a preferred area in which the government invests equally for domestic and the incoming patients. This is evident in both government and private hospitals, which know not only how to build facilities, but how to promote them (Laurence and Ilan, 2019). Korea has invested a great deal in marketing and establishing government offices to welcome medical travelers, but there is still the lack of a fully developed welcoming culture that appeals to patients outside the immediate region. The same applies to Turkey, a country that has invested heavily in its medical sector, but which stands the risk of losing out due to political and civil unrest. That said, all these countries have invested in brand equity for the long run and are committed to the process.

The most important thing, then, is to make a commitment to build a brand. Brand building takes time and never stops. It is not about brochures, websites, or new hospitals: marketing is the second phase, not the first. It is about the commitment of an entire healthcare community to build what it takes to be a competitor, to offer value, and to establish a point of difference or unique selling point.

This in itself involves all the brand touchpoints (Laurence and Ilan, 2019). So, never use scrupulous facilitators; have an airline partner that could offer real medical support; have airport facilities that can expedite services to medical patients; have a visa process that makes everything easy for medical visitors; have a complete cultural understanding of the patients you intend to accommodate, make sure they have the food they like, make sure their families have comfortable and easy accommodation while their relatives are being treated; prepare an easy release process with clear instructions for follow-up; and make sure all billing issues are transparent and clear for all.

There are many other steps along the way, but the idea here is to demonstrate that building a brand in medical travel does not consist only of the medical treatment. It involves the entire journey and involves many different entities that need to work together to provide an appealing package.

One thing to avoid is politicizing medical travel. As many governments get involved in granting budgets, tax legislation, and similar activities, in connection with medical travel, they have a tendency to try and influence the personality and the marketing strategy of such brands. That is a mistake. Politicians come and go, but the actual brand will stay there forever.

As we have seen, an example of how politics and MT brand-building does not mix occurred in Turkey, where the prime minister shuffled his cabinet and released a health minister who was a great champion of the medical travel industry in that country. As a result, advanced negotiations to create an exciting medical travel brand in Turkey came to a standstill. Thus, a country's overall branding process should be a continuous effort, executed by the direct stakeholders of the brand and the professionals who can assist them in building a long-term, lasting brand.

The second issue is allocating marketing budgets for medical travel (https://www.imtj.com/blog/how-do-you-spend-your-marketing-budget/). We all know that commercial entities allocate anything between five and 10% of their revenue to marketing spend. Currently, the medical travel field generally does not even come close to this. To succeed in a competitive market (and this is a very competitive market) countries, destinations, and hospitals must realize that spending on marketing is as important as building a new wing at the hospital or buying the latest DaVinci robotic surgery equipment.

Very few countries allow professional marketers to get involved in consulting and helping them build a brand, the way it should be done. As a result, marketing efforts are often amateurish at best. No one wants to seek treatment from an amateurish doctor or hospital, so marketing efforts and materials need to be highly professional.

The most crucial and effective marketing tool for those who want to survive in the competitive MT market, then, is to build a 'country' brand first, 'destination' brand next, followed by 'doctor' and 'hospital' brands (Laurence and Ilan, 2019).

The success of any brand, though, is the result of the unified efforts of many people-namely, the brand's stakeholders. To achieve success, a brand must be understood and passionately supported not only by its internal stakeholders, but also by its external stakeholders (Laurence and Ilan, 2019). So, to create a successful MT marketing brand, countries need the support of their government, individual hospitals, airlines, approved facilitators, tourism boards, professional marketers, and so on. And they need to work together to establish the unique selling point or points of its healthcare sector-whether that be certain specialist treatments that are available only in that country- or only a few countries; the natural beauty afforded by the country for recuperation; multiple transportation links; centers of excellence; or perhaps value for money. A brand that succeeds in having all its stakeholders speaking the same language, and cooperating with each other, is a brand with a competitive advantage- and in marketing, that is what we are always looking for.

11.6 EXTENDING YOUR HEALTHCARE BRAND, AN OLD CONCEPT, BUT A GOOD ONE (IF YOU KNOW HOW TO DO IT)

Around 10 years ago, my branding/marketing/ad agency was in charge of the Rehabilitation Institute of Chicago (RIC) account. Among the many things we did for them, was the branding of this extremely distinguished rehabilitation institute. To most of you, when I say "branding," you will immediately think that we have designed the logo and everything else that comes after that, from printed materials to online presence (mind you, that was before social media was trendy).

And you are actually wrong. We did not design the logo, the logo existed, but not much more. We designed the graphic identity manual and went on by designing every single brand touchpoint the institute has, including in 22 clinics around the Chicago metro area. We also collaborated with the institute on many other internal aspects, such as communication tone and manner, addressing the public, accreditation process, new business, recruitment, PR, and education (Laurence and Ilan, 2019).

One of the most interesting aspects in controlling the brand identity was the fact that they continuously win the U.S. News and World Report No. 1 Hospital Ranking in the rehabilitation category, for 30 years!

It not only allowed them to crow about this achievement in an annual ad campaign, but also attracted some other hospitals, which do not offer credible rehabilitation services. That attraction helps a lot in extending the brand.

Being affiliated with another successful brand is always good, and for some of the smaller, regional hospitals in the Midwest, having the RIC name on the door was a great opportunity. Extending your brand by making it an endorser to others could be a great strategy. And so, the concept of brand extension via strategic alliance became very relevant for RIC.

(Just as a side note: During our time as the brand guardians, we preached against using the acronym RIC, only the full name Rehabilitation Institute of Chicago).

Hospitals such as Alexian Brothers (AMITA), Advocate Illinois at Masonic Medical Center, Herrin Hospital, RML Specialty Hospital, Southern Illinois Healthcare, Franciscan St. Anthony Health, Silver Cross Hospital, and others were/are proud to display the RIC brand as an extension of theirs. Obviously, this affiliation acts as a brand endorser for these local/regional hospitals and elevates them to national status when it comes to rehabilitation.

It goes without saying that each of these hospitals follows the absolute lead and guidance of RIC in all matters concerning rehabilitation. A true brand extension can never be useful if the guiding principles of the original, authentic brand are not followed. The benefits are clear, but the risks are there as well.

While this model is limited to a very specific specialty, rehabilitation, other brand extension benefits both sides, as an endorsement and strategic alliance.

The financial arrangements behind the scene are kept private and are actually of no interest to consumers. But opaque as they are, the fact remains that consumers (patients) are brand savvy enough to recognize the brand equity added to their local healthcare provider who is associated with a national brand.

In the case of RIC, another obvious brand extension is having 22 outpatient locations throughout Chicago and the metro area.

These locations represented a significant branding opportunity as well as challenges at the time we handled the business.

The external opportunity, physical customer experience via brand touchpoints, as well as *the internal opportunity,* primarily manifested through an extremely diverse real estate spaces, plus clientele and staff demographics.

Ideally, any brand that faces such situation of having to extend the brand to many locations can solve certain problems by having a licensing agreement, or franchise contracts. That could immediately solve the uniformity issues.

But healthcare is not exactly fast food…In our case, the outpatient clinics were located in office buildings, as stand-alone structures, or mixed-use properties such as residential/commercial high-rises.

The clinics had different functions such as: Spine and Sports Rehabilitation Center, Center for Pain Management, Pediatric Outpatient Therapy, and Health and Fitness, etc.

Maintaining the same customer experience in vastly different spaces was a challenge, and we faced it by applying simple design solutions like easily recognizable signage for driving into the facility and navigating to it while facing confusing similar structures in suburban office malls. Needless to say, that each clinic resides in a different municipal entity, with a confusing set of rules for commercial signage.

Moving into the internal space, we found that none of the reception areas were similar. We had discussions with several clinic managers and guided them in matters such as interior ambiance and atmospherics.

That included minute details such as getting rid of "self-made" announcements, "homemade creative collateral," and other graphic displays that did not conform to the brand guidelines.

In my marketing classes, I teach that "Brand Extension" extends the current brand name and experience to new or modified product in any new category. That is all true and valid when it comes to a Product, usually FMCG.

But in the healthcare space, a brand extension, be it local, regional, national, or global, the product is irrelevant. It is all about experience and expertise.

11.7 DOES ACCREDITATION HELP YOUR HEALTHCARE BRAND EXTENSION?

During one of my many hospitals FAM visits in Asia, I could not help but notice a huge JCI logo mounted at the entrance. No one could have ignored that huge logo; it was literally covering the entire wall, almost floor to ceiling, as we walked into the entrance of the emergency ramp.

As you all know what JCI is, and probably what it stands for, there are still few questions burning in my mind:

- Does the logo of JCI represent "The Brand" for the patient?
- Or is the hospital brand more important?
- How does that JCI logo help the patients who are already about to enter the hospital? Any consumer benefits? Any brand enhancement?
- Any last chance to change your mind because you don't like JCI?

Obviously not. The fact is, very few, if at all, patients know the meaning of this logo. In awarding the Golden Seal to as many hospitals around the world, JCI is diluting the special meaning of the accreditation.

Further, it creates a "uniform" for some hospitals and prevents them from extending their own brand to other locations.

The brand association could become "JCI something…" VS. the hospital brand. Being a commercial entity, there is no doubt that JCI is trying to make a profit, but the ubiquitous nature of the accreditation prevents it from being a differentiating point for the consumer.

And my main concern is this: The patients (consumers) do not really know what JCI means. The values of JCI are not communicated to patients at all.

The reasons to choose a JCI accreditation that are being sold to *the hospitals*, not the *patients*, are:

- A deep commitment to high-quality care;
- A culture of safety for patients, visitors, and staff;A willingness to undergo rigorous preparation and a survey;
- Care delivery based on leading, evidence-based practices;
- Leadership in continuous compliance with exacting standards.

None of the reasons to choose a JCI accredited hospital mentions the comfort, both cultural and physical, or friendly and human treatment by the medical institution.

When we talk about MT, those are major consumer (patient) benefits. And my problem with this accreditation has nothing to do with the actual process. The process itself is good and necessary. It points to one portion of building a good brand, which is competence.

My problem is that healthcare institutions use the accreditation to extend their brand from one location to another and use it in marketing materials while the patients are oblivious to the real value in the specific location they chose to visit.

What many in the healthcare business forget, is the fact that the human touch is the deciding factor in choosing and preferring medical treatments. The CEO of Kaiser Permanente, Bernard Tyson, said in an interview: "We are heavy into how technology will enable us to do a better job in taking care of each individual and millions of people, but not at the expense of the human touch, the human touch will always be the deciding factor of how people feel about the care that they are getting" (Fierce Healthcare, 2016).

There is a great difference between the attempts to extend a brand by using graphic symbols and icons, award certificates, or any other achievement paraphernalia, and the actual reputation consumers seek.

I claim that certification alone is not a reason for anybody to go to a multiple JCI certified hospital in various locations. The brand extension is not created or assisted by having the JCI logo displayed in the hospital lobby and corridors or printed on the business cards of medical and administrative staff.

On many occasions, a positive brand image and reputation can be achieved by events such as the one that took place recently in Chicago.

The University of Chicago Medicine recruited a top transplant surgeon from Cleveland Clinic (https://www.modernhealthcare.com/article/20160909/

NEWS/309099995/university-of-chicago-medicine-snares-top-transplant-surgeon). We heard in the news that Dr. John Fung, formerly head of general surgery at the Cleveland Clinic, will be returning to Chicago to become director of the university's newly created Transplantation Institute.

That alone can elevate the brand image and reputation of University of Chicago Medicine to a new level and give them the opportunity to extend the brand into a new medical expertise, one they did not possess before. Accreditation is not playing a part in this move.

11.8 DOES THE BEST HOSPITAL RANKING BY U.S. NEWS AND WORLD REPORT MATTER TO YOUR BRAND EXTENSION?

Every August, many of us in the healthcare business in the U.S. wait breathlessly for the Best Hospital ranking issue.

It is the best hospital Superbowl! The report started around 1990, and in it one can find the names of almost all American Hospital Association members, who garnered reputation in 16 specific categories:

- Cancer;
- Cardiology and heart surgery;
- Diabetes and endocrinology;
- Ear, nose, and throat;
- Gastroenterology and GI surgery;
- Geriatrics;
- Gynecology;
- Nephrology;
- Neurology and neurosurgery;
- Ophthalmology;
- Orthopedics;
- Pulmonology;
- Psychiatry;
- Rehabilitation;
- Rheumatology;
- Urology.

But rankings in Ophthalmology, Psychiatry, Rehabilitation, and Rheumatology do not depend on hard data, for example. Survival outcomes are taken very seriously in this ranking, and the reputation is based on physicians' surveys.

Some hospitals that do make it to the top 20 in six categories are considered to be members of the Honor Roll, a very special list indeed.

Overall, the last number of participating hospitals is just under 5000. The methodology and the serious consideration given to the entire process made this list a very coveted and desired target by many.

I have interviewed heads of departments in hospitals who were extremely upset their department did not make it, while other hospital departments were recognized. Others are extremely unhappy about the hospital name not included or mentioned at all.

The simple equivalent to this list is another famous one: The Michelin Restaurant Guide, and the 3 Star Award (https://guide.michelin.com/en/restaurants?lat=42.1130024&lon=-87.8072818).

The simple difference could be a good life or a good meal… But the main difference is that while the Michelin Star is a global brand, the Best Hospital Ranking is a strictly U.S. recognition brand.

The problem with using the Best Hospital title, reputation, and shield image, is the fact that it is not transferable from one hospital physical facility to another.

Just like the Michelin Star is not transferable from one restaurant branch to another location. However, in corporate marketing communications, there are ways to deliver the message and create the "aura" of the transferable quality by association.

For example, the RIC has many affiliates in the Midwest, who operate from other regional hospitals. While the hospital itself is not a Best Hospital award winner, but they can use the reputation of the RIC brand in their marketing materials like this: "Our partnership with the RIC, *the #1 ranked rehabilitation hospital in the U.S. according to U.S. News and World Report,* strengthens our commitment to clinical excellence and provides a continuum of care for patients needing physical, occupational, and speech therapy."

In 2017, RIC changed its name and moved to a new building. Following a generous donation, this famed institute moved to the new building named after the donors: Shirley Ryan Ability Lab (https://www.chicagobusiness.com/article/20170217/NEWS03/170219883/first-look-ric-shirley-ryan-abilitylab).

I doubt that a Best Hospital winner here can extend that kind of brand reputation to another country. That is the limitation of the Best Hospital ranking award, since it is relevant only in the U.S. market, and might have no meaning at all in other international markets (https://health.usnews.com/best-hospitals/rankings).

Some say that many patients today are starting a search for medical treatment online. As a matter of fact, in the U.S. 90% of 18- to 24-year-old say they would trust medical information shared by others on social media networks. Around 41% say that social media would affect their choice of a specific doctor, hospital, or medical facility.

Around 60% of doctors say social media improves the quality of care that patients receive. In Google's results to searching "The best hospitals in the world," the U.S. News Best Hospital Rankings is not a factor.

Instead, we find the site called "ranking web of hospitals." That ranking system features 100 hospitals around the world, and no surprise, 52 of those are U.S. based hospitals.

Cleveland Clinic is on top of that ranking and placed as number two on the U.S. Best Hospital list. They must be doing something right to be placed at the top 3 in almost every ranking system worldwide.

No surprise, Cleveland Clinic is about to open a hospital in London in the near future. That will add another international location to a health system that includes a 165-acre main campus near downtown Cleveland, nine regional hospitals throughout Northeast Ohio, and facilities in Weston, Fla.; Las Vegas, Nev.; Toronto, Canada; and Abu Dhabi, UAE.

This health system has done so many right things to make the brand expand globally, that the name "Cleveland Clinic" is enough to open doors without the need to rely on accreditations, rankings, or any other accolades that the industry craves.

They have done it from within first and foremost. Something all of us should aspire to learn and understand. Brands are being built from within first and utilize external marketing methods second (Laurence and Ilan, 2019).

KEYWORDS

- **brand platforms**
- **global industry**
- **Malaysia Healthcare Travel Council**
- **medical services**
- **medical tourism**
- **Rehabilitation Institute of Chicago**

REFERENCES

Fierce Healthcare, (2016). https://www.fiercehealthcare.com/ (accessed on 14 July 2021).

http://global-mtm.com/medical-tourism-industry-valued-at-439-billion-poised-for-25-annual-growth/ (accessed on 14 July 2021).

https://guide.michelin.com/en/restaurants?lat=42.1130024&lon=-87.8072818 (accessed on 14 July 2021).

https://health.usnews.com/best-hospitals/rankings (accessed on 14 July 2021).

https://mtqua.org/top-10-worlds-best-hospitals-for-medical-tourists-list/ (accessed on 14 July 2021).

https://www.academia.edu/9144718/Medical_Tourism_Consumers_in_Search_of_Value_Produced_by_the_Deloitte_Center_for_Health_Solutions (accessed on 14 July 2021).

https://www.chicagobusiness.com/article/20170217/NEWS03/170219883/first-look-ric-shirley-ryan-abilitylab (accessed on 14 July 2021).

https://www.imtj.com/blog/how-do-you-spend-your-marketing-budget/ (accessed on 14 July 2021).

https://www.modernhealthcare.com/article/20160909/NEWS/309099995/university-of-chicago-medicine-snares-top-transplant-surgeon (accessed on 14 July 2021).

Joseph, W., (2008). *Patients Beyond Borders*. Healthy travel publication.

Laurence M., & Ilan, G., (2019). *Global Brand Management*. Kogan page publishing.

CHAPTER 12

HOW TO ATTRACT MORE PATIENTS, INCREASE PROFIT, AND BUILD AN IRRESISTIBLE MEDICAL TOURISM BRAND

JANET BRUNO

Arizona State University, Phoenix, Arizona, USA

ABSTRACT

The author of this chapter went from the bedside to the boardroom to become an international advisor and speaker in the global healthcare arena. In this chapter, she introduces the need for a global solution in healthcare and then lays out very specific strategies that can propel a medical tourism (MT) business to rapidly attract more patients, increase profit, and build an irresistible brand and become a leading MT destination.

12.1 INTRODUCTION

The need for healthcare is one of those basic realities in life. The old saying is that two things are guaranteed: death and taxes. Well, the authors propose that the healthcare which precedes death (and hopefully extends it much farther into the future) is also one of the things guaranteed in life. It is something that everyone, no matter where they live, should have quality access to. However, often that is tragically not the case.

This chapter is all about how you can attract an audience who are patients who want exactly what you offer, no matter where they live. Then you will be able to show them the absolute best performance and experience imaginable. Next, you can nurture them to be so loyal that they want to return to you for more services. Lastly, they will be so excited about their experience with you that they bring their family, friends, and co-workers with them as well. This is all possible to you if you embrace and act upon the blueprint that you will receive here.

You will notice that the concepts that the authors share are drawn from four worlds that the authors know very well. First, as a medical doctor, the authors draw upon my deep knowledge and experiences as a clinician. Secondly, as a seasoned Fortune 25 physician executive, the authors draw upon the lessons learned, wisdom, and network achieved during that environment. Third, as an experienced business and marketing strategist, who for years has been a behind-the-scenes manager for successful eCommerce businesses, the authors draw upon the marketing and technology wisdom gained from these valuable experiences. Lastly, as one who has lived and worked in many countries, achieved conversational fluency in six languages, and achieved a graduate degree in international affairs and economics, the authors draw upon all this cross-cultural knowledge and experience.

Solving the healthcare crisis in a myopic way, thinking only of your one country, can never deliver you to success. It has failed in the past and will continue to fail. No matter how big or powerful your country is which you call home, you will not succeed with this myopic vision. Our current world has become a global village. The solution simply must arise from this perspective. The problems of healthcare exist globally. The solution simply must arise from global collaboration and connectivity.

The early adopters are those visionaries who have medical centers who already welcome international patients or patients who have traveled from afar within one's own country. Those early adopters are going to win. That could be you. You have already seen the opportunity and are therefore nowhere reading this book. You already recognize that the global patient is the patient of the future. It is still very early in the game. True, it has already become a multi-billion-dollar industry, but it still is early.

12.2 TACKLING THE NEW ERA OF HEALTHCARE

We are entering a new era of medicine. Today's health care arena is now a global opportunity and challenge.

Statistics show the global healthcare market is anywhere from 35 to 70 billion dollars right now. It has been increasing from 15 to 25% per year. Experts predict that is expected to continue at least through 2025. These are the currently available statistics; however, it is very difficult to know how accurate they are. The market potential could even be greater. But what is absolutely true is that this market, the Global Healthcare Arena, is going to dramatically increase year after year.

It is the future of medicine absolutely. People all over the globe want and need quality healthcare. They have very specific wants and needs, just as diverse as their own unique fingerprint. Many times, that healthcare is not available locally. Or, perhaps it is not accessible locally. Or, in many locations, it is simply not affordable. So that is where you come in. Why not have your medical destination be the place that people consider when their local offerings are not a solution?

Even though the field of medical tourism (MT) is clearly growing at a rapid pace, there are many challenges. Many people who have gone into the MT arena have stumbled or have failed to attract the number and quality of patients that they want and need. The reason why they have not grown at the speed they wanted, or perhaps not achieved the profitability they wanted, is because they did not have a clear, smart marketing strategy. Or, they really did not understand who they were as a business and what their real assets are. Without a clear roadmap, a medical center could spend a lot of money with little to no profitable results.

Before we dive into the nitty-gritty details, it is important to address the belief "but healthcare is different." What I mean by this, is that too often, healthcare professionals take command and do not grasp the reality that healthcare is a real business. As such, it is governed by the universal principles of business that apply to every single business, no matter what industry you are in. So many health professionals all around the world have wonderful intentions of helping patients. They have the medical knowledge and they have excellent technical skills. But if they do not have solid business and marketing know-how, they will not succeed.

What I am about to say may sound outlandish to you, but if you do not learn another thing from our time together, please remember this...

You are not in the MT business... You are in the Marketing business. Your immediate response may be, "What?!" Stay with me here, for this is crucially important.

Although doctors, and frankly most healthcare personnel, cringe at the thought of being in marketing, or sales, that mindset represents one of the biggest obstacles to succeeding in this business. For, if you really do have a fantastic, high quality health service that you offer (which I surely hope you do!), then it is really your duty and obligation to do all you can to bring people to you and 'sell' them on coming to you for the services they need.

If you fail to make this mindset shift, it truly will stump your growth. Many people say they get this idea intellectually, but I have found that it simply stays intellectual and never moves down into a deeper understanding that then transforms actions and behavior. To be successful in this arena, this one mindset shift must be complete and comprehensive.

Bottom line: In the pursuit to build out a successful global healthcare business, just as it requires a strategy for successful practice design, it is equally important to match that with a proven marketing strategy to ensure the total package success that you desire. With that, let us dive in.

12.3 DIRECT MARKETING IN HEALTHCARE

Now that we understand the incredible opportunity and potential of MT, as well as the challenges, in this section, we will dive into what is the most powerful approach to attracting patients and growing your MT business.

That approach is called direct marketing. Or it is sometimes called direct-response marketing. They are both the same thing. A basic definition of direct marketing is: taking specific strategic steps and actions that can directly produce a measurable result. The key phrase is 'directly produce a measurable result.'

It is the approach to marketing where there is a cause of action that you take that produces a specific effect.

The reason why I am using very specific and precise language to define this is because so many people, even so-called "marketing specialists," market in a radically different way. They approach marketing with these big, broad brush-strokes with the intention to "blanket the message to as many people as possible." They do TV spots or radio spots and simply declare their generic message to all people. However, when doing it this way, they cannot measure the results.

To contrast this generalized approach with an effective direct-response approach, if a high-performing radio ad was created, it would direct a listener to call a specific phone number, or visit a specific website, to collect a promised free gift. It could be a downloadable gift, or a booklet sent in the mail. It would be a valuable piece of education that the prospective patient is really interested in. (This is key!) Ideally, it should have a high perceived value. Whether the proposed action is a website click or a phone call, either way the response is documented and "tagged" to that radio advertisement. The number or responses or actions that a single radio ad produces are exactly how that radio show ad can now measure its effectiveness. This is what "smart marketing" looks like.

Although this is a brief definition of direct marketing, this is the approach that I recommend strongly that a growing MT and global healthcare business should take. Businesses in every industry utilize this approach extremely profitably and successfully.

So, the basic approach is: one targeted action creates one specific response. It must be trackable and measurable. This allows you to actually calculate and measure your return on investment (ROI). With this strategy you can discover, for example, that for every $1 put into a certain marketing campaign, you get $1.30 in return. Or $3.50 in return. Some of the clients I have worked with have gotten returns much higher than that.

If you are using the faulty generalized approach to marketing, there is really no way to accurately calculate ROI. Make no mistake, the ability to measure ROI is critically important. It is what separates super successful businesses from those who declare bankruptcy.

12.4 THE 3 M'S OF SUCCESSFUL HEALTH MARKETING

Now that you understand the high-level view of direct response marketing, let us go a bit deeper. There are three critically important lessons to best understand this direct marketing approach. A marketing mentor of mine, Mr. D. Kennedy, taught me that there are three components to all successful direct marketing. He calls them the 3 M's.

They are:

- Message;
- Market; and
- Media.

Let us begin first with message—the message is the story you tell in all of your advertising, whether it is online or offline, through direct mail, or even person-to-person. This is the permanent foundational message that you want to deliver about your product your service you are offering. It is all about you, your team, how you are unique, and why your patients love what you do.

Your entire message is centered on what is known as your Unique Selling Proposition, or USP. Your USP is really the theme or signature of your business or your product or your service. It is what makes you extremely unique. It is what makes you the "go to." It is what makes you the only logical choice. It is what differentiates you from everyone else. It is the essence of what you offer to the world that separates you from the pack.

Determining a company's USP takes time and effort. I have a whole process I run through with my clients. But, for the sake of brevity in this chapter, let us just jump to the next critical component.

The next M that is important to address is the market. This is also something that is critical to get dialed in with 100% accuracy. The market refers to who are you targeting with your services. It is known in the marketing world as "target marketing." Or, thinking of it another way, who would you like to target if you had to communicate your message just to one particular person or group of people? Who would that person be? What qualities, characteristics, thoughts, and fears would this person have? These are all essential components of effectively identifying your market.

At this point, you may be wondering why this is so important to drill down to so much detail. Trust me, the deeper you can drill, the more powerful your marketing will be. For, whether you realize it or not, most of marketing is based solidly on psychology. Why people take certain actions or do certain behaviors is often determined by how they feel. How they feel is determined by a litany of factors such as their educational level, where they live, which type of friends they hang out with, what hobbies they tend to enjoy, what religious or political beliefs they hold, and the list goes on and on.

The more you can understand your target market, the better you are able to perfectly match your message to your market. Just like tuning in a radio station to a clear frequency, you want to aim toward your message completely 'resonating' with a prospective patient so much that all of the static disappears. So, they can now see you as the true, clear voice and solution that you are.

The third critical M is Media. The media is actually the way you communicate what you are offering. The word media does not just imply what you typically think of, such as TV and radio. That is part of it, but media in this context is HOW your message is communicated. In today's modern and technologically advanced arena, the media can involve online video, email communication, downloadable ebook, a podcast, a clever piece of direct mail, or any of the traditional avenues of advertising like TV, radio, newspaper, or billboard.

Whatever avenue you choose, you will want to strategically create it so matter if it is a TV spot or email series or a book, each communication must be designed to reinforce your desired message and perception. That means you expertly communicate your unique selling proposition. You want it to communicate one specific purpose and action that you want that prospective patient to take.

This is a brief summary of what direct response marketing is. It is a clear message, delivered to the right market, in the most appropriate and effective media. There is so much involved in this one section that I really encourage you to read it over again and again. I touched upon these concepts at a high level; however, I spend full consulting days or deliver 90-minute keynotes on this one topic alone. There is so much more I could teach on this topic, but this would be an entire book on its own. However, it is very important that you get the basics and understand the true power of direct marketing.

There is so much more involved in successful healthcare marketing. Advanced strategies include: rolling out precise action-focused campaigns, creating funnels, developing customer/patient loyalty mechanisms, initiating referral incentives, fueling viral marketing, and so much more. However, these sophisticated topics must be reserved for another place and time.

For now, if the concepts of direct marketing are fully understood and applied efficiently, effectively, and strategically to MT, the potential success is beyond your wildest dreams. I hope our time together has illuminated your understanding of successful marketing in healthcare. Very few healthcare businesses embrace direct marketing effectively, so there is a very real opportunity for you to stand out and set the gold standard.

12.5 CONCLUSION

In closing this chapter, I realize we have covered a lot of territories. Some of it may have been totally foreign to you, or some may have been semi-familiar content shared in a very unique way. In any case, it is my goal that I have instilled in you a hope that you can win in this MT game.

My biggest challenge for you is to please treat your MT business as a REAL business. Treat it with respect it deserves. It will function for you beautifully if you apply scientific, reproducible direct marketing approaches. If you do that, your business will grow like calculated clockwork. It has been a genuine honor to share with you. I wish you all the very best!

KEYWORDS

- **fueling viral marketing**
- **global village**
- **health marketing**
- **healthcare**
- **loyalty mechanisms**
- **medical tourism**
- **return on investment**

CHAPTER 13

DIMENSIONS OF MEDICAL TOURISM AND DESTINATION CAPACITY ANALYSIS: A TOOLKIT

ALI A. POORANI

Alfred Lerner College of Business and Economics, Department of Hospitality and Sport Business Management, University of Delaware, 14 W., Main Street, Newark, DE–19716, USA, E-mail: poorani@udel.edu

ABSTRACT

This chapter provides a toolkit to simplify analyzing the destination's readiness to meet and or exceed guest expectations. The script provides detailed elements for the feasibility of introducing new tourism products to the destination, including inbound medical tourism (MT). However, a robust inbound MT development must also carefully address the challenges presented by political figures, advocacy groups, media, health providers, and patients. This chapter addresses a practical approach to achieving a well-designed plan.

13.1 TOOLKIT TO ANALYZE DESTINATION'S READINESS

13.1.1 INTRODUCTION

Worldwide medical tourism (MT) destinations are being established using a combination of public and private funds. These destinations not only focus on the quality of healthcare delivery, but also include direct and indirect related services, such as medical spas, alternative medicine, health, and wellness assets, hotels, restaurants, travel, transportation, and entertainment that patients and their families need and expect while they are visiting the destination.

The innovative and emerging medical destinations face many competitions from different corners of the world today. This planning guideline explores the strategic paths helping the destination to achieve its prominence.

The first part of the toolkit addresses the destination's tourism readiness by taking stock of the destination's recreational assets that are fundamental for planners, stakeholders, and the management of the destination, as well as the destination's desire to expand to new territories, such as medical, health, and wellness tourism. The last part of the toolkit stresses planning process and steps necessary for the implementation of strategies.

The toolkit provided in this chapter will assist you in analyzing the feasibility of the destination, whether the destination is a city, region, or the country. Please note, not all the items listed are applicable to every destination. While assessing the needs, please choose all that apply.

In the first part of the chapter, we start with auditing the destinations four important dimensions that mainly addresses the existing infrastructure important to the brand. This includes four dimensions of medical tourism assets:

- product dimension;
- performance dimension;
- promotion dimension; and
- medical tourism dimension.

Collectively, we can use the acronym of PPP+MT to address these dimensions. In the second part of the chapter, we address planning process and strategies as how to build a brand. This planning process begins with establishing the purpose and guiding principles, critical for leading the destination to build a right brand. Then we highlight how to engage all stakeholders in planning a strategy and acquire buy-in from various groups to achieve a well-designed brand strategy.

13.1.2 FOUR DIMENSIONS OF MEDICAL TOURISM ASSETS: PRODUCT DIMENSION

Product dimension includes the following subsections.

13.1.2.1 DESTINATION'S CORE ASSETS (ATTRACTIONS)

In this section, you will review the destination's out-door, in-door, and nature attractions; shopping; cuisine and culinary experience; entertainments including concert venues, nightlife, and theater. Other entertainments such as sports, boating, and cruising are also included.

13.1.2.1.1 Outdoor, Indoor, and Nature Attractions

List all the outdoor and nature attractions, something for every taste, all ranked highly by visitors or presented in social media reviews (Table 13.1).

TABLE 13.1 Product Dimension: Outdoor and Indoor Attractions

Attractions	Insert a Brief Highlight for Each	Insert Ratings for Each (%)
Gardens		
National parks		
Harbors		
Aquariums		
Museums		
Galleries		
Art centers		
Other		

13.1.2.1.2 Shopping

Identify and list a variety of shopping activities from fancy upscale shops and boutiques to budget and discount stores. Also, remember shoppers may engage in other activities as well (Table 13.2).

TABLE 13.2 Product Dimension: Shopping

Shopping	Insert a Brief Highlight for Each	Tourists' Reasons for Visiting (%)
Famous shopping districts		
Jewelry		
Luxury fashion stores		
Ethnic shopping districts		
Mass retails		
Future plans		

13.1.2.1.3 Cuisine and Culinary Experience

In this segment, you will identify world-renowned dining places, great microbreweries, international cuisines, and annual food festivals, locally grown cultural foods, and socially and environmentally sustainable restaurants. In addition, you will identify homegrown foods, community gardens, food markets, and food deliveries (Table 13.3).

TABLE 13.3 Product Dimension: Cuisine and Culinary Experience

Cuisines and Culinary Experience	Insert a Brief Highlight for Each	Tourists' Reasons for Visiting (%)
World-renowned dining places		
Breweries and microbreweries		
International cuisines		
Annual food festivals		
Popular in local foods		
Locally grown cultural foods		
Socially and environmentally sustainable restaurants		
Other		

13.1.2.1.4 Entertainment

Identify an array of entertainment including live music, small music venues, concerts, casinos, and bars, etc. (Table 13.4).

TABLE 13.4 Product Dimension: Entertainment

Entertainment	Insert a Brief Highlight for Each	Tourists' Reasons for Visiting (%)
Concert venues		
Music festivals		
Nightlife		
Popular bars		
Entertainment districts		
Dance venues		
Theatres		
Other		

13.1.2.1.5 Sports

A number of cities promote sport tourism as an economic development initiative at the community level. In this segment, identify services including number of national and provincial sports; multi-sport and major games organizations; and variety of other sport and tourism industry partners; including, sharing best practices, measuring the economic impact of sport events, and enhancing the profile of sport tourism. These events provide year-long excitement and active grounds for sport fans (Table 13.5).

TABLE 13.5 Product Dimension: Sports

Sports and Activities	Insert a Brief Highlight for Each Include: Locations, Dates, etc.	Tourists' Reasons for Visiting (%)
Major teams		
Professional sport teams		
Sport entertainments		
Home-games		
Game organizers and industry partners		
Economic impact of sport tourism		
Other		

13.1.2.1.6 Boating and Cruising

Waterfronts are considered a great source of entertainment and tourist attraction. A well-planned expansion can contribute to environmental improvements, economic activity, and overall enhancement of quality of life through the development of a designated waterfront area, providing a potential for additional visitor attraction. At the same time, poorly planned development can have negative effect on marine life, noise levels, traffic, pollution, etc. Therefore, economic, health, social, and environmental impacts all need to be considered. Commitment to social responsibility is critical (Table 13.6).

13.1.2.2 QUALITY OF TOURISM SERVICES

In this segment visitor satisfaction, tourist information services, signage, tour guide quality, taxi services, and Wi-Fi must be assessed. These activities directly impact visitor experience at the destination.

TABLE 13.6 Product Dimension: Boating and Cruising

Boating and Cruising	Insert a Brief Highlight for Each	Tourists' Reasons for Visiting (%)
Guided tours		
Marine life viewing		
Major water activities		
Boat rentals		
Variety of other tours and activities		

13.1.2.2.1 Visitor Satisfaction

Historical *Visitor Satisfaction Scores* provide valuable information about how welcoming and friendly the destination is. It clearly summarizes customer perceptions. Beyond the scores, you might ask: Does the destination have a philosophy or a slogan? Does the destination offer a free training program for frontline employees? Does the destination have partnerships to support visitor experience? The goal is to make the destination a welcoming place (Table 13.7).

TABLE 13.7 Product Dimension: Visitor Satisfaction

Visitor Satisfaction: Collect and Analyze Visitor Satisfaction Scores	Insert a Brief Highlight for Each	Tourists' Response Rate from Highly Satisfied to Not Satisfied at All (%)
Satisfaction with visitor services		
Satisfaction with customer service representatives		
Staff ability to provide complete and accurate information		
Overall satisfaction with the visit		
Other		

13.1.2.2.2 Information Services

Visitor Information Services, such as things to do, information desks and mobile apps are essential for a visitor-friendly destination, particularly for visitors from abroad. Also, information desks planning visitor trips, marketing, and advertisement the destination are important (Table 13.8).

TABLE 13.8 Product Dimension: Information Services

Information Services Identify and List Destination's Information Services	Insert a Brief Highlight for Each	Availability or Future Plans
Information desks in various touchpoints, airports, train stations, major tourist centers, etc.		
Mobile visitor information		
Free apps for transportation		
Free apps for street foods		
Free apps for food delivery		
Mobile apps for scheduled popular events and attractions		
Free entertainment magazine		
Other		

13.1.2.2.3 Signage

Beyond directional street signs, modular signage systems, such as digital way finders can help improve visitor comfort and get visitors to their destinations day and night (24/7). In addition, in most popular places universal modular signs are important to accommodate visitors with disabilities (Table 13.9).

TABLE 13.9 Product Dimension: Signage

Signage Identify Important Destinations and Provide Modular Signage and Digital Way Finders to Guide Visitors to:	Insert a Brief Highlight for Each	Availability or Future Plans
Hospitals, clinics		
Important public spaces		
Major office buildings		
Parks, and walking trails		
College and university campuses		
Universal modular signs to accommodate people with disabilities		
Other		

13.1.2.2.4 Tour Guide Quality

The quality of tour guides is essential to visitor perception of the destination and satisfaction rates. The following items are essential to consider (Table 13.10).

TABLE 13.10 Product Dimension: Tour Guide Quality

Tour Guide Quality	Insert a Brief Highlight for Each	Availability or Future Plans
Self-guided tour apps		
Self-guided tours: Walking, cycling, cultural, other		
Provide tour suggestions		
Provide maps		
Provide instructions and directions		
Trained tour guides		
Multilingual services		
Other		

13.1.2.2.5 Taxi Services and Uber

Taxi services with additional training and improvement. An analysis of Yelp comments or other reviews can help to improve taxi services. Uber now have the listing of most popular visits.

13.1.2.2.6 Wi-Fi Access

Evaluate offering free Wi-Fi in parks and public spaces, such as subways, airport, City Hall, City Center, and public and publicly accessible places. In addition, free Wi-Fi map helps locating hot spots in the city.

13.1.2.3 VALUE PROPOSITIONING

By creating new partnerships and innovating products, such as tour packages and city passes, you can make the destination economically competitive and offer value to visitors. Partnership may include hospitality and tourism, labor organizations, medical profession, and government. Major plans, such as properly designing the destination for medical tourism special economic zones (MTSEZs). More information about MTSEZs is provided in Chapter 6:

1. **Partnerships:** Cross-promotions and packages within the destination and other destinations are very important for value creation. Tour

packages can be created combining the destination with neighboring attractions.

2. **Tour Packages:** Presents a comprehensive website listing accommodation, attractions, restaurants, tour operators/tourism partners, and transportation. This is a valuable source of information for visitors.
3. **Ticket Access:** Make ticket access available online, via phone or in person. City passes are a great way of skipping long lines at various attractions. You may package them for groups of visitors, bundled attractions (Table 13.11).

TABLE 13.11 Product Dimension: Value Propositioning

Value Propositioning	Insert Brief Highlights for Each	Availability or Future Plans
Partnerships		
Tour packages		
Ticket access		
Other		

13.1.2.4 DESTINATION ACCESSIBILITY

Easily accessible destinations by air, train, and bus with various price ranges are a great way to attract visitors. Connectivity among airports, train, and bus stations are also added value. Plans also be made for a well-run public transportation system (Table 13.12).

TABLE 13.12 Product Dimension: Destination Accessibility

Destination Accessibility Identify and Evaluate Each One of the Following Means of Transportation	Insert a Brief Highlight for Each	Availability or Future Plans
Airports serving number of destinations		
Express services to and from airport		
Highways connecting surrounding cities and neighboring countries		
Rail stations and connections		
Within destination transportation network		

13.1.2.5 DISABILITY ACCOMMODATIONS

Wheelchair accessibility is a necessary feature for MT purposes. A dedicated and well-publicized Website must be developed providing customized experience for patients and international travelers who need wheelchair or vision and hearing assistance.

13.1.3 PERFORMANCE DIMENSION

Performance dimension include the following subsections.

13.1.3.1 LODGING ACCOMMODATIONS

In this section, we will cover:

i. Lodging accommodation including number of visits and visitor profile;
ii. Lodging yield-occupancy, ADR, and length of stay;
iii. Critical acclaim.

13.1.3.2 NUMBER OF VISITS AND VISITOR PROFILE

Accounting for a number of visits and visitor profile in destination will help estimating supply and demand for lodging and growth rates. It will guide the destination's current readiness and future planning (Table 13.13).

TABLE 13.13 Product Dimension: Number of Visits and Visitor Profile

Number of Visits and Visitor Profile	Data Analysis Insert Brief Highlights for Each	Data and Graphs
Number of overnight visitors for the past 3 or 5-years		
Forecasts for the next year		
Profile of visitors by region: (Domestic and overseas)		
Profile of visitors by country of origin		
Expenditures by region (domestic and overseas)		
Expenditures by country of origin		

13.1.3.3 LODGING YIELD (OCCUPANCY, AVERAGE DAILY RATES, AND LENGTH OF STAY)

A clear understanding of revenues and taxes generated are essential for the growth and the economy of the destination. In this segment you will highlight the following data (Table 13.14).

TABLE 13.14 Product Dimension: Lodging Yield

Lodging Yield: Occupancy, ADR, and Length of Stay	Insert Important Highlights Here	Included Data and Graphs
Retrieve monthly occupancy rates		
Retrieve average daily rates		
Number of room nights sold for the period under calculation		
Calculate lodging revenues, monthly, and yearly		
Estimate revenues generated by country of origin		
Diagram lodging revenues generated by destinations in various districts, downtown, etc.		

13.1.3.4 CRITICAL ACCLAIM

Scan the destination for any number of awards, recognitions, achievements, rankings, and popular events. Destination may greatly benefit from including these in all forms of communications (Table 13.15).

TABLE 13.15 Product Dimension: Critical Acclaim

Critical Acclaim	Insert Important Highlights Here	Available or Plans
Sustainability initiatives, such as sustainable conventions, trade shows, conferences		
Green initiatives		
Acclaimed health resorts, monuments, landscapes, parks, buildings, etc.		
Celebrity chefs, acclaimed entertainment places, etc.		
Destination's world rankings, such as most accessible city, best quality of living, world-renown health, and wellness destination		
Other		

13.1.4 PROMOTION DIMENSION: SUSTAINABILITY, SUPPORT, AND MAINTENANCE

i. Destination marketing;
ii. Product renewals;
iii. Carrying capacity.

13.1.4.1 DESTINATION MARKETING

Does the destination have an official tourism focuses to promote and sell the region or destination as a remarkable destination for tourists, MT, convention delegates and business travelers? (Table 13.16)

TABLE 13.16 Product Dimension: Destination Marketing

Destination Marketing	Insert Important Highlights Here	Available or Plans
Does the destination have a tourism office to market destination?		
Does the destination have other offices in popular countries or cities of origin?		
Services offered by the tourism office		
Does the destination have any tourism association?		
Services offered by the association, such as member support, advocating the destination, education, or communication		
Other		

13.1.4.2 PRODUCT RENEWALS

Destination renewals provides an economic engine for future development and offers many promises, such as relocation of businesses, new business segments, and other investments. Continued commitment to invest in lodging and attractions is a sign of the destination's commitment to meeting and exceeding visitor expectations. In this segment, you will list and describe city projects, lodging, and food service industry renewals and other attraction plans already underway or planned for future development. Obviously, continued investments in tourism assets can add to the desirability of the destination for visitors (Table 13.17).

TABLE 13.17 Product Dimension: Product Renewals

Product Renewals	Insert Important Highlights Here	Available Now or Plans
Lodging projects currently underway		
Lodging project future plans		
Restaurant and entertainment projects underway		
Restaurant and entertainment projects future plans		
Other attractions underway or in the future plans for renewal		
Other		

13.1.4.3 CARRYING CAPACITY

Many customers today appreciate and demand the destinations proactively engage in greening their cultural, natural, and other resources, such as energy and businesses. Table 13.18 provides some ideas towards managing the destination's carrying capacity.

TABLE 13.18 Product Dimension: Product Carrying Capacity

Carrying Capacity	Insert Important Highlights Here	Available Now or Plans
Website describing the destination's green initiatives		
Green tourism guidebooks		
Apps and maps providing information on sustainable attraction		
Green restaurants		
Green hotels and eco-lodges		
Destination's other green initiatives		

13.1.5 MEDICAL TOURISM (MT) DIMENSION

MT dimension include: pool and push factors which are discussed in subsections (Alnakhi, et al, 2019).

13.1.5.1 PUSH FACTORS

Push factors are most applicable in understanding the motivational factors of medical travelers. Included are:

- Patients' characteristics;
- Case complexity;
- Financial related;
- Health services related to home country.

13.1.5.2 PULL FACTORS

The pull factors are the factors that attracted patients to the country of destination as they are perceived by the patient. Included are:

- Patients' previous experience;
- Source of information;
- Information related to destination country;
- Health services related to destination country.

13.2 PLANNING AND BUILDING A BRAND STRATEGY

Planning: A robust inbound MT development must also carefully address the challenges presented by political figures, advocacy groups, media, health providers, and local patients. In Part 2, we address the purpose, guiding principles, and steps to take to acquire *buy-in* from various groups to achieve a well-designed brand strategy.

13.2.1 THE PURPOSE

As stated, and confirmed, by MT experts in this textbook, establishing MT just because, '*it is a nice thing to do*,' or embarking on it because '*everyone else is doing it*' is a major pitfall. To achieve a successful MT destination, the overarching goal and purpose must be clear; shared by major players and the community; and follow vibrant guiding principles.

13.2.1.1 GUIDING PRINCIPLES

The success of any program to become a leading MT destination requires at least the following guiding principles that incorporates: patient-first philosophy; partnerships; impact; diversity; engaged community; and active government participation (PP IDEA):

1. **Patients First:** All stages of the project must assure that the focus is on enhancing patient experience at all touchpoints and that guests receive individualized attention and service.
2. **Partnerships:** No one institution, alone, can provide all services that traveling patients and their family's demand. The destination will need to create partnerships with the community in order to create innovative MT assets, economic development programs, infrastructure, and educational programs that will address the critiques, critical needs of patients, and concerns of local communities at all touchpoints. Leaders from *Influential media, lodging industry, food service, entertainment, sport, and travel and transportation authorities* are all critical in this partnership.
   i. Media must be a partner. Leaders from influential media can discuss the pros and cons of establishing an effective MT program. Media buy-in is critical for political reasons and convincing the oppositions. In addition, media can help publicizing the destination and her achievements.
   ii. Leaders from the lodging industry and hotel operators can help to find innovative designs and ways to make patient accommodations a pleasant, uncomplicated, and trouble-free experience both esthetically (please refer to Chapter 17 for more details) and service-wise, including *building-modifications, amenities-modifications, and service-modifications.*
   iii. Engage foodservice leaders, associations, and entrepreneurs to formulate plans to help build a healthy, innovative, and sustainable gastronomic destination.
   iv. Partnerships with entertainment, sport, and hospitality segments can facilitate entertainment and activities to satisfy the needs of diverse visitors, patients, and their families while waiting for their care.
   v. Involve travel and transportation authorities and businesses to address time-critical issues and needs of various patients at

different times and locations. Configure a transportation network that accommodates various budgets and specific needs of patients and visitors.

3. **Impact:** It is essential for the destination to realize that this endeavor is a *path* that requires continuous nourishing, renewal, and support. To make significant differences, you must establish specific goals, measure successes; and *communicate and publicize* your achievement stories. Creating a signature medical destination requires bringing intellectual, physical, financial, and human resources (HR) together. The most important aspects of this path are:
   i. Develop innovative and *impactful* community and research initiatives.
   ii. Build a world-class medical destination infrastructure in key areas identified in the Toolkit.
   iii. Measure success of all programs.
   iv. Communicate achievement stories.
4. **Diversity:** One of the characteristics of a globally-renown destination is the inclusion of diversity of mind, socioeconomic, ethnic, geographic, and physical abilities of its inhabitants. The community and partners must commit to fostering a robust diversity education that all people from around the world feel welcome.
5. **Engaged Community:** The community will engage medical groups, research institutions, students, and management and associates in all touchpoints by addressing the most important matters identified subsequent to a comprehensive study (outlined in Toolkit) while, considering care and wellbeing of all visitors in a healthy and sustainable manner. To maintain the mission of a world-class medical destination, community values and engagement are essential to avoid opposition. The followings are some useful recommendations:
   i. Involves major players in world-renown research, seminars, and conferences in medical destination management.
   ii. Engage health community in continuous innovations.
   iii. Inspire residents to exhibit welcoming gestures to all visitors showcasing the community a world-class destination.
6. **Active Government Support:** The government buy-in is needed for a smooth transition to a defined MT destination. An economic impact analysis will help to drive this strategy.

➢ **Key Steps:**
- Work with partners and develop: *Implementation plans, Action Steps, and Timetables.*
- Initiate focus groups to address recruiting, training, guest services and other resources, such as diversity initiatives.
- Identify the future needs and the path to a prominent Destination Medical Center.

13.2.1.2 BUILDING A BRAND STRATEGY

The following recommendation is offered to develop a brand strategy for the region that stimulates tourism growth, creates a new vision for the area, and enhances its local, regional, and global awareness to include two phases as described below:

➢ **Phase 1: Identify Existing Brand Image and Future Aspirations:**
- Identify the goals of branding (results to achieve);
- Identify participants;
- Identify core values of the region-what the region stands for (See the Toolkit);
- Identify the region's attributes (current image) to internal/external constituents;
- Identify the region's future aspirations.

➢ **Phase 2: Develop Region's Brand Strategy:**
- **Step 1:**
    - o **Develop Brand Image:** Identity and involve targeted sponsors and organizations to determine what the region stands for, what makes it different from other places, and what are the core values that can sustain and transform from generation to generation. A detail analysis using the Toolkit is very helpful.
- **Step 2:**
    - o **Build up the Region's Brand Positioning Strategy:** This involves how to move from current brand identity to aspiration identity for each region, and segment of the industry (Focus Groups).

- **Step 3:**
  - o **Establish Priorities:** For region's touchpoints where the target audiences encounter with each segment, based on their impact and importance, such as train stations, roads, signage, museums, art centers, nightlife, etc. This will include pre-visit, during-visit, and post-visit decisions.
  - o **Assist to Execute and Communicate the Brand (PR):** It takes a few years to change the perceptions of the region. Therefore, specific value propositioning and targeted messages, as well as a media strategy will be developed for each group to change perceptions.
  - o **Provide Criteria for Measuring Results:** Develop metrics that would help to measure if the original goals are materialized.

13.2.1.3 BENEFITS OF BRANDING

An honest, credible, and sustainable brand identity and positioning will help region to:

- Transform the perception of the region by internal and external constituents;
- Enhance its local, regional, and global positioning;
- Involve community and international partners to proactively participate in the vision and brand;
- Determine specific touchpoints and that deliver ideal value and experience for all constituents.

13.3 CONCLUSION

We divided the chapter into two important parts. Part 1 addressed the four dimensions involved in assessing the destination's tourism assets that included: *Product Dimension, Performance Dimension, Promotion Dimension, and Medical Tourism Dimension* (PPP MT). We used a detailed Toolkit (PPP MT) and itemized each dimension to make it easy for constituents to navigate the toolkit. We also noted that not all items in the toolkit are equally applicable to all destinations. Users can evaluate each item for their pertinence and importance to the destination's MT goals and aspirations.

In Part 2, we discussed the importance of planning, establishing guiding principles, and the key steps for engaging the community. Finally, we outlined necessary steps for building a brand strategy and gaining buy-in from various stakeholders and the community. We hope that this process will help to evaluate your destination for incorporating any tourism product health, wellness, or MT to your destination.

KEYWORDS

- **average daily rates**
- **medical tourism**
- **medical tourism special economic zone**
- **nature attractions**
- **planning guideline**
- **product dimension**
- **wellness assets**

REFERENCES

Alnakhi, W., Segal, J., Frick, K., Ahmed, S., & Morlock, L. (2019). Motivational factors for choosing treatment destinations among the patients treated overseas from the United Arab Emirates: results from the knowledge, attitudes and perceptions survey 2012, Retrieved from https://www.researchgate.net/publication/335901198_Motivational_factors_for_choosing_treatment_destinations_among_the_patients_treated_overseas_from_the_United_Arab_Emirates_results_from_the_knowledge_attitudes_and_perceptions_survey_2012/figures?lo=1.

PART V
Medical Travel Facilitators

CHAPTER 14

SUCCESS FACTORS AND BEST PRACTICES FOR MEDICAL TRAVEL FACILITATORS: TIPS FOR BENCHMARKING AND STARTUPS

ALI A. POORANI and YUSHU FU

Alfred Lerner College of Business and Economics, Department of Hospitality and Sport Business Management, University of Delaware, 14 W., Main Street, Newark, DE–19716, USA, E-mail: poorani@udel.edu (A. A. Poorani)

ABSTRACT

Medical tourism facilitators (MTFs) play a crucial role in the landscape of medical travel. As an emerging industry, medical tourism is unfamiliar to many patients, leading to the need for specialists to assist their medical travel abroad. In this chapter based on the survey of global medical tourism facilitators, we address sources of their business, their sales and marketing, most beneficial promotion, types of services they offer, and other aspects of operation of Medical Travel Intermediaries. Finally, we address their motivation entering this business and most important challenges they face.

14.1 INTRODUCTION

Brokers, medical travel agents, medical travel intermediaries, medical travel facilitators, and medical tourism (MT) companies are all terms used to describe the types of companies that have emerged to act as links between the medical tourists and MT products (Haarhoff and Mokoena, 2017). For the purpose of this chapter, the terms medical tourism intermediaries (MTIs) and medical tourism facilitators (MTFs) will be used interchangeably.

Gran and Frederick (2018) identified four types of facilitators:

- **Facilitator 1:** Self-facilitate (on my own);
- **Facilitator 2:** Medical tourism agents;
- **Facilitator 3:** Third-party payers (Insurers and employers);
- **Facilitator 4:** Domestic healthcare providers or physicians.

Although these four partake in medical travel decisions in one form or another, this study focuses more specifically on the MT agents or facilitator types, as medical travel intermediaries.

MTFs play a crucial role in the landscape of medical travel. As an emerging industry, as well as a non-routine patient practice, MT is unfamiliar to many patients, leading to the need for specialists to assist their medical travel abroad. Beyond tourism-related activities, traveling internationally to receive medical services faces many legal issues, as well as obtaining medical care information, all of which require essential services of MTFs (Gan, 2011; Snyder et al., 2011). The study of various literature reveals the following main reasons tourists choose services of intermediaries:

Convenience	Intermediaries are familiar with local medical services and information, thus can provide conveniences of one-stop shopping.
Experience	Experienced MTIs have experience in identifying and networking with healthcare providers and other travel arrangements.
Logistics	Make travel arrangements, including air travel, visas, accommodations, etc.
Savings	They can better negotiate packages and deals.
Pre-and-post travel services	They assist in follow-ups with any questions arising before and after travel.

Interestingly, little is known about the economics of MTIs and their business practices (Cook, 2010; Jones, 2011; Solomon, 2011). American hospitals more and more are forming relationships with MT facilitator startups to reverse the lost MT market to South East Asia, Germany, Israel, Finland, and Switzerland (Frank and McCarthy, 2013). Though the partnership takes longer to show returns, it allows the two parties to align strategies, improve case selection and screening, which in turn can reduce financial risks and costs (Frank and McCarthy, 2013).

This study's objectives are to investigate the operational issues and practices of MT facilitator business.

1. **Why is This Study Important?** As mentioned earlier, this segment of the MT is relatively new. Additionally, their role as an intermediary is significantly important to patients traveling abroad (Cohen, 2012; Jagyasi, 2010); and more importantly, according to the MT Association (Stephano, 2013) nearly 36% of all travelers used services of medical travel facilitators; more than 50% of the patients surveyed in India and UAE reported that they had used MTIs for their seeking treatment abroad (Alsharif et al., 2010).
2. **Entering to MTF Business:** Currently, to our knowledge, there are no objective and reliable sources of information for new entrepreneurs to use as a blueprint to enter this market; or the existing facilitators to benchmark their operations; and reduce prevailing skepticism in this segment. This study seeks to explore the economic viability of the MT facilitator business and investigate the ecosystem of MT on a macro level. More specifically, the study reports data on the following areas: Sources of business; most beneficial advertising methods and advertising expenditures; purpose of patient travel; number of travel agencies they work with; location issues, such as proximity to providers; combining tourism activities with treatment; average length of stay; where their customers come from and which destinations do better; types of plans, services; charges, fees, discounts, and collection methods; customer feedback mechanisms; and, reasons for entering this business, their success factors and challenges. We hope that this information will help aspiring intermediaries to jump start their businesses and current facilitators to benchmark their practices.

14.2 BACKGROUND

In the minds of many, the idea of MT may be of cosmetic surgeries on a tropical island; but in fact, the destinations and types of procedures are far more varied (Horowitz, 2007; Moody, 2007), complicating the manner in which such services are selected, performed, and/or followed up. Therefore, the professional services of intermediaries/facilitators can help to ease the process. To help medical travelers, a number of established MT destination hospitals, such as Thailand's Bumrungrad International Hospital, already provides hospitality services to visiting patients, including accommodations in luxury hotels, managed recovery, and therapy sessions (Boyd, 2011; Simon, 2006); or India's Apollo Hospitals Group that has attracted patients from around the world (Apollo Hospitals Group, 2010) provide patient services to international travelers. These practices denote the travelers' need for convenience, logistics, and other services listed earlier. No matter what kind of services travelers seek, or the providers offer, MTFs play a significant role in the landscape of MT.

Patients seeking treatment abroad are generally faced with two major options: (i) a patient might conduct a personal research of medical care facility abroad; (ii) employ the services of a MT facilitator (Cattaneo, 2009; Herrick, 2007; Turner, 2007). Option one requires that a patient have some knowledge of MT ecosystem facilitated by efficient customer services provided by the destination hospital. In the absence of such reliable information, services of MTFs can fill the gap. A MT facilitator (intermediary) functions much like specialized travel agency, such as MedRetreat that offers prospective patients 183 medical procedures in seven countries (med retreat) or Arriva Costa Rica that allows patients to comparison shop by listing procedures and prices on their website (Arriva Costa Rica Website, 2014). Intermediaries for medical travel will often employ certified doctors and nurses to analyze treatments, assess the quality of care in destination, and assist in their decision-making process (Alsever, 2006; Fetscherin, 2016). But are medial facilitators' services reliable?

14.2.1 FACILITATORS AND QUALITY ISSUES

Concerns about quality of care are of prominent importance to anyone making decisions about their personal health. Hospitals around the world that wish to attract medical travelers and seek to improve their reputations globally, have distanced themselves from the view of third world medicine mainly by employing western-trained doctors from renowned medical schools and or through partnering with world's most prestigious hospitals, such as the partnership between Wockhardt with Harvard Medical School and Apollo Hospitals Group with Johns Hopkins (Turner, 2007). Studies, such as Project Hope conducted at a number of international hospitals that regularly treat western patients, show that Bumrungrad, Apollo, and Wockhardt (another popular destination hospital in India) all are accredited by the Joint Commission International (JCI) and experience mortality rates of less than 1% in coronary artery bypass graft (CABG) surgeries. These findings are comparable to renowned American hospitals like Johns Hopkins and Cleveland Clinic (Milestein, 2007).

Beyond the remarkable competitions among MT providers striving for quality and reputation, some studies have expressed concerns about the ***Triage Role*** intermediaries play in the selection of healthcare facilities and providing patients with information and advice about which doctors to consult for particular illnesses (Chee, Whittaker, and Hong Por, 2017). In this role, intermediaries act as medical professionals without actual medical training. However, their study concluded that using an assemblage model, intermediaries play a stabilizing role at three levels: (i) individual level; (ii) the organizational (hospital) level, determined as the relations between MTFs and hospital staff; and (iii) national and international levels to facilitate medical travel from one country to the other. We suggest intermediaries to employ the services of doctors, nurses or other clinical professionals with specialist knowledge, quality control, and support services, thus offering something unique and valuable to their clients. In the following section, we address the facilitator professionalism and ethical issues that address some of the concerns raised above.

14.2.2 FACILITATOR PROFESSIONALISM AND ETHICAL ISSUES

Some researchers have expressed those intermediaries emphasize the benefits of MT more than its risks (Lee et al., 2014). For example, Maguire et al. (2016), studied 27 MT agency websites in Europe and concluded that facilitators do not regularly provide a balanced information on their websites. Since most travelers get their information from facilitators' websites, risks, regulatory issues, and oversight, emergency arrangements or financial ties in their websites are generally absent. Although their study has some merits, we think that holding intermediaries to such high standards are too harsh and an unusual practice. Imagine expecting the same standards from the foodservice industry, hotels, travel agents, or other similar businesses to alarm their customers on their websites about the risks of eating certain foods or traveling to various destinations. We suggest it is the customers' responsibility to seek such valuable and necessary information regarding risks, regulations, etc., when they are consulting with intermediaries privately.

Others have expressed concerns about attractive brochures and tourist packages, which are often found as a result of an information search on the Internet (Haarhoff and Mokoena, 2017). In this widely unregulated industry, ultimately, it is upon the facilitators to differentiate themselves in the marketplace and to satisfy the needs of growing number of medical tourists. Obviously, those who do not provide high quality service or an excellent customer satisfaction rating will find it more and more difficult to compete in an increasingly competitive marketplace.

Milbank (2018) interviewed 23 professional intermediaries including (11 women and 12 men), 8 were based within Australia, while the rest were located in the USA ($n = 6$), the UK, Canada, and Greece ($n = 2$ for each), and Thailand, South Africa, and Israel ($n = 1$ for each). Based on her study she concluded that there was a strong sense of 'embodied' expertise and claims to ethical practice based upon that expertise, particularly in the form of specialist knowledge, quality control, and support services, which are hardly offered by other providers; thus, offering something unique and valuable to their clients. In addition, some facilitators defined their role as a form of labor of love to help people and children, especially in cross-border assisted reproduction.

The study also recommended some industry-specific regulation or certification to be in place for improving and safeguarding the field from unscrupulous players, particularly in the trans-national context. We concur that some laws regarding MTFs are necessary. In Chapter 15, we address this issue in more detail.

Currently, there are a number of organizations providing certification, courses, or training in MT, such as, to name a few (not in a particular order):

- Certified medical tourism professional (https://www.medicaltourism.com/cmtp);
- Medical tourism training (https://www.medicaltourismtraining.com/);
- Medvarsity (https://www.medvarsity.com/courses/certificate-course-in-medical-tourism/); and
- Medical travel quality alliance (https://mtqua.org/certification/).

In additions, a number of colleges and universities around the world offer courses or degrees in medical tourism.

We hope the information provided in this chapter will serve to answer to the growing needs of this emerging industry, provide benchmarking opportunities for entrepreneurs entering this market, and further maximize their strategic and operational endeavors.

14.2.3 MEDICAL TOURISM FACILITATORS (MTFS) AND LEGAL ISSUES

As patients from the Middle East are traveling more and more often to South East Asia, American hospitals are seeking to regain their position as a popular MT destination. A major hurdle facing referrals to the United States (U.S.) is the facilitators' payment through commissions paid by the hospital (Frank and McCarthy, 2013). Stark Law and other anti-kickback legislation prohibit American medical providers from paying commissions to referral agents, a practice that is accepted in most other medical tourist destinations. In an effort to conform to U.S. laws, some American hospitals are forming relationships with MT facilitator startups developing new compensation schemes, providing collection services, and taking contingency fees as well as guaranteeing payment on behalf of patients which allows the facilitator to contract with hospitals as a payer (Frank and McCarthy, 2013). More specific legal issues regarding legal issues involving MTFs is discussed in Chapter 15.

14.3 RESEARCH METHODOLOGY

A lot needs to be explored about the business of MTFs. In addition, routine practices of intermediaries and their knowledge of the field can reveal valuable information about the sources of business, services demanded by their customers, and the ways they market their businesses and other success factors. In this study, we explored the following specific topics:

- Lead generators for their business;
- Most beneficial advertising methods and advertising expenditures;
- Purpose of patient travel;
- Number of travel agencies they work with;
- Location: Proximity to providers;
- Combining tourism activities with treatment;
- Average length of stay;
- Where the customers come from;
- Which destinations did better;
- Types of plans offered;
- Types of services offered;
- Charges, fees, discounts, and collection methods;
- Customer feedback mechanisms;
- Reasons for entering this business;
- Success factors and challenges.

14.3.1 RESEARCH DESIGN AND INSTRUMENTATION

The research design utilized for this empirical study was a survey method. In order to establish the validity of the survey instrument, the researchers conducted a pilot study and interviewed 20 MTIs during the 6th World Medical Tourism and Global Healthcare Congress in Miami, Florida. Based on the feedback from the participants, the final questionnaire was developed and distributed to survey participants.

14.3.2 DATA COLLECTION

The final instrument was then distributed via the mailing list of MT Association. Within the MT Association, there is a MT facilitator group, which is made up of MTIs from all over the world. The MT intermediary group is relatively small, and this was the most productive way to approach this group. A total of 77 responses were collected using the URL provided in the mailing. The questionnaire clarified that the responses will be kept anonymous and only aggregate results will be published. Figures 14.1 and 14.2 exhibit demographics of participants.

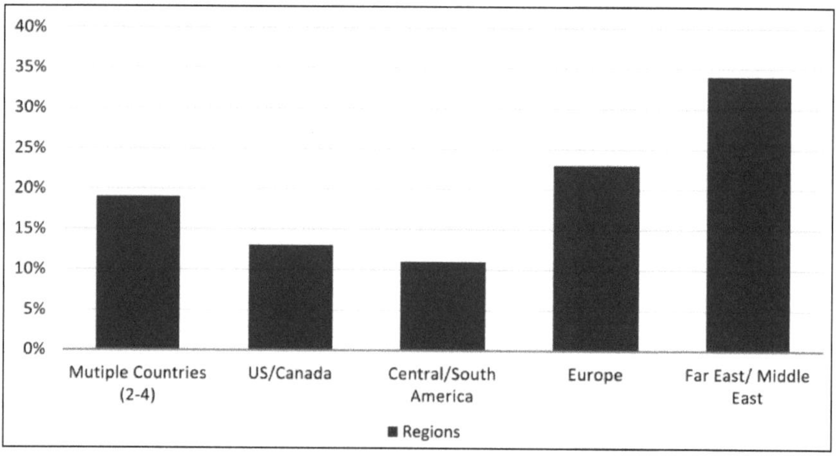

FIGURE 14.1 Where are your operations located?

14.4 DATA ANALYSIS

14.4.1 LEAD GENERATORS

MTFs used a wide variety of lead generators. According to Figure 14.3, among the most frequently mentioned and top choices were: Sales and marketing staff (45%), company website (42%), advertising (41%), and word of mouth (40%); followed by doctors and hospital referrals (37%), social media (37%), repeat guests and MT directories.

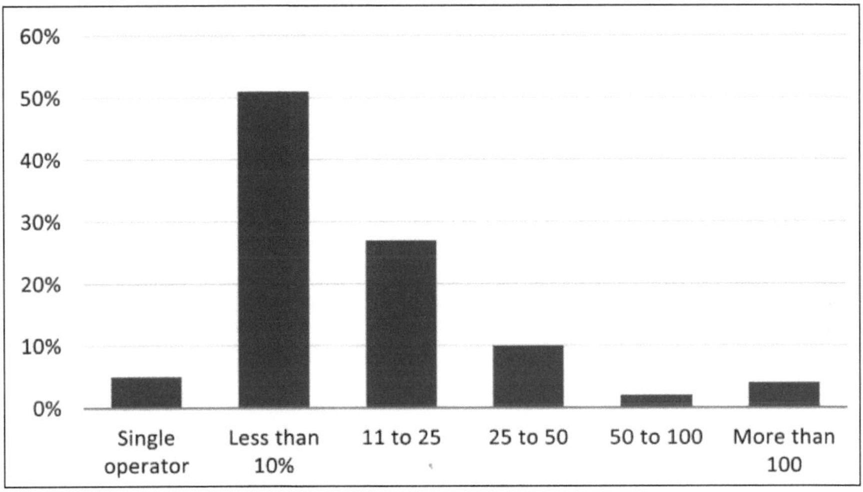

FIGURE 14.2 Number of employees in your company?

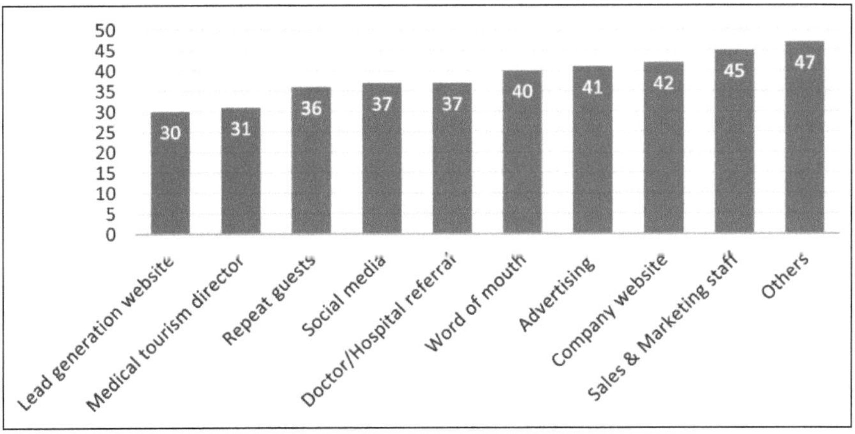

FIGURE 14.3 Lead generators (sources of business).

14.4.2 MOST BENEFICIAL ADVERTISING METHODS AND ADVERTISING EXPENDITURES

When companies were pushed to choose their top three most beneficial advertising methods, search engines, such as Google, Bing, etc., were the number one most beneficial advertising method (70%), followed by social media (60%), and tourism bureaus (50%). As shown in Figure 14.4, these companies spent between $0.00 to $50,000 dollars in advertising during the last 12-months of the study.

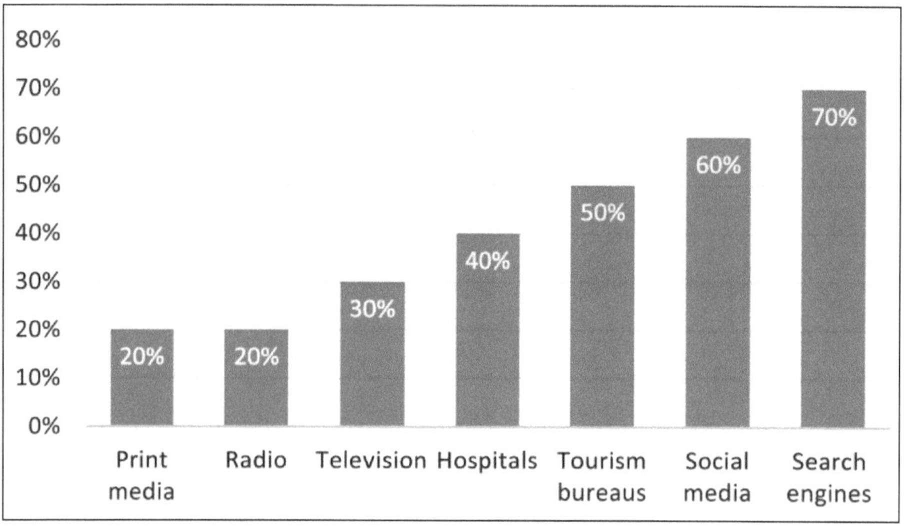

FIGURE 14.4 Most beneficial advertising (Spent: $0.00 to $50,000).

14.4.3 PURPOSE OF TRAVEL

Based on the assessments of MTFs, medical tourists traveled to receive a wide variety of treatments, including plastic surgery, oncology, cardio-vascular, dental, and other forms of surgery (see Figure 14.5). Over 56% of patients sought orthopedic surgery followed by plastic surgery (38%). Surprisingly, Weightless surgery was the lowest reported (22%). Other forms of treatment, such as stem cells, stress reduction, neurology, and urology were also mentioned.

14.4.4 NUMBER OF TRAVEL AGENTS THEY WORK WITH

Based on this study, as shown in Figure 14.6, over 40% reported working with 1–3 travel agencies, 20% worked with 4–6, and 10% reported working with more than 6 travel agencies, respectively.

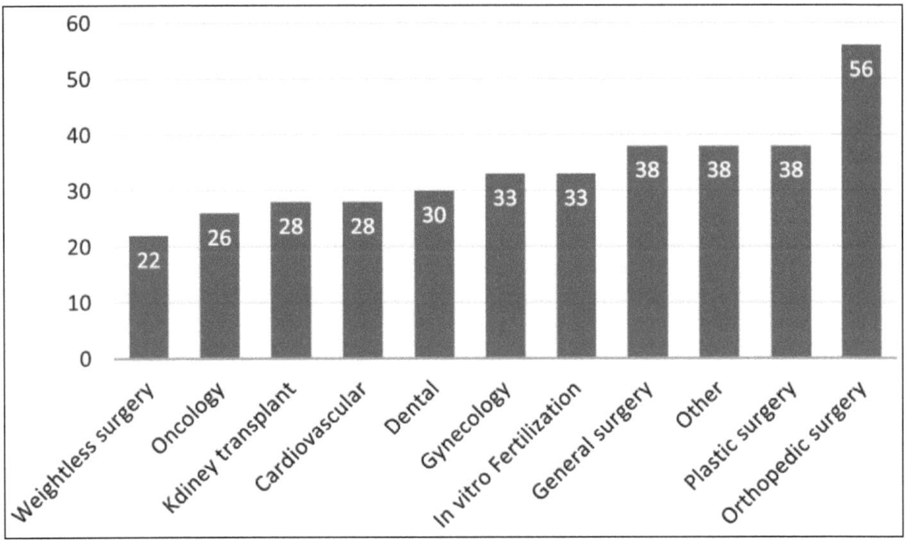

FIGURE 14.5 Purpose of travel.

Note: Other: Stem cells; Stress; Neurology; Urology.

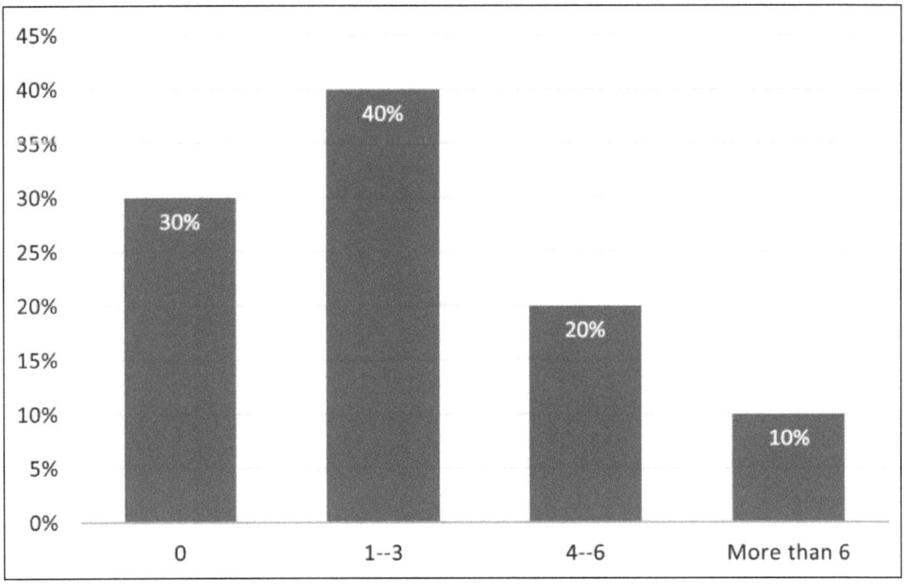

FIGURE 14.6 Number of travel agencies you work with.

14.4.5 LOCATION-PROXIMITY TO PROVIDERS

Nearly 68% of respondents reported that proximity to providers was *very important* or *important* to the success of their business. Only 18% indicated that proximity to the provider was not important (see Table 14.1 for more details).

TABLE 14.1 Location-Proximity to Providers?

Location Proximity	Importance
Not important	18%
Slightly important	14%
Important	20%
Very important	48%
Total	100%

14.4.6 COMBINING TOURISM ACTIVITIES WITH TREATMENT

A remarkable finding was that about ½ of medical travelers combine tourism activities with treatment. As shown in Figure 14.7, another interesting finding, as reported by MTFs, is that outbound travel outperforms inbound travel with a ratio of 2 to 1.

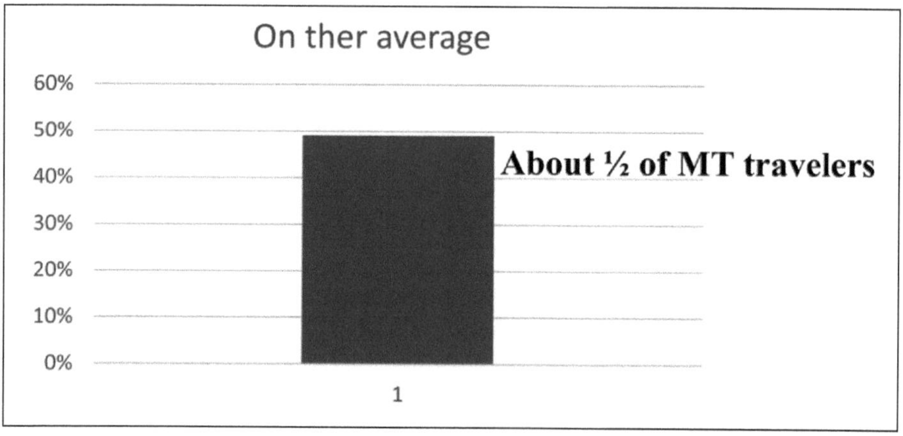

FIGURE 14.7 Combining tourism activities with treatment?

Note: Outbound travel outperforms inbound travel with a ratio of 2 to 1.

14.4.7 AVERAGE LENGTH OF STAY

MTFs generally have a direct insight regarding the length of stay. As depicted, MT travelers' time spent in destination widely varied between the ranges of a minimum of 3-days and maximum of 30-days (see Table 14.2). The average hospital stay reported was 7-days; average tourism activities was also 7-days; with an overall average stay of 13.13-days.

TABLE 14.2 Average Length of Stay?

Average Length of Stay	Number of Days
Hospital stay	7 days
Tourism activities	7 days
Overall average	13.13 days

Note: Min: 3 days and Max: 30 days.

14.4.8 WHERE THE CUSTOMERS COME FROM?

As shown in Figure 14.8, travelers from Asia and the U.S. top the figure of MT travelers (40%), followed by Africa, the Middle East, and Europe, 37%, 35%, and 33%, respectively. Other up-and-coming countries mentioned were Ukraine, Georgia, and Azerbaijan.

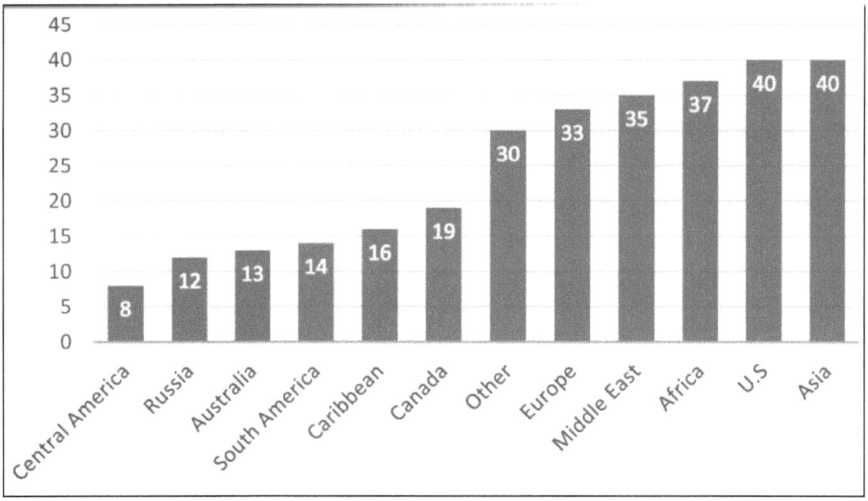

FIGURE 14.8 Percent customers come from?

Note: Other: Ukraine, Georgia, Azerbaijan.

14.4.9 WHICH DESTINATIONS DID BETTER?

MTFs were asked to rate their customer satisfaction with various destinations. As shown in Figure 14.9, South Africa, Israel, and Brazil topped the figure with a rating of 2–2.5 in a Likert Scale of 1–3, with 3 being the highest; surprisingly, the U.S. occupied the last place with a rating of 1.25 in the same scale.

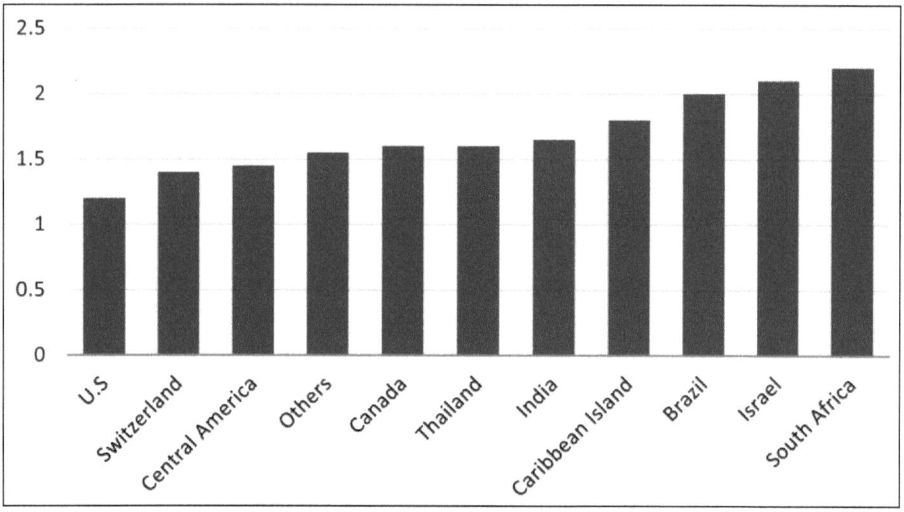

FIGURE 14.9 Which destinations did better?

14.4.10 TYPES OF PLANS OFFERED

The majority of MTFs (51%) offered both custom plans and standard plans. Nearly, 41% offered custom plans and 8% offered only standard plans (see Figure 14.10 for details).

14.4.11 TYPES OF SERVICES OFFERED

MTFs offer a wide variety of services. For detail, please refer to Table 14.3. Among the most common services are booking hotels, contacting doctors, translation, airplane, and other transportation, assist patients and families in getting visas, personal assistance in touring and shopping, and searching for international medical insurance. Nearly 2/3rd offer 24-hour customer service, daily patient care, and private nursing on request. Other services included:

- Medical reports;
- Attention on airport;
- Attending to personal and medical problems;
- Special cuisines.

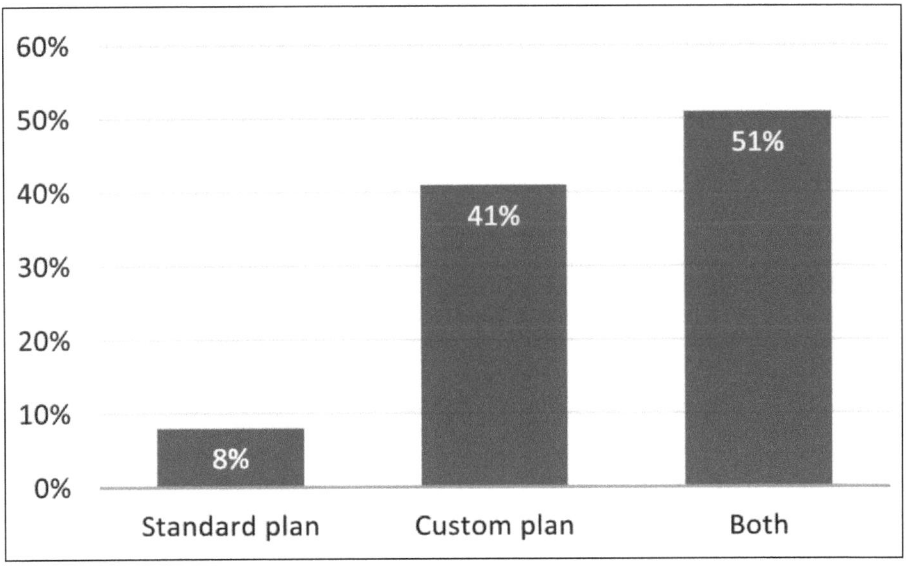

FIGURE 14.10 Types of plans offered.

TABLE 14.3 Types of Services Offered

Types of Services Offered	Percentage (%)
Booking hotels	95%
Contact doctors	86%
Translator	84%
Booking airplane tickets	84%
Transportation	73%
24/7 customer service support	70%
Daily patient check	70%
Private nursing on request	68%
Assist patients and their families in getting a visa	62%
Personal assistant for touring and shopping	62%
Search for international medical insurance for patients	46%
Others innovative services	19%

14.4.12 CHARGES, FEES, DISCOUNTS, AND COLLECTION METHODS

As shown in Table 14.4, 51% of MTFs charge commission on both medical procedures and travel/tourism; 47% charge commission on medical procedures alone; and, other 45% reported charging fixed rates. More than 80% offer some sort of discount ranging from customer spending to repeat business. Only 10% of respondents do not offer any discounts (Figure 14.11). The majority reported collecting fees upfront (85%) or after procedures (53%). Only 30% reported accepting installments (Table 14.5).

TABLE 14.4 Type of Charges

Types of Charges	Percentage (%)
Commission on medical procedure and travel/tourism	51%
Commission on medical procedure	47%
Fixed-rate	45%

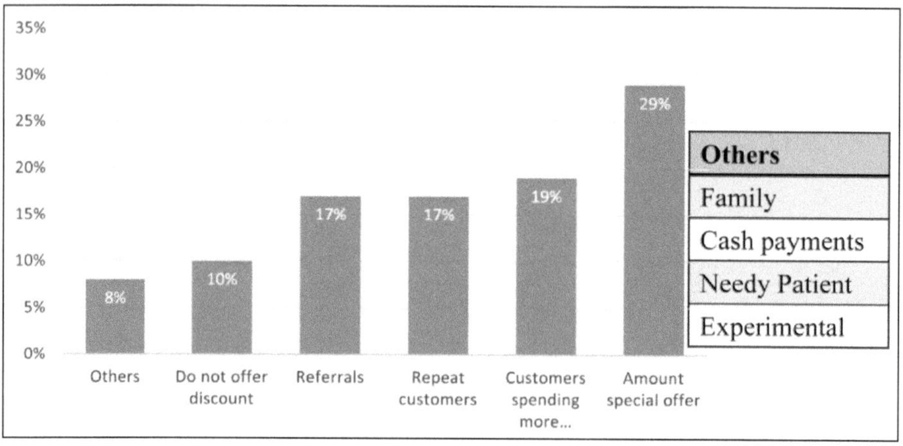

FIGURE 14.11 Percentage offer discounts and special offers.

TABLE 14.5 Collection Methods

Collection Methods	Percentage (%)
Up-front	85%
After the procedure	53%
Installments	30%

14.4.13 CUSTOMER FEEDBACK MECHANISM

Table 14.6 depicts customer feedback methods used by facilitators. Written customer feedback leads the figure by (88%) followed by phone calls (56%), online reviews (50%), and social media (46%).

TABLE 14.6 Customer Feedback Methods

Customer Feedback Methods	Percentage (%)
Written customer feedback	88%
Phone calls	56%
Online reviews	50%
Social media (medical tourism online reviews, likes, etc.)	46%
Others: Greeting cards, personal interaction, survey	8%

14.4.14 REASONS FOR ENTERING THIS BUSINESS

Nearly half of the survey participants (46%) reported they entered this business to own their business; 35% were a part of their company's business; 13% transferred from a travel agency; and 7% reported that they entered this business after retirement (see Figure 14.12 for more details).

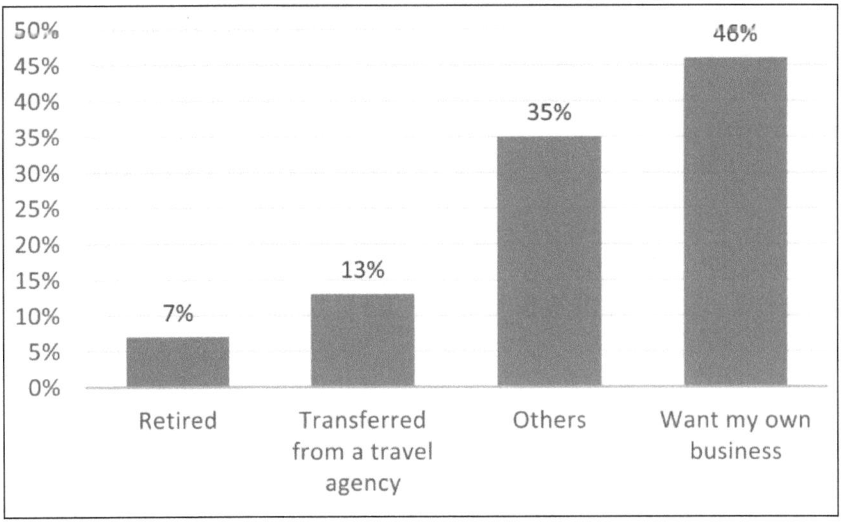

FIGURE 14.12 Why have you chosen this business?
Note: Other: Part of the company's business.

14.4.15 SUCCESS FACTORS AND CHALLENGES

Participants reported a long list of success factors for MT facilitator business, chief among them were quality, trust, and safety. Other success factors were offering better price, effective marketing and branding, relationship with providers and location.

This emerging business is also facing many challenges. The most frequently mentioned challenges were a number of unqualified operators entering this business, unsafe facilities, international exposure, currency, familiarity, regulations, and competition (see Table 14.7).

TABLE 14.7 Success Factors and Challenges

Success Factors:	Challenges:
• Quality/trust/safety	• Unqualified operators
• Price	• Unsafe facilitators
• Marketing/branding	• International exposure
• Relation with providers	• Currency/familiarity
• Location	• Regulation
	• Competition

14.5 FINDINGS

MTFs enter the industry for a variety of reasons. Chief among them is the entrepreneurial spirit, a goal of operating their own business. Others had been part of a travel agency and began specializing in the field. A small segment reported being a part of a related business, or a post-retirement second career for a small minority.

Traditional advertising and marketing staff are among the top source of lead generation to provide personalized services to MT clients. Though slightly less productive than the above top lead generators, company web site, social media, and "Word-of-mouth" were substantial sources of leads. Like any other businesses, company websites and social media play a significant role in providing sources of business for MTFs. Interestingly, but not surprising, though traditional advertising is still being a part of advertising scheme; MTFs are more and more relying on online sources to take advantage of video capabilities, such as step-by-step walkthroughs; to benefit from the low cost of their use; and, reaching greater numbers.

MTFs tend to specialize not just in a geographical region but also in the types of procedures they patients demand. Some are specialized exclusively in one type, such as dental surgery, others offer a variety of procedures ranging from orthopedic surgery, cardiology, oncology, gynecology, and in-vitro fertilization procedures, general surgery to plastic surgery, to name a few.

With respect to location of MTFs, for the purpose of fostering a relationship with a destination hospital, most respondents affirmed that it is important to be close to their patient's destination. This also allows the MT facilitator to meet their client in person and offer other services rather than simply introducing them to the hospital and doctors. We also heard a number of respondents noting that it was unimportant to be near the destination medical treatment center. This might be explained by the variety of business models and practices throughout the industry.

The process of reservation of travel arrangements, it appears to be a fairly streamlined process, with most using 1–3 booking agencies. This is likely due to the upfront price quoting, common in the industry. Being able to quote travel prices as well as medical procedure costs, allows this option to be sold easily. This transparency constitutes a major strength for MT facilitation industry.

On the average, half of all patients engaging in MT, partake in some type of tourism activity. This number is an average. Some respondents answered that all of their patients engage in touristic activities while some had a very few that engaged in any tourism activity (as few as 5%).

We uncovered that there is a large variation in the length of stay among patients. Some trips lasting only three days and some as long as a month. Many include at least some leisurely activities outside of the realm of a hospital stay.

U.S. was the most common country of origin for patients, perhaps due to medical costs being prohibitively expensive in this country. The analysis also suggested that perhaps there exist a degree of regional specialization in the MT facilitator industry. As numerous facilitators claimed that their clients exclusively or almost exclusively were from a particular region.

Further reflecting on the wide variety of business models used by MTFs, the services they offer are many and varied.

Payment for services tends to follow either commission-based models, fixed rates, or some combination of the two. Many earn money by the way of commission from the hospitals and doctors, tourist destinations, or from a combination of these sources. Such practices have created a disadvantage for the U.S. facilitators that are prohibited by law to charge doctors and hospitals for a commission. This has prompted new business models to circumvent these unfair practices.

Costs vary greatly depending upon the procedure. Typical patient expenses ranged from $5,000 USD to $100,000. Unlike receiving treatment in the U.S., patients will typically know ahead of time what their medical expenses will be as there is a great deal of transparency in the foreign medical marketplace, which the U.S. medical system is advised to follow.

Some facilitators will have special promotional offers for a variety of reasons to gain customers and their loyalty.

Payment systems are generally flexible, offering some relief to patients. Secure bank transfers, credit cards, cash, personal check, and travelers' check or a combination of those are accepted. Most facilitators require at least partial payment upfront, and receive some payment following the procedure, while a small number of facilitators will bill patients in installments, not unlike the medical industry in the U.S.

Customer feedback appears to be appreciated by MTFs. Most respondents are utilizing some form or another, and many MTFs seek feedback from more than one method. Written customer feedback, phone calls, online reviews, social media all were used to gauge customer satisfaction.

When exploring the factors which make a MT facilitator successful, it was surprising to see neither price nor time be the top response. Although these were named among the most important, the top response was safety. Safety concerns and quality of health care have often been sighted as the top factors that may dissuade potential patients from seeking care abroad. The finding of this study supports the notion that MTFs give top priority for professionalism and their reputation in the market. Trust was also sighted as an important factor for success. Perhaps building on this is the recognition that the relationship between facilitator and medial provider must be strong.

14.6 RECOMMENDATIONS FOR FUTURE RESEARCH

There are considerable variations between patient satisfaction ratings of destinations. This may be due, in large part, to the regional travel nature of the industry and its facilitators. Future investigations into particular aspects of patient experience could gain the desired insights. The responses seem to hint that MTFs tend to work exclusively with either inbound or outbound patients. This could be explored further. Finally, challenges faced by facilitators are widely varied. Price is often sighted as a hurdle, as is patient safety.

The customer experience is also mentioned several different ways. Through providing better service and being able to protect their privacy rights. All these challenges could be explored further.

14.7 LIMITATIONS

MT is a global phenomenon making the sample size an important requirement. A small sample size of this study limited further breakdown of data for particular regions with unique characteristics, businesses with small, medium, or large sizes, and different business models.

KEYWORDS

- **Joint Commission International**
- **medical tourism facilitators**
- **medical tourism intermediaries**
- **medical tourists**
- **medical travel agents**
- **medical travel intermediaries**

REFERENCES

Alsever, J., (2006). *Basking on the Beach, or Maybe the Operating Table.* New York Times. Retrieved from: http://www.nytimes.com/2006/10/15/business/yourmoney/15care.html?pagewanted=1&r=1 (accessed on 14 July 2021).

Alsharif, M. J., Labonté, R., & Lu, Z., (2010). Patients beyond borders: A study of medical tourists in four countries. *Global Social Policy, 10*(3), 315–335.

Apollo Hospitals Group. (2010). Available at: http://www.apollohospitals.com/ (accessed on 14 July 2021).

Arriva Costa Rica, (2014). Available at: http://www.arrivacostarica.com (accessed on 14 July 2021).

Arunanondchia, J., & Fink, C., (2007). Trade-in health services in the ASEAN region. *Health Promotion International, 21*(S1), 59–66.

Berger, S., (2005). *Analyzing Your Hospital's Labor Productivity.* Healthcare financial management. Retrieved from: http://www.allbusiness.com/humanresources/employee-development-employee-productivity/393722-1.html (accessed on 14 July 2021).

Boyd, J., McGrath, M., & Maa, J., (2011). *Emerging Trends in the Outsourcing of Medical and Surgical Care.* American Medical Association. Retrieved from: http://archsurg.ama-assn.org/cgi/reprint/146/1/107 (accessed on 14 July 2021).

Cattaneo, O., (2009). *Trade-in Health Services: What's in it for Developing Countries? (Working Paper)* (pp. 2–30). The World Bank International Trade Department.

Chee, H. L., Whittaker, A., & Hong, P. H., (2017). Medical travel facilitators, private hospitals, and international medical travel in assemblage. *Asia Pacific Viewpoint, 58*(2), 242–254.

Cohen, I. G., (2012). How to regulate medical tourism (and why it matters for bioethics). *Developing World Bioethics, 12*(1), 9–20.

Connell, J., (2013). Contemporary medical tourism: Conceptualization, culture, and commodification. *Tourism Management, 34*, 1–13.

Cook, P. S., (2010). Constructions and experiences of authenticity in medical tourism: The performances of places, spaces, practices, objects, and bodies. *Tourist Studies, 10*(2), 135–153.

Deloitte, (2008). *Medical Tourism, Consumers in Search of Value*. Deloitte Center for Health Solutions, Washington, DC.

Deloitte, (2009). *Medical Tourism, Updates, and Implications*. Deloitte Center for Health Solutions, Washington DC.

Fetscherin, M., & Stephano, R. M., (2016). The medical tourism index: Scale development and validation. *Tourism Management, 52*, 539–556.

Fox, M., (2010). *Malpractice Liability Costs the U.S. $55.6 Billion: Study.* Reuters. Retrieved from: https://www.reuters.com/article/us-malpractice-usa/malpractice-liability-costs-u-s-55-6-billion-study-idUSTRE6860KN20100907 (accessed on 14 July 2021).

Gan, L. L., & Frederick, J. R., (2011). Medical tourism facilitators: Patterns of service differentiation. *Journal of Vacation Marketing, 17*(3), 165–183.

Gan, L. L., & Frederick, J. R., (2018). The choice of facilitators in medical tourism. *Health Marketing Quarterly, 35*(1), 65–83.

Gary, H., (2008). Medical tourism: Crossing borders to access health care. *Kennedy Institute of Ethics Journal, 18*(2), 193–201

Haarhoff, R., & Mokoena, L. G., (2017). The market place of medical tourism facilitators in South Africa. *African Journal of Hospitality, Tourism and Leisure, 6*(3), 1–17.

Herrick, D. M., (2007). Medical tourism: Global competition in health care. *National Center for Policy Analysis, 304*, 01–35.

Horowitz, M. D., Rosensweig, J. A., & Jones, C. A., (2007). Medical tourism: Globalization of the healthcare marketplace. *Medscape General Medicine, 9*(4), 33.

Jagyasi, P., (2010). *Medical Tourism: Research & Survey Report*. Dr Prem.

Kavilanz, P. B., (2009). *Biggest Medical Mystery: The Bill.* Retrieved from.

Lancaster, J., (2004). *Surgeries, Side Trips for 'Medical Tourists* (p. A01). Washington Post.

Lazzaro, V., (2011). How global case rates differentiate medical travel facilitators. *Employee Benefit Plan Review*, 13, 14.

Lee, H., Wright, K. B., O'Conner, M., & Wombacher, K., (2014). Framing medical tourism: An analysis of persuasive appeals, risks and benefits, and new media features of medical tourism broker websites. *Health Commun., 29*, 637–645.

Maquire, A., Bussmann, S., Kocker, C. M. Z., Verra, S. E., Giurgi, L. A., & Ruggeri, K., (2016). Raising Concerns about the information provided on medical travel agency websites: A place for policy. *Health Policy and Technology, 5*, 414–422.

McCarthy, E., & Frank, P., (2013). U.S. Academic. *Medical Centers and Medical Facilitators: A Changing Relationship*. Partners Health International.

Med retreat. Retrieved from http://www.medretreat.com/ (accessed 22 September 2021).
Milbank, J., (2018). The role of professional facilitators in cross-border assisted reproduction. *Reproductive BioMedicine and Society Online, 6*, 60–71.
Milstein, A., & Smith, M., (2007). Will the surgical world become flat? *Health Affairs, 26*(1), Retrieved from: https://www.healthaffairs.org/doi/10.1377/hlthaff.26.1.137 (accessed on 14 July 2021).
Moody, M., (2007). Medical tourism: Employers can save significant healthcare dollars by having employees seek overseas options. *Rough Notes, 150*(11), 114–116.
Roth, M., (2006). *Surgery Abroad an Option for Those with Minimal Health Coverage*. The Pittsburgh Post-gazette. Retrieved from: https://www.post-gazette.com/life/travel/2006/09/10/Surgery-abroad-an-option-for-those-with-minimal-health-coverage/stories/200609100214 (accessed on 14 July 2021).
Simon, B., (2006). *Luxurious Faraway Hospitals [Television Series Episode]*. In (Jeff Fager), 60 minutes. New York, NY: CBS.
Singh, S. N., & Wachter, R. M., (2008). Perspectives on medical outsourcing and telemedicine: Rough edges in a flat world? *New England Journal of Medicine, 358*, 1622–1627.
Snyder, J., Crooks, V. A., Adams, K., Kingsbury, P., & Johnston, R., (2011). The 'patient's physician one-step removed': The evolving roles of medical tourism facilitators. *Journal of Medical Ethics*, *37*(9), 530–534.
Solomon, H., (2011). Affective journeys: The emotional structuring of medical tourism in India. *Anthropology and Medicine, 18*(1), 105–118.
Stephani, R., (2013). *Medical Tourism Facilitators: The Good, the Bad-the Unknown*. Retrieved from: https://medicaltourismassociation.com/upload/file/Medical-Tourism-Facilitators-the-Good-the-Bad-the-Unknown-26.pdf (accessed on 14 July 2021).
Tolliver, K., (2011). *Proliferation of Cross Border Trade in Health Services* (pp. 1–42). University of Tennessee Honors Thesis Projects.
Turner, L., (2007). First world health care at third-world prices': Globalization, Bioethics and Medical Tourism. *Bio Societies, 2*, 303–325.
York, D., (2008). Medical tourism: The trend toward outsourcing medical procedures to foreign countries. *Journal of Continuing Education in the Health Professions, 28*(2), 99–102.

CHAPTER 15

LEGAL ISSUES IN MEDICAL TRAVEL FACILITATION

STEPHEN M. WEINER

Founding Chair, Heath Law Practice Mintz, Levin, Cohn, Ferris, Glovsky and Popeo, P.C, Boston, Massachusetts

ABSTRACT

This chapter of legal analysis of medical travel focuses on relationships, and rights, responsibilities, and potential liabilities that may arise from relationships. The world of medical travel is filled with relationships. This chapter focuses on the medical facilitator and their role in understanding healthcare and risk. Medical travel is a healthcare business. An expertise in travel alone is not sufficient. The medical facilitator's role should not entail the direct provision of services, the facilitator is arranging for healthcare services for a person in need of them, and therefore must understand the medical and social context in which a client seeks care from a destination provider. Historically facilitators generally came to a large extent from a travel background. Some medical facilitators, destination providers, clients are described in this chapter along with relevant legal analysis. Disclaimer, though: This chapter is descriptive and is not intended to provide legal advice, only macro analysis of issues that arise from some medical travel relationships.

In order to keep this chapter chapter-length rather than book-length, we will narrow our focus to legal issues associated with the role of the "middleperson" in the medical travel arena, whether we call that person (individual or entity) the facilitator, the broker, the expediter, or whatever.

The author would like to thank Mack Ramsden, Yale Law School Class of 2023, who, as a Project Analyst at Mintz Levin, served as a research assistant to the author in preparing this chapter.

We will use facilitator for ease of convenience. The facilitator makes the arrangements by which the person who is to travel to receive medical care (the "client") receives care from a provider in another jurisdiction (the "destination provider"). The scope of its services may be limited or broad, depending on the facilitator's business model and the needs of its clients. The services may include any of the following: (i) arranging for travel for medical purposes but also travel incident to the receipt of medical services, including excursions in the destination country if the client is fit for that; (ii) advising on or helping to select the destination provider; (iii) arranging for or facilitating the transmission of medical data from the client's providers in her place of residence (the "local provider") to the destination provider; (iv) arranging for or facilitating payment of the destination provider's fees; (v) arranging for or facilitating aftercare when the patient returns home; and/or (vi) facilitating administration of an ERISA plan's medical option, as described below. Each of these functions raises legal and other issues that are briefly described in this chapter. The facilitator may also have a formal affiliation with a destination provider. That aspect of the facilitator's business model is not addressed in this chapter.

Some of the concepts described below in this chapter would be relevant regardless of what localities-the client's place of residence or the location to which the client is to travel-are involved, but much of law is local in its fashion, i.e., based on specific jurisdictions, and so there will always be differences in legal analysis and application depending on what localities are involved in a particular medical travel situation. For the present purposes, then, we will assume that: (i) the facilitator is U.S.-based; (ii) the client is domiciled in the U.S.; and (iii) the client is leaving the U.S. for care in a "foreign," i.e., non-U.S. destination. We will not focus, other than on a very limited basis, on the destination provider, other than to highlight issues related to the provider that affect the facilitator's responsibilities and the client's rights.

In considering the role of the facilitator, we need first to know something about the client. Most specifically, we need to know how the client intends to pay for the services or have them paid for. Many clients will be self-pay patients, that is, they plan to pay out of pocket for their care. Some clients may have insurance coverage that covers some of the medical travel benefits, but it may not cover some or any of the care the client will be specifically seeking. The increasing popularity of high deductible health plans in the U.S. means that some clients with insurance may nonetheless choose to engage in medical travel in order to reduce their out-of-pocket costs. Some clients will be covered by a self-insured employer plan subject to ERISA in the U.S. ERISA does not prevent employers from offering a medical travel option as part of the health benefit they provide employees, but the scope and terms of an optional medical travel benefit must be clearly spelled out. For example, the benefit could be limited to a certain scope of procedures provided only at specified destination providers and no more. The medical travel benefit terms must be clear in the plan documents so that covered employees know the nature of the benefit before they access it. Further, employers may provide financial incentives for employees to seek care through the medical travel option they offer. But they cannot require or force ("coerce") employees to use medical travel to receive benefits under an ERISA plan.

When considering legal issues associated with any economic actor, including facilitators, questions of risk, liability, likelihood of being sued, and likelihood that a lawsuit would be successful inevitably arise, generally quite quickly. It is important to differentiate these issues. Risk may not necessarily lead to actual liability. Risk is a combination of legal and business judgment, and, where the law is not 100% prescriptive, which is most times, then judgment is needed as to how much risk a party is prepared to accept. Risk may be mitigated generally by actions that would in fact seem to be based on simple common sense. Liability means that something has gone wrong in a way that gives another party a claim against the facilitator. Even the best risk mitigation strategies may not avoid liability, but liability can be mitigated, to some extent, by contract. Likelihood of litigation is in some respects the most difficult to predict. It may be a truism to say, certainly in the litigious culture of the U.S., that, since risk rarely can be reduced to zero, the risk of being sued can commensurately rarely be reduced to zero. The key question is then often not whether a party will avoid a lawsuit but how likely it is that the lawsuit might have a successful outcome for the plaintiff, and whether the defending party has insurance to cover the litigation costs it might bear regardless of outcome. In considering the legal context in which a facilitator functions, strategies for assessment, allocation, and management of risk are necessary, and good legal documents that can define and limit the scope of potential liability are important elements of risk mitigation.

Before discussing risk mitigation in more detail, however, another general introductory observation should be kept in mind in assessing legal risk. If a facilitator is brought to court by a client, the legal outcome may to a degree depend on the judge's "view" of the client. The judge may need to determine if the client appears to be an informed consumer or if the client appears to be a party in need of protection by the court. There is no necessary predetermined outcome of such a weighing. The judge will, if swayed by either of these possibilities, consider a variety of facts and circumstances. The important point for a facilitator is that the clearer information it provides to a client and the more appropriate the choices are that it offers to a client, the more likely the client will be seen as an informed consumer who understood the risks she was taking. This perception of the client is helpful in defending the facilitator. This observation should therefore help frame some of the risk mitigation factors described below.

As a further generalization, the facilitator who wishes to reduce the likelihood of suit from a client should be clear about its responsibilities vis-à-vis the client's responsibilities about traveling for medical care and the risk attendant to such activity. The facilitator should engage in clear and well-documented communications with the client about the nature of the risks and the allocation of responsibilities. This information should be documented, that is, written in clear and plain language that the client can be expected to comprehend. The information should also be communicated via a personal connection between the facilitator and the client, that is, through what the facilitator or its staff *says* to the client and how. The information should be consistent between the written and oral communications, of course. This is an area in which it is especially important that the facilitator understand that it is engaged in providing support for healthcare services, not simply making travel arrangements.

With the foregoing as background, we should recognize that, in defining their relationships with clients, facilitators should adhere to a set of risk reduction principles that guide the written documentation they use with clients. It should also guide the personal interactions that their staff has with clients and, as applicable, clients' families or guardians. Some of the key principles for risk reduction may be summarized as follows, a list that is of course not intended to be exhaustive:

- **Principle 1:** The facilitator should understand healthcare. Medical travel is a healthcare business. An expertise in travel alone is not enough. Even though the facilitator's role should not entail the direct provision of services, the facilitator is arranging for healthcare services for a person in need of them, and therefore must understand the medical and social context in which a client seeks care from a destination provider. Historically facilitators generally came to a large extent from a travel background. The "medical" in medical travel is far more important to reducing risk for a facilitator than is the "travel" component.

- **Principle 2:** The facilitator should be sure that the client knows the right questions to ask, even if the facilitator is not necessarily the source for providing the answers. These questions include, but are not limited to, the following:
  - The nature of the client's condition.
  - The risks associated with travel for care.
  - Information about the destination provider.
  - How will the client's medical data be transmitted between local providers and destination providers?
  - What privacy rules apply to the client's medical data when it is outside the U.S., and how will those privacy rules be honored?
  - What happens if unexpected emergencies arise while traveling or while under the care of the destination provider?
  - What if something goes wrong that could be considered malpractice in the destination provider's jurisdiction?
  - How will the provider be paid? and
  - What happens when the client returns home and needs aftercare?
- **Principle 3:** Be sure that the documentation used with clients is complete, that it accurately and clearly addresses respective responsibilities, has the proper representations (discussed below), and is written in a lucid and understandable language. If the client is not fluent in English, nonetheless, make sure that he or she understands the provisions of each of the agreements the facilitator uses. No small type or unintelligible legalese.
- **Principle 4:** Do not rely just on documentation. The facilitator should communicate with the client orally as well as in writing to be sure that the client has received, and understands all the relevant information. It may be useful to have an informal standard script that will be followed by facilitator staff in discussions with a client about what is involved with medical travel and what the documentation says and means. A "script" would allow the facilitator to maintain a record of the information provided to clients orally and would provide assurances that the information that is given to clients is complete, consistent with the written documents, and is being provided consistently for all clients.
- **Principle 5:** Risk reduction is correlated with information. Information must be communicated clearly and, where recommendations are involved, as for instance about what destination providers to use, the information should be based on validated and objective information

to the maximum extent possible. It is generally better to provide choices, not a single recommendation. There is, however, a delicate balance between providing useful information that a client may act upon, being sure it is complete and accurate, and assuming responsibility for the information and the actions taken by the client. It is generally better to caution the client about the need to obtain her own independent assessment of the information.

While we emphasize the importance of clear and comprehensible communications between facilitator and client, we have not yet addressed in any detail the specifics, that is, what should the facilitator be communicating. Here are some points; their significance should be obvious, but some will be further explained:

- The facilitator must be clear that it is not the agent of the destination provider. In some cases, the facilitator entity may in fact be affiliated with a destination provider, and so the distinction between its responsibilities, and those of the affiliated provider, must be carefully delineated, as provided below. Separation of the entities would be important, although any such affiliation, and especially if it is one where the facilitator is owned by the destination provider, increases the level of risk that the client may seek to hold the facilitator liable for actions that properly reside with the provider.
- The facilitator should be clear that it is not engaging in the practice of medicine and that it is not making medical decisions.
- The client should be advised to consult with her own caregivers locally, both before and following travel for medical care. The facilitator should seek assurances that such consultation has taken place.
- The facilitator should also seek assurances that the client has been fully advised by an appropriate clinician or clinicians about the risk of travel in light of her medical condition.
- The facilitator should clearly define the scope of services that it will arrange or provide itself, both medical and non-medical. For example, will it arrange for transportation to and from the destination provider, both internationally and in the local jurisdiction? Will it handle travel that is not directly associated with the medical care, such as touring that the patient may be able to undertake while in the foreign jurisdiction? Will the facilitator arrange for appropriate amenities at the site of the destination provider, such as internet or translation services?

With respect to the specification of services to be provided, quite importantly, the facilitator should be clear about whether it will have responsibilities related to payment for the destination provider's medical services and for the transmission of the client's medical data from domestic providers to the destination provider. Each of these issues is discussed in more detail below.

The facilitator should be cautious about endorsing providers. As a generalization, it is better than the client selects her own destination provider, in consultation with her local caregiver(s), but that is not always, or perhaps even generally, the case. There is often an expectation that the facilitator will play some role in the selection of the destination provider. So, when asked for recommendations, the facilitator should recognize that offering choices is better than making a single recommendation; and that, in making any recommendations, the facilitator should emphasize the client's independent responsibility for validating the quality and qualifications of the destination provider, preferably, as noted above, in consultation with her own domestic provider(s). The facilitator should recognize that in making any recommendations, notwithstanding appropriate qualifications and caveats, it could be seen as endorsing the quality of the providers it recommends, and so should take care not to be "foolish" in recommending providers of known questionable quality. Some facilitators that do not have a formal affiliation with a provider system develop a network of preferred providers as a mechanism for offering choice to clients. If that approach is used, the facilitator should be clear as to the standards it uses to include a provider in its network; and those standards used for selection should be reasonably objective. It is desirable for the facilitator to look to independent certification, accreditation, or licensure bodies that have given the provider an imprimatur that can stand as a proxy for quality. If possible, to support its recommendation, the facilitator should arrange for its staff to visit providers it is considering recommending, and to undertake background checks from available information on the provider's clinical and managerial staff.

One of the key considerations in selecting a destination provider is of course the cost differential between the client's receiving the service from that provider and her receiving it closer to home. In selecting providers to recommend, the facilitator will of necessity be cognizant of cost differences, but that should not be the sole driver for selecting providers to recommend. In addition to the types of proxy quality standards described above, the facilitator should have some understanding of the overall sophistication of the medical system in the jurisdiction in which the destination provider is located, and could also give some consideration to what is essentially an ethical issue: is the destination provider the only high quality provider in the jurisdiction and is it open equally to local residents of the jurisdiction as it is to patients imported from elsewhere, like the U.S.

Other considerations that may serve as selection criteria include the extent to which the client will feel she is in "familiar surroundings" to her own home location, and whether (assuming the client is an English-speaking American resident) English is commonly spoken in the area and by the medical staff of the destination provider. One additional consideration, which is relevant to the discussion on malpractice below and is of great interest, especially to U.S. residents going abroad, is whether the jurisdiction of the destination provider has a transparent legal system and what the rules of that jurisdiction are with respect to malpractice actions and recoveries.

As noted earlier, risk reduction for a facilitator means both oral and written communications that are clear and comprehensible for the client. In addition to the communications described above, much of which can be conveyed orally, the facilitator should have a written agreement with the client that, without being overly legalistic and off-putting, does make sure that the client has had to think about some of the key points that the facilitator wishes to communicate. Any such agreement should include written acknowledgments by the client that:

- She has consulted with and received medical advice from a local caregiver who knows her;
- She has assessed with a local caregiver the risks of travel and the treatment options available to her;
- She recognizes the risk associated with her travel and her seeking care in another jurisdiction;
- She, directly or through her local caregiver(s), will be providing the destination provider with complete and accurate medical records sufficient for undertaking the treatment being sought;
- She understands that the scope of services she anticipates receiving may be modified upon arrival at the destination provider based on assessment or emergencies that may arise; and
- She has identified what additional services she may wish to receive while at the destination provider.

A facilitator must also decide how it intends to handle three critical areas, and the choices it makes must be clear to the client in written documents and oral communications. These are: (i) client fee arrangements; (ii) provisions for payment to destination providers; and (iii) management of medical records. Each encompasses a variety of legal issues that are described briefly in further sections.

15.1 FACILITATOR FEE ARRANGEMENTS

Facilitation being a service business, the facilitator naturally expects to be paid for the services it provides. There are choices as to the source for the payment and the structure of the payment. Whatever approach is agreed upon between the facilitator and the client must be accurately and clearly recorded, with any reasonably anticipated contingencies accounted for in advance, to the extent possible. Each choice has certain possible legal ramifications.

If the client is to be responsible for paying the facilitator's fee, the payment could take a number of different forms, including a lump sum payment covering all services, an a la carte payment for each discrete subset of services, or a fee based on a percentage of the estimated destination provider fees. The last of these seems least tied to the actual resources used by the facilitator, but is an available option. However, it should accommodate to the types of changes in provider fees for which the client will be responsible beyond the originally estimated amounts, as described below in the discussion on provider payments. A facilitator fee based on provider charges that accommodates such changes, however, may require the facilitator to bill the client for multiple fees as circumstances change, which can be frustrating to clients who seek certainty and predictability as to their financial obligations. Where the fee arrangement is between the facilitator and the client, the terms must be clear as to how the fee is calculated, when the payment is due and what form or forms the payment may take. If, for example, the facilitator accepts payment via credit card, it will need to be aware of its own responsibilities vis-à-vis credit card issuers, the financial regulatory structure that surrounds credit card payments, and the fees for which it will be responsible by accepting that method of payment.

The fee arrangement for facilitator services may be between the facilitator and the destination provider, as distinct from the facilitator and the client. This makes the fee less transparent to the client while at the same time raising potentially more legal issues. As with fees paid by the client, the fees paid by the provider could also be a flat amount per procedure or incident of care, or a percentage of the total fees to be paid to the provider for the client's services. Payment could be structured as an add-on to the charges that the client would otherwise be responsible for (the client pays both the destination provider's established charges and an additional amount for the facilitator fee, and the destination provider remits the fee amount to the facilitator); or it could be a discount from the provider's charges that is retained by the facilitator, who assesses the client for the destination provider's full undiscounted charges, and not passed back to the client (the client pays the facilitator for the provider's fees based on an established provider charge schedule but the amount actually remitted by the facilitator to the provider is discounted).

Regardless of the structure, where the provider is directly or indirectly covering the facilitator's fee, the payment could be characterized as a referral fee and might be illegal in the jurisdiction in which the provider is located (While not directly within the purview of this chapter, it should be noted that, if a facilitator arranges for care on an inter-state basis within the United States (U.S.), it must be aware of both the source of the client's coverage for the care (e.g., self-pay, private insurance) and whether the particular state in which the destination provider is located has anti-kickback-type statutes that apply to care paid for by the particular payer analogous to those that apply under the Medicare or Medicaid programs nationally.).

15.2 PROVIDER FEE ARRANGEMENTS

The facilitator also has choices with regard to how the payment to the destination provider is to flow from the client to the provider. Under any of the usual models, the facilitator should be expected to arrange for the client to have an estimate of the cost of the services she will be receiving prior to arrival at the destination provider. The estimate would be based, preferably following consultation between the destination provider and the client's local provider(s) and after a review of the client's medical records, on the proposed scope of services to be provided and a pre-determined fee schedule, which could be, for example, the destination provider's usual charges, or some discount from those charges. Depending on the agreement between the facilitator and the destination provider, the client may be obliged to pay a certain percentage of the estimated charges within a certain time period before the client arrives at the destination provider's locale, such as 50% at least 30 days beforehand. Further, generally the destination provider will expect full payment of the estimate to be made by the time services commence. The facilitator's client agreement, therefore, must also provide for what happens in the event the client is unable to arrive at the scheduled time or the services are not able to commence on time. In addition, the agreement should contemplate what happens in the event the scope of services changes once the client arrives, or if something unexpected happens while she is undergoing treatment with the destination provider. While not all scenarios can be predicted, several of the more obvious ones should be spelled out for the client, with written understandings in place as to how they will be handled if they arise. The client should be aware that the estimated amount she is responsible for prior to commencement of services is not necessarily the full amount for which she will have ultimate liability. As noted above, where the facilitator fee is based on the provider's charges (estimated or as adjusted), this will need to be clear to the client as well.

A further issue the facilitator must determine is whether it will be a conduit for the funds to be paid to the destination provider. The facilitator could provide for the client to pay the destination provider directly, on the schedule agreed to between the facilitator and the provider. Or the client could remit the payments to the facilitator, who then itself undertakes the responsibility for providing the fees to the destination provider. If the latter approach is used, it can be limited just to the payment of the estimated fees, which would typically be due before services commence, with the client being directly responsible for payment of any additional charges that may arise once she arrives at the destination provider. There are pros and cons associated with any approach taken to how the destination provider is to be paid. For example, if the funds pass through the facilitator, there is a higher likelihood of assuring the provider is paid timely. If the facilitator's own fee is being paid by the provider, the facilitator could withhold the fee from the amounts remitted to the provider to be sure it receives its payment as well. On the other hand, the facilitator may not wish to have the legal liability to the destination provider of assuring payment where the client may not be forthcoming in a timely matter, especially where the amount of fees may change based on events occurring once the client arrives at the destination provider. Whatever approach is used, to restate points made earlier, the payment arrangements for the destination provider must be clearly spelled out, the facilitator must be sure the client fully understands the terms, and, as much as is reasonably possible, contingencies are identified and accommodated in advance. Further, the facilitator will need to build into the agreements with the client and with the destination provider protection for itself if the client fails to meet her obligations for timely provider payment.

15.3 MEDICAL RECORDS

It is critical for medical travel that full and accurate medical records are made available to the destination provider prior to it providing an estimate of the services to the client. Earlier in this chapter, we discuss the importance of the facilitator clarifying with the client that the client, working with her local provider(s), has the responsibility for assuring this. The facilitator should not take responsibility for the completeness or accuracy of the medical records. In addition, while the facilitator may arrange for the destination provider to provide records of its services to the client and the client's local provider(s) as appropriate (for example, for follow-up care purposes), the facilitator should not take responsibility for assuring that the destination provider meets this obligation. These precautions are inherent in the principle that the facilitator is not in the business of providing medical services.

There is a separate question, however, as to how medical records are transmitted from the client or her local provider to the destination provider. The facilitator again has a choice, of whether to require the client to be responsible for the direct transmission of her medical records, so the facilitator plays no role other than to secure assurance from the client that this is being or will be timely done; or the facilitator takes responsibility for the communication (but not the content) of the medical records. Again, there are pros and cons. If the facilitator acts as the intermediary for the transmission of the records, then it has assurance that the destination provider has received the relevant records as it determines the services it will provide and establishes the estimate of fees to be remitted. At the same time, though, the facilitator will need to comply with the privacy and security requirements that apply in the jurisdiction in which it receives the records (in the U.S. that would be HIPAA and any applicable state privacy law requirements) and in the jurisdiction to which the records are sent, that is, the jurisdiction in which the destination provider is located. Compliance will be critical to reduce the risks associated with medical travel facilitation, and it could entail potentially significant additional costs for the facilitator to implement the systems necessary to assure compliance. In any event, the facilitator will need to determine what role it wishes to play in the transmission of medical records and the extent to which it protects itself either as a compliance matter or in being sure that the client has adhered to its contractual obligations regarding that transmission.

15.4 MALPRACTICE

A question that a facilitator inevitably faces in working with clients is what happens if something goes wrong while the destination provider is providing care. Clients who reside in the U.S. would generally prefer to bring a malpractice action against the destination provider in the U.S., where malpractice jurisprudence is more "generous" toward plaintiffs than in many other jurisdictions. To a large extent, the issue of handling malpractice claims is between the client and the destination provider, not the facilitator, but a responsible facilitator should be sure that its clients are aware of what recourse they may have in the event errors in treatment occur. This would include general information about the malpractice jurisprudence where the destination provider is located; the availability of civil proceedings and the transparency of the judicial system in that jurisdiction; and the nature of the consent forms that the destination provider is likely to request the client to sign as a condition of treatment. Such forms are likely to contain agreements that disputes, including overtreatment, would be subject to the destination provider's local jurisdiction and an agreement by the client not to seek recourse against the provider in any other jurisdiction. The client will likely wish to know if she could bring an action against the destination provider in a court in the U.S. This raises complex issues of law that entail questions of whether the U.S. court would honor a consent form limiting actions to the destination provider's jurisdiction, whether the U.S. court would take jurisdiction based on theories of whether the destination provider does or does not have sufficient contacts within the U.S. to justify asserting that jurisdiction; and whether, even if the U.S. court takes jurisdiction, the court should apply local U.S. law or the law of the jurisdiction in which the destination provider is located. These are complex areas that tend to turn on a variety of "facts and circumstances" considerations, so it may be difficult for a facilitator to give concrete guidance to a client, and, indeed, it may be just as well for the facilitator to suggest that the client should consult with counsel in advance of treatment to understand what might happen in the event of an error, and not take responsibility for providing the information itself.

An alternative for dispute resolution is not to rely on the judicial process but to use arbitration. The facilitator could work with both the destination provider and the client to have any agreements the client enters with the provider call for this. The venue would again be an issue but, by using this approach, the client may avoid being subject to malpractice jurisprudence that is not especially patient-friendly.

It is also, of course, possible for the client to seek recourse against the facilitator in the event an error occurs in treatment. While theoretically, in the U.S., anyone may seek to sue anyone else when something goes wrong, the facilitator can minimize its exposure to a malpractice action by being clear, as is discussed above, about the limits of its responsibility for the provision of medical care the client is seeking. There could be a liability for matters for which the facilitator is responsible depending on the arrangements used by the facilitator, such as the transmission of medical records. But with respect to the actual provision of care, the facilitator should be able to protect itself by proper documentation and by otherwise being quite clear that it is not taking responsibility for the care of the patient (and preferably not directing the patient to the particular destination provider but giving a patient choice in making that decision).

15.5 CONCLUSION

As recourse to medical travel may increase, especially for U.S. residents seeking less expensive choices when they have financial liability for the healthcare services they need, the business opportunity for a facilitator may outweigh the risks attendant on the business model. However, facilitating medical travel encompasses a web of relationships that have substantial legal implications. Identifying potential sources of liability and managing the risk attendant thereto requires thoughtful attention to the details and mechanics of the facilitation/client relationship, clarity as to the scope of respective responsibilities, clear communication, and good protective documentation, especially in the form of contracts with clients and destination providers.

KEYWORDS

- **facilitator's business model**
- **good protective documentation**
- **liability**
- **medical travel**
- **risk mitigation**

BIBLIOGRAPHY

Cohen, I. G., (2010). *Medical Tourism: The View from 10,000 Feet.* The Hastings center report.

Cohen, I. G., (2010). Protecting Patients with Passports: Medical tourism and the patient-protective argument. *Iowa Law Review, 95*, 1467–1567.

Cortez, N., (2008). Patients without Borders: The emerging global market for patients and the evolution of modern healthcare. *Indiana Law Journal, 83*, 71–132.

Cortez, N., (2010). Recalibrating the legal risks of cross-border healthcare. *Yale Journal of Health Policy, Law & Ethics, 10*, 1–89.

Cortez, N., (2012). Into the void: The legal ambiguities of an unregulated medical tourism market. In: Hodges et al., *Risks and Challenges in Medical Tourism* (pp. 187–206). Praeger Publishing; Chapter 9.

Crooks et al., (2013). Ethical and legal implications of the risks of medical tourism for patients: A qualitative study of Canadian health and safety representatives' perspectives. *BMJ Open,* 3

Crooks et al., (2015). Inbound medical tourism to Barbados: A qualitative examination of local lawyers' prospective legal and regulatory concerns. *BMC Health Services Research, 15*, 291

Johnston, R., et al., (2010). What is known about the effects of medical tourism in destination and departure countries? A scoping review. *International Journal for Equity in Health, 9*, 24–36.

Kassim, P. N. J., (2009). Medicine beyond borders: The legal and ethical challenges. *Medicine and Law, 28*, 439–450

Parsiyar, A., (2009). Medical tourism: The commodification of healthcare in Latin America. *Law and Business Review of the Americas, 15*, 379–404.

Steklof, C., (2010). Medical tourism and the legal impediments to recovery in cases of medical malpractice. *Washington University Global Studies Law Review, 9*, 721–742.

Terry, N., (2007). Under-regulated healthcare phenomena in a flat world: Medical tourism and outsourcing. *Western New England Law Review, 29*, 421–472.

Williams, H., (2011). Fighting fire with fire: Reforming the healthcare system through a market-based approach to medical tourism. *North Carolina Law Review, 89*, 607–684.

PART VI

Patient-Centric Approaches: Managing Medical Traveler Expectations

CHAPTER 16

PATIENTS RATE HOTEL-LIKE HOSPITALS BETTER THAN STATE AND NATIONAL AVERAGE IN PATIENT SATISFACTION

ALI A. POORANI and RITA FEARON

Alfred Lerner College of Business & Economics, Department of Hospitality and Sport Business Management, University of Delaware. 14 W. Main Street, Newark, DE 19716, USA, E-mail: poorani@udel.edu (Ali A. Poorani) E-mail wittyreta@yahoo.com (Rita Fearon)

ABSTRACT

Healthcare providers are contending with the trend of moving from basic pay-for-services to value-based systems as patients are demanding first rate services and are shopping for superior experience. The purpose of the study was to explore how patients rate hotel-like hospitals as compared to state and national averages as measured by hospital consumer assessment of healthcare providers and systems (HCAHPS). The findings indicate that the majority of hotel-like hospitals did better than state and national averages with regards to: (i) patient satisfaction scores; and (ii) recommending the hospital to friends and family.

16.1 INTRODUCTION

Is there anything that hospitality can offer to hospitals? Service and experience, certainly. How about facilities design? Yamaguchi (2015) in *Harvard Business Review* suggested, "Better Healing from Better Design." But, what about, "Better Satisfaction from Better Design?" That is the primary goal of this chapter.

It has been long acknowledged that the physical environment can have an immediate effect on the attitudes and behaviors of customers and employees (Bitner, 1992). In this chapter, we imperially explore the effect hotel-like or hotel-inspired hospital design on customer satisfaction.

Due to new government CMS (Center for Medicare and Medical Services) and HCAHPS (hospital consumer assessment of healthcare providers and systems) requirements, as well as intense industry competition, the effectiveness of healthcare delivery is being measured by the quality of related direct and indirect services, including quality of accommodations, communication, and other auxiliary services, which impact hospitals financially, as reimbursement rates are based on both medical and guest experience measures (Chee et al., 2016).

Yamaguchi (2015), in his article, featured in *Harvard Business Review*, entitled, Better Healing from Better Hospital Design, stated that, "Innovative new hospital designs have changed patients' experiences and expectations of what a hospital should be. These sleek buildings have design features found in many high-end hotels—and they are built with the express intention not just of increasing patient satisfaction but of reducing costs and improving outcomes." Additionally, hospitals strive for hotel-quality service, borrowing the fundamentals from the hospitality industry to improve patient experience and customer service (Binder, 2015).

Though very compelling argument for the trend, hospital CEO Barry Rabner (2014) acknowledged the difficulty of knowing where to invest in facility design, since there is little research available on the amenities that most improve health outcomes, he asked, "...where's the evidence that it works?" Additionally, the design of a hospital facility, along with its equipment and technology providers have not historically considered their impact on the quality and safety of patients (Reiling, Hughes, and Murphy, 2008). Saddiqui et al. (2015) studied the new patient-centered design features incorporated into the new clinical building. The results showed improved room- and visitor-related satisfaction without significant improvement in satisfaction with clinical providers, ancillary staff, and only 1 of 4 measures of overall satisfaction improved. The most prominent increase was with pleasantness of décor (33.6% vs. 64.8%) and visitor accommodation and comfort (50.0% vs. 70.3%). Our review of literature supports the above notion and found little evidence whether hotel-like hospitals actually increase patient satisfaction scores.

The objective of this study was to address this void and investigate if hotel-like hospitals deliver higher patient satisfaction as measured by HCAHPS. HCAHPS (CMS.gov), is a standardized patient experience survey instrument for all hospitals in the U.S. that enables customers to compare hospitals and for the hospitals to hear the voices of their patients and their families.

Discerning the facts and statistics around the benefits of hotel-inspired hospital designs is critical for both healthcare organizations and developers. Just in 2015, there were over $100 billion in construction of mixed projects across the U.S., including apartments, retail space, hotels, and patient care facilities (Drummer, 2015). According to the Economic Research Data published by the Federal Reserve Bank of St. Louis, healthcare construction spending reached nearly 41 billion in 2017; and, it is expected to grow about 2% per year until 2020.

Additionally, due to the increasingly competitive healthcare market, healthcare consumers have more options for the type of care they select; thus, healthcare providers must attempt to build environments that encourage repeat visits and increase patient satisfaction and facility recommendations (Fottler et al., 2000).

16.2 BACKGROUND

As related to the healthcare, until early 2000s, the role of the health care physical or tangible environment, including the facility, was essentially an unstudied area (Hutton and Richardson, 1995). Swan, Richardson, and Hutton (2003), in an experimental study, concluded that patients in well-decorated and well-appointed hotel-like rooms rated their attending physicians, housekeeping, and food-service staff, the food, and the hospital better than patients in standard rooms in the same hospital. Also, they had stronger intentions to use the hospital again and would recommend the hospital to others. Douglas and Douglas (2004) highlighted that the need for personal space, a homey welcoming atmosphere, areas for visitors, access to external surroundings, and the provision of facilities for recreation and leisure were important. Ulrich and Zimring (2004), based on the study of literature found strong evidence for the link between the physical environment and patient and staff outcomes in four areas: (i) Reduced staff stress and fatigue and increased effectiveness in delivering care; (ii) improved patient safety; (iii) reduced stress and improved outcomes; and (iv) improved overall healthcare quality. This and similar studies (Nelson, West, and Goodman, 2005) led to the popularity of evidence-based design (EBD), denoting that healthcare organizations out to acknowledge the critical role that the healthcare facility plays in improved patient and staff outcomes.

Along the same line, Becker and Douglas (2008) concluded that there is a positive correlation between more attractive environments and higher levels of perceived quality, and satisfaction as well as a reduction of patient anxiety. Goldman and Romley (2010) studied amenities at hospitals in the greater Los Angeles area and analyzed the choice behavior of Medicare pneumonia patients in this market. They concluded that amenities such as good food, attentive staff, and pleasant surroundings played an important role in hospital demand. They concluded that one standard deviation increase in a hospital's amenities, raised its demand by 38.5% on average.

While the above studies were focused on EBD and patient outcomes, a new trend was at work that derived on the patient-centric design. For example, in an article entitled, "Many hospitals look to hotels for patient satisfaction tactics," published in the *New York Times,* Weed (2016) stated that, "While clinical care is the focus of any medical center, hospitals have many incentives to move toward hotel-inspired features, services, and staff training." The main driver for this trend, according to the article, is hospitals' interest in marketing to patients with private insurance who have a choice in where they receive care and encouraging word-of-mouth recommendations.

The new trend, also known as (guest-focused care), which started about a decade ago, more regards patient as a 'guest' rather than a 'patients.' The logic behind this assertion is that, patients were (are) generally perceived as something to be processed, while guests are to be treated, and are with you for a short time (*Healthcare Design Magazine*, 2010). In sum, the idea of guest-focused care is about introducing design attitudes that focus on the guest experience, which affects not just finished design but operations, efficiency, and staff interaction. The result is a renewed approach to the concept of "hospitality meets healthcare design" that goes further than what has been done in the past. It shapes everything from check-in to check-out and is based on experiential design. For example, more recently, Suess, Mody, and Guarracino (2017), in a case study based on the theoretical foundations from previous research, reframed the '*healthscapes model,*' that examines the infusion of hospitality attributes in the healthcare environments. They examined patients' preferences for hotel-like attributes using conjoint analysis, and then assessed their perceived well-being and willingness to pay higher out-of-pocket expenses. Their findings supported patient preferences for hospitality features in hospital environments; patients will come back; recommend the hospital; interior design consistently came out on top; and, the service atmosphere was highly desired.

Finally, the need for a patient-centric design, or guest-focused care is more exacerbated by patient experience becoming a top priority for hospital design, and the Center for Medicare and Medicaid Services (CMS) by linking reimbursements to the HCAHPS scores, since 2012. HCAHPS is the first standardized, publicly reported survey of patients' perspectives regarding their hospital stay (CMS.gov).

16.2.1 CUSTOMER/PATIENT PERCEPTION, EXPERIENCE, AND SATISFACTION

Perception is defined as the process of recognizing and interpreting sensory stimuli (2017). Omachonu (1990) asserted that the quality of care is perceived through a patient's own experience and perceptions of that experience. Hemmington (2007) argued that perceptions are important because the memorable experience is what the patient does take home from the hospital.

As a concern for the healthcare organization, Powers and Bendall-Lyon (2003) asserted that there were two key components of patient perception of quality of care: (i) The patient perception of the physical environment; and (ii) patient interactions with staff members. To account for the physical environment, we embarked on comparing patient satisfaction scores of hospitals with well-known hotel-inspired features with ordinary hospitals. The goal was to explore if hotel-like hospitals have a better patient satisfaction scores as compared to the average hospital at the state and national levels.

16.3 METHODOLOGY

16.3.1 RESEARCH DESIGN

The study was designed to investigate and compare the patient satisfaction scores of the hotel-inspired hospitals and compare the results with the state and national scores. More specifically, the study compared patient satisfaction scores as measured by HCAHPS scores from hotel-inspired hospitals with the HCAHPS scores of statewide and national scores.

16.3.2 INSTRUMENTATION

The data sets were made available from the Centers for Medicare and Medicaid Services (CMS) website. For the purpose of this study, we selected the last two items in the HCAHPS survey that measure: overall rating of a hospital:

"please answer the following questions about your stay at the hospital named on the cover letter. Do not include any other hospital stays in your answers."

- Using any number from 0 to 10, where 0 is the worst hospital possible and 10 is the best hospital possible, what number would you use to rate this hospital during your stay?
- Would you recommend this hospital to your friends and family?

Note: For this study, we analyzed the patient satisfaction results based on the highest scores that hospitals strive to achieve:

- Patients who gave their hospital a rating of 9 or 10 on a scale of 0–10.
- Patients who reported YES, they would definitely recommend the hospital.

16.3.2.1 ASSUMPTION

We acknowledge that overall satisfaction scores may mean different things to different patients. However, given a large number of subjects (n = 12,563 in 2017 study) and (n = 9,959 in 2020 study), it is safe to assume that satisfaction scores in this study represent a collective perception of patients about their experience in their hospital stays.

16.3.2.2 ADJUSTMENT OF HCAHPS LINEAR SCORES

One may question the differences in population mix and characteristics using a hotel-like hospital versus other varieties. To answer this question, CMS applies adjustments for the effects of patient mix, survey mode, and quarterly weighting through a series of steps, as follows.

First, CMS applies patient-mix adjustment (PMA) to quarterly HCAHPS scores to account for the tendency of certain patient sub-groups to respond more positively or negatively to the HCAHPS Survey. PMA enables fair comparisons across hospitals by adjusting hospital scores as if all hospitals had an identical mix of patient characteristics. (For more information about the application of the HCAHPS PMAs, including the definition of the PMA factors, please refer to the documents on the HCAHPS On-Line Web site, www.hcahpsonline.org, under the "Mode and Patient-Mix Adj" button).

Second, HCAHPS quarterly scores are adjusted for the effect of mode of survey administration (mail, telephone, mixed-mode, or Interactive Voice

Response). CMS derived the survey mode adjustments from a large-scale, randomized mode experiment. Similar to PMA, these adjustments are applied on the original scale shown on the HCAHPS survey instrument for each measure. For detailed information about the mode experiment and survey mode adjustment, please see www.hcahpsonline.org. If the measure is a multi-item composite, averaging of items within composites takes place at this point. Third, regarding rescaling of HCAHPS linear adjusted scores, each adjusted HCAHPS linear measure score on the original scale is then transformed into a 0–100 linear-scaled score.

16.3.3 SAMPLING

We compared the sample population (hotel-like hospitals) in two intervals, January 2020 and in year 2017. The idea was to see if there is a degree of consistency in the results. Population for the study in year 2017 were patients (n = 12,569) who had utilized the services of the hotel-inspired hospitals as provided in Hospital Compare's data set. The population in 2020 were (n = 9,959). The hotel-inspired hospitals were selected based on literature review, the Internet search, and word associations of hospitals having hotel-like features. This non-probability sampling technique was the only way to select a particular group of hospitals based on specific characteristics believed to be representative of the sample (Gay et al., 2009; Hair, Babin, Money, and Samouel, 2003).

After the final analysis, 9 hospitals appeared to possess hotel-inspired featured published about them:

- Cedars-Sinai Medical Center;
- Emory Johns Creek Hospital;
- Henry Ford West Bloomfield Hospital;Medical City Dallas;
- Mount Sinai Hospital, NY;
- Pennsylvania Hospital Philadelphia;
- Ronald Reagan UCLA Medical Center, Los Angeles;St. Dominic-Jackson Memorial Hospital;
- Woman's Hospital of Texas.

The data collected were self-reported data based on the patients' experiences, satisfaction as well as their perceived quality of care at the hotel-inspired hospitals. The data published by CMS (Centers for Medicare and Medicaid Services) are both quality data, interpretable, and credible as suggested by Bloomberg and Volpe (2008) necessary for any reliable study.

16.3.4 DATA ANALYSIS

Data analysis for this study was straightforward. We wanted to compare average patient satisfaction scores for the sample, hotel-like hospitals, with the averages of other hospitals in the state and also with the national averages. We also looked at the scores of individual hotel-like hospitals and compared them individually with the state and national averages. Additionally, we ran a t-test: two-sample assuming unequal variances to find out if the differences were significant.

Lived experience of hospitality within the hotel-like hospital settings was critical to the study. As such, it was important that participants had been admitted to the hospital for a period that enabled them to experience the hotel-infused amenities at a hospital.

We analyzed data based on scores of 9 and 10 only, the highest numbers on a Likert-Scale of 0–10. These levels of satisfaction are very stringent; those best hospitals pursue; and, very hard to achieve. Then we calculated the average scores for the two following patient satisfaction variables for the years 2017 and 2020:

- **Variable 1:** Patients who gave their hospital a rating of 9 or 10 on a scale of 0–10.
- **Variable 2:** Patients who reported YES would definitely recommend the hospital a rating of 9 or 10 on a scale of 0–10.

16.4 FINDINGS

Collectively, on both variables, hotel-inspired hospitals received better ratings as compared to state and national averages. This result was pretty consistent based on both 2017 and 2020 observations. Actually, hotel-inspired hospitals did slightly better in 2020 observations than in 2017. In addition, the (P Values) using t-test: two-sample assuming unequal variances, showed significant difference (PV= 0.003) for both sets (Tables 16.1 and 16.2; Figures 16.1 and 16.2).

- ➢ **Detail Analysis 2017:** To further analyze the findings, we looked at the performance of each hospital individually as well. In 2017 observation, with respect to variable 1 (Patients who gave their hospital a rating of 9 or 10 on a scale of 0–10); 6 out of the 9 cases (67%) in the sample did better than state averages. Three hospitals performed less than the state averages, but the means we're very close (mean difference: –2.3%). As compared to the national averages, 7 hospitals out of 9 (68%) in the sample had patient ratings higher than national averages. For the 2 hospitals that were rated slightly lower than the national average, the mean difference was (–3.5%) (Table 16.3).

TABLE 16.1 Mean Scores Year 2017

Variables	Hotel-Like Hospitals (Mean Scores)	State Averages (Mean Scores)	National Averages (Mean Scores)
Patients who gave their hospital a rating of 9 or 10 on a scale of 0–10.	77%	72%	73%
Patients who reported YES would definitely recommend the hospital	78%	72%	73%

TABLE 16.2 Mean Scores Year 2020

Variables	Hotel-Like Hospitals (Mean Scores)	State Averages (Mean Scores)	National Averages (Mean Scores)
Patients who gave their hospital a rating of 9 or 10 on a scale of 0–10.	76%	71%	73%
Patients who reported YES would definitely recommend the hospital	78%	70%	72%

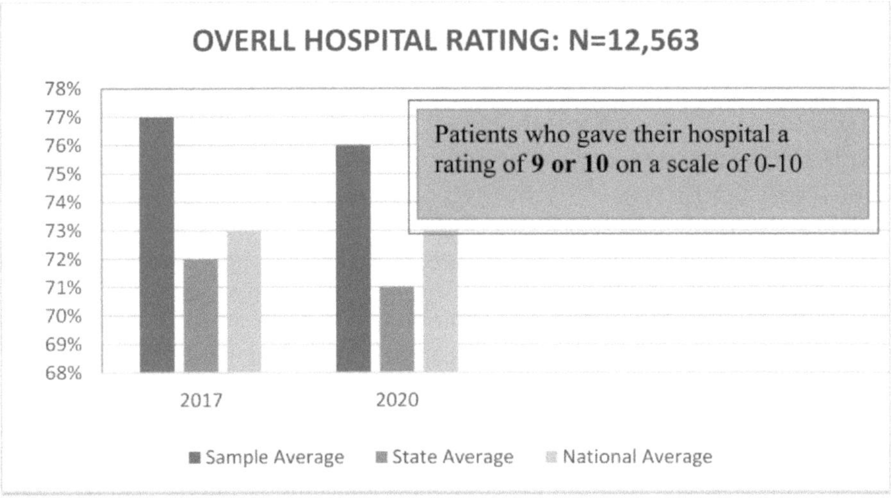

FIGURE 16.1 Hospital rating 2017 and 2020.

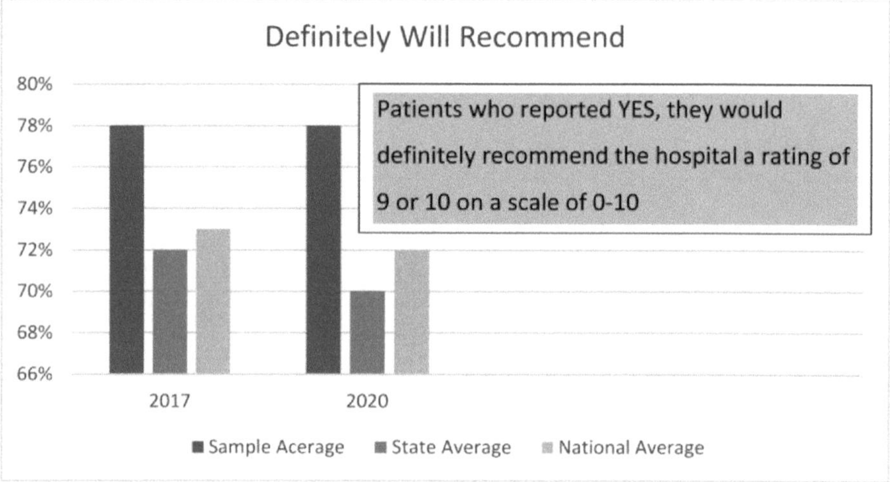

FIGURE 16.2 Definitely will recommend 2017 and 2020 ratings.

TABLE 16.3 Patients Who Gave Their Hospital a Rating of 9 or 10 on a Scale of 0–10

Hospital Name	Hotel-Like Hospital Scores	State Averages	National Averages	Number Responded N = 12,563
Cedars-Sinai Medical Center	80%	69%	73%	2,778
Emory Johns Creek Hospital	74%	72%	73%	644
Henry Ford West Bloomfield	80%	74%	72%	739
Medical City of Dallas	75%	76%	73%	3,494
Mount Sinai Hospital, NY	68%	66%	73%	970
Pennsylvania Hospital, Philadelphia	79%	71%	73%	424
Ronal Regan UCLA Medical Center	86%	69%	73%	565
ST Dominic-Jackson Memorial	83%	73%	73%	1,692
Woman's Hospital of Texas	71%	76%	73%	1,267
Mean Scores (Year 2017)	**77%**	**72%**	**73%**	**N = 12,563**

With regards to Variable 2 (Patients who reported YES, they would definitely recommend the hospital), a nearly similar pattern emerged. Six hospitals out of 9 (67%) reached better patient satisfaction scores than state averages, one was equal, and the other two hospitals had an average mean difference of –2%. As compared to national averages, six out of the nine (67%) of hospitals performed better than the national averages. The other three were slightly underrated with a mean average of –2.3% (Table 16.4).

TABLE 16.4 Patients Who Reported YES, They Would Definitely Recommend the Hospital

Hospital Name	Hotel-Like Hospitals Scores	State Averages	National Averages	Number Responded N = 12,563
Cedars-Sinai Medical Center	77%	76%	80%	2,778
Emory Johns Creek Hospital	78%	71%	72%	644
Henry Ford West Bloomfield	81%	72%	72%	739
Medical City of Dallas	75%	75%	72%	3,494
Mount Sinai Hospital, NY	71%	66%	72%	970
Pennsylvania Hospital, Philadelphia	77%	70%	72%	414
Ronal Regan UCLA Medical Center	86%	70%	72%	565
ST Dominic-Jackson Memorial	82%	71%	72%	1,692
Woman's Hospital of Texas	73%	75%	72%	1,267
Mean Scores (Year 2017)	**78%**	**72%**	**73%**	**N = 12,563**

➢ **Detail Analysis 2020:** To further analyze we looked at the performance of each hospital individually as well. According to the 2020 observations, the results were slightly better. With respect to variable 1 (Patients who gave their hospital a rating of 9 or 10 on a scale of 0–10); 6 out of the 9 cases (67%) in the sample did better than state averages. Three hospitals performed equally as compared to state averages. As compared to the national averages, 7 hospitals out of 9 (68%) in the sample had patient ratings higher than national averages. And, 2 hospitals were rated slightly lower than the national average (Table 16.5).

TABLE 16.5 Patients Who Gave Their Hospital a Rating of 9 or 10 on a Scale of 0–10

Hospital Name	Hotel-Like Hospital Scores	State Averages	National Averages	Number Responded N = 9,959
Cedars-Sinai Medical Center	81%	70%	73%	2,072
Emory Johns Creek Hospital	71%	71%	73%	450
Henry Ford West Bloomfield	80%	73%	73%	533
Medical City of Dallas	76%	75%	73%	3,191
Mount Sinai Hospital, NY	66%	66%	73%	970
Pennsylvania Hospital, Philadelphia	75%	71%	73%	432
Ronal Regan UCLA Medical Center	84%	70%	73%	458
ST Dominic-Jackson Memorial	75%	71%	73%	707
Woman's Hospital of Texas	75%	75%	73%	1,146
Mean Scores (Year 2020)	**76%**	**71%**	**73%**	**N = 9,959**

With regards to Variable 2 (Patients who reported YES, they would definitely recommend the hospital), the results were somewhat better. All hospitals in sample 9 (100%) reached better patient satisfaction scores than state averages. As compared to national averages, 8 out of the 9 (89%) of the sample hospitals performed better than the national averages. Only one hospital had a rating slightly less than national average –2% (Table 16.6).

TABLE 16.6 Patients Who Reported YES, They Would Definitely Recommend the Hospital

Hospital Name	Hotel-Like Hospitals Scores	State Averages	National Averages	Number Responded N = 9,959
Cedars-Sinai Medical Center	85%	70%	72%	2,072
Emory Johns Creek Hospital	77%	70%	72%	450
Henry Ford West Bloomfield	80%	71%	72%	533
Medical City of Dallas	78%	75%	72%	3,191
Mount Sinai Hospital, NY	70%	66%	72%	970
Pennsylvania Hospital, Philadelphia	78%	70%	72%	432
Ronal Regan UCLA Medical Center	84%	70%	72%	458
ST Dominic-Jackson Memorial	78%	71%	72%	707
Woman's Hospital of Texas	76%	75%	72%	1,146
Mean Scores (Year 2020)	**78%**	**70%**	**72%**	**N = 9,959**

16.5 CONCLUSION

The study suggests that overall, hospitals can benefit from hotel-like (hotel-inspired) design as measured by patient satisfaction scores. The findings were not a direct result of asking if subjects preferred hotel-like hospitals over others, but the study provided some circumstantial, subliminal, or in-direct evidence that patient satisfaction in the hotel-like hospitals was more favorable as compared to experiences of other patients statewide and or nationally during the same periods.

The study has also shown that the favorable ratings have been relatively stable and consistent by comparing the (year 2017) scores and the results on the latest observation in (January 2020). As for recommending the hotel-like hospital to friends and family, the results were very strong. Finally, the study may favor the notion that hospitals can use a guest-centric approach in designing healthcare to achieve customer satisfaction and patient recommendation and eventually generate more demand.

16.5.1 MANAGERIAL IMPLICATIONS

Hospitals could develop their strategy and future investment plans that involve redesigning and restructuring the hospitals for overall hotel-inspired comfort and hotel-like features those patients expect and demand. Based on the findings of the study, it is very likely to make hospitals truly hospitable and prepare current and future healthcare providers, doctors, nurses, healthcare scientists, students, and allied healthcare professionals to deeply connect to the emotional side of patient care and experience.

The findings are also a great compliment and endorsement for the hotel industry that has achieved tremendous success, making hotel-like hospitals the hospitals of the future. This movement also strengthen the idea and practice of hospitals employing hospitality students and professionals for operating and managing some aspects of their operations, in positions such as Patient Experience Officers and staff.

16.5.2 LIMITATIONS

First, the subject hospitals, hotel-like hospitals, were selected based on literature review. If a hotel-like hospital did not the on the literature review, was not included in the study. Second, the patients were not directly asked if their experience was solely as a result of hotel-like features. Their overall satisfaction could include many other aspects of the hospital, such as providers communication, staff, and auxiliary services, the control of which was not possible in this study. However, the findings shed some light on the subtle impact of hotel-inspired hospitals on customer satisfaction and their perceived quality of care. Also, this study did not extend its scope to other services offered by hotels that hospitals share, such as revenue management, food service, concierge, and other aspects of hospitality in each touchpoint.

From the research design perspective, future studies are encouraged to: (i) examine the data using a more sophisticated methods to control other variables; (ii) repeat the entire process at a later time to examine the role of time in this dynamic and ever-changing industry; and (iii) focus on utilizing a larger and more diverse sample or apply a global approach to achieve stronger generalizations.

KEYWORDS

- **evidence-based design**
- **hospital design trends**
- **hotel-inspired attributes**
- **hotel-like hospitals**
- **patient satisfaction**

REFERENCES

Advisory Board, (2014). *Redesigning Hospital Rooms to Help Patients Heal*. The Daily Briefing. News for Healthcare Executives. Retrieved from: https://www.advisory.com/daily-briefing (accessed on 14 July 2021).

Becker, F., & Douglass, S. J., (2008). The ecology of the patient visit: Physical attractiveness, waiting times, and perceived quality of care. *Journal of Ambulatory Care Management, 31*(2), 124–137.

Beryl Institute, (2018). Retrieved from: http://www.theberylinstitute.org/?page=definingpatientexp (accessed on 14 July 2021).

Binder, L., (2015). *Hospitals Strive for Hotel Quality Service*. Forbs. Retrieved from: https://www.forbes.com/sites/leahbinder/2015/05/06/hospitals-strive-for-hotel-quality-service/#3de3a8b952cf (accessed on 14 July 2021).

Bitner, M. J., (1992). Services capes: The impact of physical surroundings on customers and employees. *Journal of Marketing, 56*(2), 57–71.

Bloomberg, L. D., & Volpe, M., (2008). *Completing Your Qualitative Dissertation: A Roadmap from Beginning to End.* Thousand Oaks, CA: Sage Publications.

Chee, T. T., Ryan, A. M., Wasfy, J. H., & Borden, W. B., (2016). Current state of value-based purchasing programs. *Circulation, 133*(22), 2197–2205. http://doi.org/10.1161/CIRCULATIONAHA.115.010268.

Douglas, C. H., & Douglas, M. R., (2004). Patient-friendly hospital environments: Exploring the patient's perspective. *Health Expectations, 71*, 61–73.

Fottler, M. D., Ford, R. C., Roberts, V., Ford, E. W., & Spears, J. D., (2000). Creating a healing environment: The importance of the service setting in the new consumer-oriented healthcare system/practitioner application. *Journal of Healthcare Management, 45*(2), 91.

Fred Economic Data, Federal Reserve of St. Louis. Retrieved from: https://fred.stlouisfed.org/series/TLHLTHCONS (accessed on 14 July 2021).

Gay, L. R., Mills, G. E., & Airasian, P., (2009). *Educational Research Competencies for Analysis and Applications* (9th edn.). Upper Saddle River, New Jersey: Pearson.

Goldman, D., & Romley, J., (2010). *Hospitals as Hotels: The Role of Patient Amenities in Hospital Demand.* National Bureau of Economic Research. Retrieved from: https://www.nber.org/papers/w14619 (accessed on 14 July 2021).

Goulding, C. R., & Chin, L., (2016). Healing havens. *Commercial Construction & Renovation, 15*(4), 116–118.

Hair, J. F., Babin, B., Money, A. H., & Samouel, P., (2003). *Essentials of Business Research Methods.* River Street, Hoboken, NJ: John Wiley & Sons Inc.

Healthcare Design Magazine, (2010). *Hospitality Meets Healthcare.* Retrieved from: https://www.healthcaredesignmagazine.com/architecture/hospitality-meets-healthcare/ (accessed on 14 July 2021).

Hutton, J. D., & Richardson, L. D., (1995). Healthscapes: The role of the facility and physical environment on consumer attitudes, satisfaction, quality assessments, and behaviors. *Health Care Management Review, 20*(2), 48–61.

Kelly, R., Losekoot, E., & Wright-StClair, V. A., (2016). Hospitality in hospitals: The importance of caring about the patient. *Hospitality & Society, 6*(2), 113–129. doi: 10.1386/hosp.6.2.113_1.

Lončarić, D., Lončarić, D., & Marković, S., (2015). Health tourism, customer satisfaction and quality of life: The role of specialty hospitals. *Tourism in Southern & Eastern Europe*, 3159–3173.

Marberry, S. O., (2006). *Improving Healthcare with Better Building Design.* Concord, CA: The Center for health design.

Medina-Muñoz, D. R., & Medina-Muñoz, R. D., (2014). The attractiveness of wellness destinations: An importance-performance-satisfaction approach. *International Journal of Tourism Research, 16*(6), 521–533. doi: 10.1002/jtr.1944.

Nelson, C., West, T., & Goodman, C., (2005). *The Hospital Built Environment: What Role Might Funders of Health Services Research Play?* No. 05-0106-EF. Rockville, MD: Agency.

Nyberg, M., & Grindland, B., (2008). The influence of the room context in the meal experience: Examples from a hospital and a nursery. *Journal of Foodservice, 19*(1), 35–43. doi: 10.1111/j.1745-4506.2007.00078.x.

Omachonu, V. K., (1990). Quality of care and the patient: New criteria for evaluation. *Health Care Management Review, 15*(4), 43–50.

Powers, T. L., & Bendall-Lyon, D. The satisfaction score. Mark Health Serv. 2003 Fall;23(3):28-32. PMID: 12958748.

Rabner, B., (2014). *Redesigning Hospital Rooms to Help Patients Heal.* Daily Briefing. Retrieved from: https://www.advisory.com/daily-briefing/2014/08/25/redesigning-hospital-rooms-to-help-patients-heal (accessed on 14 July 2021).

Reiling, J., Hughes, R. G., & Murphy, M. R., (2008). The impact of facility design on patient safety. In: Hughes, R. G., (ed.) *Patient Safety and Quality: An Evidence-Based Handbook for Nurses.* Rockville (MD): Agency for Healthcare Research and Quality (US); Chapter 28. Available from: https://www.ncbi.nlm.nih.gov/books/NBK2633/ (accessed on 14 July 2021).

Rice, G., Ingram, J., & Mizan, J., (2008). Enhancing a primary care environment: A case study of effects on patients and staff in a single general practice. *Br. J. Gen. Pract., 58*(552), 465–470. doi: 10.3399/bjgp08X319422.

Siddiqui, Z., Zuccarelli, R., Durkin, N., Wu, A., & Brotman, D. J., (2015). *Changes in Patient Satisfaction Related to Hospital Renovation: Experience with a New Clinical Building.* Retrieved from: https://www.ncbi.nlm.nih.gov/pubmed/25652720?report=abstract (accessed on 14 July 2021).

Suess, C., & Mody, M., (2017). Hospitality healthscapes: A conjoint analysis approach to understanding patient responses to hotel-like hospital rooms. *International Journal of Hospitality Management,* 6159–6172. doi: 10.1016/j.ijhm.2016.11.004.

Swan, J. E., Richardson, L. D., & Hutton, J. D., (2003). Do appealing hospital rooms increase patient evaluations of physicians, nurses, and hospital services? *Health Care Management Review, 28*(3), 254.

Ulrich, R., & Zimring, C., (2004). *The Role of the Physical Environment in the Hospital of the 21st Century.* Concord, CA: The Center for Health Design.

Watson, C. A., (2005). Integration of technology and facility design: Implications for nursing administration. *Journal of Nursing Administration, 35*(5), 217–219.

Weed, J., (2016). *With Room Service and More, Hospitals Borrow from Hotels.* New York Times. Retrieved from: https://www.nytimes.com/2016/08/02/business/making-hospitals-more-like-hotels.html?rref=collection%2Fsectioncollection%2Fhealth&action=click&contentCollection=health&region=stream&module=stream_unit&version=latest&contentPlacement=7&pgtype=sectionfront&_r=1 (accessed on 14 July 2021).

CHAPTER 17

FROM ANTISEPTIC TO AESTHETIC: EXPERIENCE AS A NEW NARRATIVE IN CURATIVE CARE

KATIE CONTRERA

Johns Hopkins University, Krieger School of Arts and Sciences, Advanced Academic Programs, Washington DC 20036, USA, E-mail: katie.contrera@gmail.com

ABSTRACT

The goal of this chapter is to investigate whether space and aesthetic can heal. You might forget you are in Cleveland, Ohio, when you walk into the sunny atrium. Even on the dreariest of days, it sparkles. Cleveland Clinic's new cancer center was designed to improve efficiencies for both patients and physicians. All administrative and clinical services orbit the patient. The space allows care teams to coalesce around types of cancer, rather than clinical disciplines, making it truly a model of patient-centered care. Exploring the shifting nature of hospital design requires an understanding of the complex relationship between people and medicine. This chapter aims to investigate how social movements have shaped the hospital aesthetic over time, how current iterations of modern design are catering to patients in inventive ways, and how shaping the patient experience has implications beyond healthcare. Modern trends in hospital design recognize patients are more than sick bodies shuffling in and out of hospitals. They have identities, they have stories to tell, and sometimes just being seen can be efficacious. In this chapter the focus is on the emerging field of neuroaesthetics that aims to understand what type of aesthetic humans might be intrinsically drawn to. To this end, we explore the Cleveland Clinic's story and how the hospital relies on a nature-inspired aesthetic to create a healing environment.

17.1 INTRODUCTION

When you walk into the main entrance of Cleveland Clinic, you are greeted in 21 different languages all at once. *Bienvenue*, *Bun Venit, Willkommen*, Welcome—all inscribed on the entrance down the main hospital corridor. You are immediately intercepted by an affable individual in a cherry red blazer. A Red Coat, they are called. As in, "If you have any questions, find someone in a red coat. They will help you." And they do. They are human way finders, directing patients and families to appointments. They are trained to detect uncertainty and gently guide visitors to their destination. They are also called caregivers. Everyone in the hospital is called a caregiver. From the treating physicians, nurses, and clinical staff, the scientists, the janitorial team, and the executives. The walls around you are all a clean, bright white, the only paint color that is permitted in patient areas. Nearly every corridor features art that has been perfectly curated by a team of professionals. The designs are intended to provide surface-level appeal with an added layer of depth, inviting patients and visitors to linger just a little longer.

This is all done to make patients and their families feel more like welcomed guests, rather than mere acquaintances. It is fitting, really, for individuals who are sharing some of their most intimate moments with the care teams working within these walls. Now, arguably more than ever before, there is a consciousness and a dialog about space, esthetic, and experience in healthcare settings. Understanding what we do next might be more important than we realize.

17.2 ESTHETIC IN THE PRIVATE

Architecture and design are full of clues that reveal insights about the goals and attitudes. Even materials can relay information about a moment in time, a cultural shift, or a yearning for the future. Some design experts argue that design trends a few years ago were more likely to favor synthetic materials, like plastics, while more modern trends favor woods and other natural or weathered materials that imitate a certain authenticity and hominess (Barco, 2017). This may suggest a desire to retreat to our homes or a preference for comfort reflected in home-based esthetics (Barco, 2017). It is prudent to pay attention to how and why certain design choices make us feel a certain way—and what that says about humanist aspirations.

Let us consider this concept beyond private life and examine it in the public sphere. Institutions like hospitals, schools, and prisons have corrective and restorative goals—to heal, to educate, and to reform. While these institutional goals have remained largely intact, arguments about *how* these goals should be approached, spatially, change over time. This analysis aims to investigate how social movements have shaped the hospital esthetic over time, how current iterations of modern design are catering to patients in inventive ways, and how shaping the patient experience has implications beyond healthcare.

17.3 MEDICINE AS A SOCIAL INSTITUTION

Exploring the shifting nature of hospital design requires an understanding of the complex relationship between people and medicine. The relationship between history and medicine can be best explained as symbiotic—two factions working together to influence the other (Waddington, 2011). The social history of medicine does not ignore the reality of disease, rather, it invites us to question where economic, social, and cultural narratives are imbued (Waddington, 2011). In the context of healthcare, we must consider how normative beliefs may shape approaches to health and wellness. As such, definitions of health and healing, and appropriate interventions, change over time (Waddington, 2011).

Language, for example, shapes many facets of life, including how medicine is perceived and practiced. In the late 1600s, *nostalgia* was a potentially deadly disease (Watts Smith, 2017). The last documented nostalgia fatality was a World War I American soldier in France (Watts Smith, 2017). *How is that possible? What changed?* Telephone and railways would transform communication and travel, opening new doors in social connectivity. Modernism placed incredible emphasis on independence, travel, and discovery (Watts Smith, 2017). Social ideals are always changing. This cannot be untangled from medicine—and nostalgia is a far from isolated example.

Interventions in mental health also demonstrate a dramatic change over the last century. It was not uncommon for postpartum woman in the late 19th century to be institutionalized after a diagnosis of puerperal insanity (Theriot, 1989). Symptoms of puerperal insanity were described as fatigue, frequent mood swings, and loss of appetite following childbirth. At one point, up to 10% of female asylum occupants were being treated for puerperal insanity (Theriot, 1989). And just like that, by early-to-mid-20th century, the diagnosis nearly vanished. (Theriot, 1989) *Was it mysteriously cured?* In a sense—yes. Social expectations about motherhood changed. Today, clinicians may diagnose those women with postpartum depression. Or maybe there is no diagnosis at all, but a recognition that motherhood can be physically, emotionally, and mentally demanding. As medical interventions and social perceptions change, standards of care change.

Figure 17.1 represents esthetic associated with psychiatric inpatient treatment over 100 years ago. Space is a nonverbal articulation that can affirm condition and empower the patient to heal, or it can reinforce harmful stereotypes about diseases. This psychological burden of illness can be a powerful determinant of one's own efficacy.

FIGURE 17.1 The National Library of Medicine (1904). Interior of the electric room in Davis House of the Adams Nervine Asylum (Boston, MA) showing equipment used for electrotherapy. *Source:* With permission via the Creative Commons Attribution License; https://creativecommons.org/publicdomain/mark/1.0/.

17.3.1 RELIGION AND MEDICINE

The investigation into human health is, arguably, as old as civilization. Until the 16th century, ancient Greek and medieval philosophies largely guided curative medicine (Waddington, 2011). Christianity was such an integral part of early modern European culture that it shaped nearly every aspect of life (Waddington, 2011). Religious orders founded the early European hospital, and operated at every level, as administrators and caregivers (Waddington, 2011). They believed the body and soul were inextricably linked. As such, disease was viewed as both a product of God's wrath, and healing, his intervention. Hospitals were, in essence, modeled after monasteries, and physicians refused to treat patients who had not received the sacrament of confession (Buklijaš, 2008). Space and architecture were an extension of this soul-body narrative. Many hospitals were even built in a cross-shaped design to reflect that sanctity of the crucifix, a symbol of Christianity and martyrdom, while interiors mirrored a monastery with biblical tableaus, altars, and Christian iconography (Buklijaš, 2008).

Today, spiritual domains still have a place in hospitals, however, they are largely positioned as an ancillary support for patients, rather than a primary intervention for care. Given the long, complex relationship between disease and spirituality, and so long as humans turn to religion to grapple with the devastation of illness, it seems unlikely they will divorce entirely.

17.3.2 A BRIEF SOCIAL HISTORY OF MEDICINE

Sociocultural movements have played a critical role in shaping transformation in hospitals. Reformation and Enlightenment periods in 17th and 18th century Europe introduced more progressive ideas about narratives in healthcare. These movements would ultimately lead to the secularization of hospital design. Physical illness and spiritual care began to slowly separate, and physicians, rather than clergy, took charge of medical institutions (Buklijas, 2008). Medieval perspectives that focused on the alignment of physical and spiritual health would lose some momentum. There was a growing interest in science, and greater skepticism toward all-encompassing religious views (Figure 17.2).

Later, industrial revolutions in Europe, and then, in the U.S., would place new emphasis on manufacturing and economic development. By the 1850s, growing urban populations, increased wealth, and a sense of social responsibility, would lead to more order and expansion for the modern hospital (Kisacky, 2017). Ornate, religious relics were largely replaced with a utilitarian esthetic. A heightened sense of health and hygiene would guide the next wave of design as initiatives to promote disease prevention became critical in the late 19th and early 20th centuries (Kisacky, 2017).

Designers decidedly focused on minimal décor for purposes of sterility, arguing that "hospital buildings were designed for cure and reform, and only contingently for beauty" (Kisacky, 2017). The pavilion ward-style building, which included several full functioning wards separated by a long, narrow hallway, would become a near-universal architectural style in hospital design (Kisacky, 2017). It was built flatly to control the spread of airborne impurities. Hospitals required that patients have direct access to sunlight, as it was integral to their care. It was believed that toxins manifested in dark spaces and that sunlight countered this (Kisacky, 2017). While germ theory would gradually refute some of these beliefs, even today, windows play an important role in care. Natural light connotes a certain positivity and connection to the natural world.

FIGURE 17.2 Wellcome Collection. The Hôtel Dieu, Paris: interior showing patients being nursed by monks and nuns (Woodcut).

Source: With permission via the Attribution 4.0 International (CC BY 4.0) license; https://wellcomecollection.org/works/qptprtrn.

The 20th century brought advances in medicine that would transform the institution's capacity for care. New discoveries in hygiene, anesthesia, and surgery led to longer inpatient visits and growing specialty areas (Kisacky,

2017). While clinical staff were more poised than ever to heal patients, many hospitals lacked an organizational structure for efficiency (Kisacky, 2017). Changes in medicine led to changes in design, and by the early to mid-20th century, many hospitals transitioned to the vertical tower, which looked similar to other civic buildings (Kisacky, 2017). Central to the new organizing principle of design was how to best coordinate the layout of care for staff and patients to maximize health outcomes. Often, this approach sacrificed patient comfort and the patient experience (Kisacky, 2017).

The hospital, once considered an integral part of a patient's therapy, become an ancillary tool to facilitate health. In other words, its structure is a host for those who come to heal (Kisacky, 2017). A closer look at 21st century hospital design reveals the hospital experience aims not only to be endured but to be enjoyed.

In the last 20 years, the emerging trend in healthcare esthetic has been to cater to the patient by making the hospital a *part* of a holistic, therapeutic experience. Worth noting, Cleveland Clinic and Mayo Clinic, two globally renowned programs, were thinking about medical esthetic long before it was in vogue, which is, perhaps, a testament to their institutional legacies as premier healthcare destinations.

17.4 TWO INSTITUTIONS: FINDING JUXTAPOSITION IN ESTHETIC GOALS

Both institutions share a similar early architectural history. In fact, they shared the same original architecture firm, Ellerbe & Co., headquartered in St. Paul, Minnesota (Ellerbe and Hayskar, 1980). In 1916, leaders of Cleveland Clinic, then known as Crile Clinic, selected the firm to model the flagship building off what Ellerbe created for the Mayos in Rochester, Minnesota. Both institutions were considered early adopters in space and design concepts that both created efficiencies in the delivery of care, a major concern at the time, and also to modernize the patient experience. For example, Cleveland Clinic was the first hospital to include private bathrooms in patient rooms (Ellerbe and Hayskar, 1980).

Dr. George Crile, a skilled surgeon and one of Cleveland Clinic's founders, is a name worth mentioning in the hospital's architectural history. Seemingly ahead of his time, he told Ellerbe he wanted a striking building that symbolized progressive medicine. Dr. Crile's vision came from a genuine desire to foster the future of medicine for patients and providers. Though at the time, such grandiosity in medicine was discouraged. Crile was deeply concerned with patients' emotions and the healing process. An illustration of his empathy, Crile has actually been credited with instituting "hospital green," the traditional color of surgical wear, because he felt white was too institutional, too clinical (Ellerbe and Hayskar, 1980).

Remarkably, he also led the charge to renovate inpatient rooms to make them look more like an inviting bedroom than a clinical space. Dr. Crile introduced colorful stained-glass windows to patient rooms, which was a novel design choice since, up until then, white was used almost exclusively in hospital rooms. George Crile's, then controversial, approach to hospital esthetic would set the hospital's legacy of design into motion.

The original building was situated on Cleveland's once iconic, "Millionaires' Row," established by industrial tycoons that put down roots in Cleveland in the late 19th century, including John D. Rockefeller. While few of the palatial mansions remain today, Cleveland Clinic has been building its own brand of towering structures. Today, the building that once stood among the likes of the Victorian-inspired "Millionaires' Row" mansions, now finds itself looking dated against the shining exteriors of the new structures. The juxtaposition of the two buildings is a sign of how far design and esthetic has come in hospitals, even for an institution that was ahead of its time.

Cleveland Clinic has remained on the cutting-edge of healthcare architecture and esthetic. Toward the end of the 20th century, it took on a decidedly more modern, earthy esthetic. Aptly, in 1986, the Crile Building, received a prestigious Honor Award from the American Institute of Architects for its innovative, pyramid-inspired design (Figure 17.3).

Interestingly, about 80 years after Crile's quest to reinvent the patient experience by introducing color to stark, clinical spaces, Dr. Delos "Toby" Cosgrove, another transformative figure in Cleveland Clinic's history would come along and reverse these design elements, but with the same goal in mind. The Cosgrove Era from 2004–2017 is known for its clean, meticulously manicured esthetic. Aside from well-placed art and sculpture, nearly every wall is a blinding white. It is breathtaking, almost jarring, the first time you walk through the hospital.

In a Cleveland Clinic press release (2017), Dr. Cosgrove writes, "Architecture and design can truly help comfort patients and their families during a difficult time at the hospital. Natural light, noise-dampening features, green spaces, accommodations, and other environmental features go a long way toward enhancing the patient experience." Drs. Crile and Cosgrove represent two very different approaches to design for their times, one reaching for the warm and personal and one, seemingly, harkening back to the antiseptic.

Paradoxically, both with the same goal in mind—to create spaces for patients to heal.

FIGURE 17.3 Cleveland Clinic (2009). Exterior of Crile Building (Cleveland, Ohio). *Source:* Cleveland Clinic. With permission via the Creative Commons Attribution 4.0 International license; https://commons.wikimedia.org/wiki/File:Crile_Building_A6110.jpg.

17.4.1 A NEW ERA: ESTHETIC AS EXPERIENTIAL CARE

The "Cosgrove Era" was one of capital expansion. During this time, the newly relocated Taussig Cancer Institute was built. In 2014, Cleveland Clinic broke ground on the project, which was proudly conceived, from the ground up, with the patient in mind. Clinical and administrative leadership embarked on a global tour of cancer buildings to assess best practices in cancer centers around the world. Cohorts of cancer specialists, oncologists, and surgeons, as well as patients and their families formed to counsel on every aspect of the esthetic experience. You might forget you are in Cleveland, Ohio, when you walk into the sunny atrium. Even on the dreariest of days, it sparkles. The art provides a balanced statement, careful not to overwhelm visitors, but offer just enough levity as patients and families navigate the space (Figure 17.4).

FIGURE 17.4 Cleveland Clinic (2016). Atrium at front entrance of Taussig Cancer Center (Cleveland, Ohio).

Source: With permission via the Cleveland Clinic Newsroom; https://newsroom.clevelandclinic.org/wp-content/uploads/sites/4/2016/12/ClevelandClinic_06.jpg.

Cleveland Clinic's new cancer center was designed to improve efficiencies for both patients and physicians. All administrative and clinical services orbit the patient. The space allows care teams to coalesce around types of cancer, rather than clinical disciplines, making it truly a model of patient-centered care. This structure was designed to eliminate wait-time for patients and streamline communication efforts within their care team.

Seemingly, every detail has been carefully considered—from the massive skylight that allows light to flood into the lower level where patients undergo radiation and imaging to the patient support services wing, which includes an in-house wig boutique and retail store with products designed specifically for patients with cancer. Even every infusion room has a garden-facing view with floor-to-ceiling windows designed to create a calm, therapeutic environment for patients.

By the mid-20th century, hospital designers chose function over form, but the 21st century hospital has, seemingly, found a place for both (Kisacky, 2017). Across the country, new projects in healthcare seem to point to one converging goal—to create a transformative experience for patients and families by taking a step out of the institutional and into the *personal.*

This is especially compelling in the case of children's hospitals, where the youngest and most vulnerable patients receive care. The Charlotte R. Bloomberg Children's Center Building at Johns Hopkins Hospital in Baltimore, Maryland, is another example of hospital architecture and design that creates an experience for patients and families. The building was recognized by *Interior Design Magazine* for its use of space and design. Michael Bloomberg, philanthropic donor of the building, asked that art play an integral role in every facet of the project (Kaiser, 2012). The children's center is full of whimsical, over-the-top charm with animated art and sculpture at every turn.

In fact, the building itself is a work of art, thanks to the artist Spencer Fitch. Mounted behind the window are colorful exterior panels with flecks of color, the "frit" pattern, modeled off of a Monet painting. It is peculiar, at a distance, to see the building standing in cheerful awareness next to mostly brick buildings. Many of the creative choices have a functional purpose as well. The frit pattern was designed to provide coverage for patients in the atrium, allowing them to feel less exposed (Kaiser, 2012). Similarly, the curved hallways in the building function to reduce noise pollution (Kaiser, 2012). The building has a remarkable ability to elicit awe and wonder from everyone who walks through it, not just children. It is that subtle, or not-so-subtle, levity that invites children and their families to feel entertained, inspired, or maybe even relax *just a little*.

Swedish Ballard Behavioral Health Center in Seattle, Washington is bringing a nature esthetic to patients through manipulated lighting (Matambanadzo, 2017). The technique is described as a shift in lighting that mimics the natural movement of the sunrise and sunset. This is done to simulate patients' natural circadian rhythms. In some areas, the lighting seems to create an effect of walking through a forest with green and blue hues seamlessly integrated into the design. The functional aspect of the design helps guide and way-find, leading patients into main corridors or activity centers.

Cleveland Clinic also relies on a nature-inspired esthetic to create a healing environment. One of the most compelling structures in the hospital is BluBerg, a 30-foot geometrically-inspired iceberg display by artist Inigo Manglano-Ovalle. It juts down from the ceiling to greet visitors walking through the Great Hall. The piece was designed to "explore(s) the allegorical and other symbolic aspects of natural phenomenon—the unknown below the surface…" This is a tragically fitting narrative for many of the patients who arrive at Cleveland Clinic in search of a treatment. The strategy to create whimsy and sophistication with art and sculpture and consciously bring elements of nature inside are undoubtedly trends in healthcare design that we have not seen before at this scale (Figure 17.5).

17.5 WHY DOES THE ESTHETIC EXPERIENCE MATTER TO US?

We might pause for a moment and ask a few pointed questions. Why are we attracted to this esthetic? Why are healthcare leaders spending so much time cultivating this experience, and does any of it have any real bearing on health and healing?

FIGURE 17.5 Cleveland Clinic (2018). Image of BluBerg, a 30-foot geometrically-inspired iceberg display by artist, Inigo Manglano-Ovalle (Cleveland, Ohio).

Source: With permission via the Cleveland Clinic Newsroom; https://newsroom.clevelandclinic.org/wp-content/uploads/sites/4/2018/02/art-Manglano-Ovalle.jpg.

There might be a greater neural attraction happening that we do not even realize. The emerging field of neuroesthetics aims to understand if there is a type of esthetic humans are intrinsically drawn to. In 2010, a museum director in Baltimore, Maryland, and a neuroscientist at Johns Hopkins University, partnered to gain insights into neural preferences of art. They duplicated different iterations of the same amorphous blob-like shape and asked museum patrons to identify which ones they liked the most. The answers, they believed, would offer new insights into the brain's esthetic preference. The results revealed that an overwhelming majority of people were attracted to the shapes that presented as softer, with greater curvatures, rather than the pointier counterpart (Hughes, 2010). As a follow up, researchers also leveraged brain imaging technology to observe the neural response to viewing these shapes. The visual cortex, part of the brain that interprets shapes, appeared more hyperactive when viewing these "softer" shapes, another indication of attraction to that shape.

While these findings were not published, continued investigations into neuroesthetics may reveal new insights into how our brains are wired and could have significant implications for better understanding our affinity for a specific esthetic.

Further, we are also increasingly drawn to consumer experiences, and that is exactly what the modern hospital is creating through space and esthetic. *The Experience Economy* describes our 21st-century desire to prioritize consumer experiences. The experience economy relies on orchestrated details not directly related to the immediate product or service (Pine and Gilmore, 1999). Experiences are marketed across many industries, and healthcare is no exception. Like any industry, the success and viability of a hospital rely on differentiating its services from those of competitors. Marketers are taking stock of the psychological impact of consumer experiences—and this applies to everything from a hamburger to a prostate exam. Waiting for your check-up in a clean, light-filled room with a featured art collection, reading material and a friendly staff makes the experience a little easier to *endure*—and perhaps even *enjoy*.

Architecture itself can also have a really interesting, almost magical effect on us. At the most granular level, some architects argue that all space is interactive (Plummer, 2016). Exerting ourselves in manmade space is so normalized that it is rendered almost meaningless. Even mundane interactions, like walking upstairs or opening a window, remind us that we are aware, we are alive, and we have more power than we might think. Space gives us the agency to be the narrators of our own story.

Whether it is how our brains are wired, how we have been socialized as consumers, or a primal need to explore and interact with architecture, healthcare leaders, now more than ever, are using space and design to directly appeal to patients to create an *experience*.

17.5.1 SPACE AS A FORM OF HEALING

Can space and esthetic heal? The answer is not straightforward. And certainly, we cannot dismiss the reality of disease, but space does provide distinct nonverbal clues about institutional motivation. If you, as a patient, feel seen, respected, empowered, your outcomes would very likely be more positive than counterparts in a less desirable setting.

Take the Athens Lunatic Asylum, located in Athens, Ohio, a small town near the state's southeastern border. The asylum was built in the late 1800s as an extension of the "moral treatment experiment" (Ziff, 2012). This was a new, progressive approach to treating the mentally ill, and the architecture of this asylum was "central to the moral treatment curative schema" (p. 21). The experiment continued for 20 years. Staff and providers remained dedicated despite many obstacles, including treating a wide spectrum of conditions with relatively few resources. However, major strides were made during this experiment. The first female asylum assistant physician in America was hired to oversee women's health. The Asylum would also become the first to secure funding from the state to fund communal dining halls for patients. This gave them the option to eat together, and on white tablecloths, rather than in their rooms (Ziff, 2012). Ultimately, overcrowding, and medical progress would dilute the experimental efforts. By the turn of the century, models for psychiatric care would become "became administrative and managerial rather than curative" (p. 164) and would eventually lead to its closing (Ziff, 2012).

This narrative offers a moment in time where moral treatment was the framework for patients who had, historically, been treated inhumanely. The experiment lasted 20 years, and there is no data that tracks outcomes linked to their improved care. By the turn of the century, there was a new charge for the asylum structure, one that was framed in medical research (note: more observation and surveillance), less compassionate care, and patient housing became commensurate with what patients could afford (Ziff, 2012).

Medical records from the asylum reveal the medical history of Alice, a young patient who was diagnosed with epilepsy. By modern standards, epilepsy is understood as a neurological disease characterized by uncontrollable seizing, but in the late 1800s, epileptic patients were viewed as insane. Medical notes document her time as a patient and her eventual recovery. Genealogical records suggest she may have married in Michigan several years later and died in Los Angeles in 1943 at the age of 80 (Ziff, 2012). We have no way of knowing whether the moral experiment directly contributed to Alice's outcomes, but we can be fairly certain that compassionate treatment can be efficacious.

In some ways, this call for compassionate treatment is not unlike the more modern emphasis on institutionalizing empathy across healthcare systems. Industry leaders have noted a need for "...more empathy in medicine and better service to meet the needs of patients and caregivers alike" (Boissy, 2017,

p. 543). In the inaugural issue of the *Journal of Patient Experience* (2014), the editor, Adrienne Boissy, MD, MA, and leader in the patient experience industry, writes that the goal of the journal is to provide content that focuses on "relationship-centered care, empathy, and patient perspectives into the care we provide." This has become a hot topic in healthcare in the last decade. The industry is paying close attention to the effects of empathy in healthcare.

17.5.2 SPACE, POWER, AND MAGICAL THINKING

Space and power are very Foucaultian concepts (Crampton and Elden, 2015). Foucault asserts that architecture itself is an argument and represents a culture perspective on surveillance (Crampton and Elden, 2015, p. 60). Prisons, schools, factories, and hospitals, underscore a power dynamic that is worth exploring. Historically, these institutions create maximum exposure of the subjects and minimal exposure of the authority figures. This creates a very real division of "superior and subordinate" and may inculcate a culture of surveillance, distrust, and disempowerment. Surveillance is at odds with free will (Hutson, 2013). If institutions want to empower individuals to heal, learn, and reform, then they will have to find a way to maximize free will and limit determinism.

There is value in examining organizational structures more closely and identifying where the culture of surveillance is at odds with a culture of compassion. It is difficult to divorce space and esthetic from the goals of the institution. The administrators at the Athens Lunatic Asylum advocated for a new dining hall with white table clothes for patients to enjoy together. It is a minor detail, but it illustrates a culture of caring, "a moral treatment." In modern institutions, there is undoubtedly a larger dialog about the role of empathy in the delivery of care.

There is "magical" value of being in the presence of something greater than oneself. In one study, one cohort stood next to a 25-foot tyrannosaurus skeleton for one minute, while the other was instructed to stare down an empty hallway. The folks who stood next to the skeleton reported feeling more optimistic and more connected to a larger social system than their hallway counterparts (Hutson, 2013). In more studies of this kind, when feelings of awe and wonder are elicited, people tend to report greater feelings of purpose and connection to humanity (Hutson, 2013).

The concept of magical thinking has some really interesting dimensions that demand a closer look. Humans are wired to draw connections and make meaning out of symbols and patterns (Hutson, 2013). It may be irrational, but it also gives us a little bit of hope when things feel hopeless. And that *is* kind of magical. In the same way that the people standing next to the dinosaur skeleton were more empowered than those standing in the bleak hallway, sometimes esthetics can make all the difference.

Institutions, like hospitals, are beginning to catch on to this. More to the point, 100 years ago, hanging a 30-foot "iceberg" sculpture from the ceiling of a busy hospital hallway might have seemed absurd. Today, it provides visual interest and dimension and contributes to an esthetic that aligns with institutional goals of healing. It invites patients to feel less isolated, less institutionalized, less supervised—and just a little more *normal.*

17.5.3 BEYOND HEALTHCARE

Keeping this in mind, it stands to reason that a student who shows up to a school in deplorable conditions might not be as motivated to learn as a student in a school with an esthetic more conducive to learning. And the long-term impact of this implicit messaging is "your access to quality education is less important than that of the kids' on the other side of town."

The same is true of prison systems. Remember that Foucault says that architecture is an argument. In theory, the goal of penal systems is to rehabilitate inmates and support a successful reintegration back into society and their communities. In practice, however, prisons often send a conflicting message of surveillance, isolation, and subpar conditions. Space and esthetic must align with institutional goals if it is going to be a successful strategy in creating a culture of compassion and promoting humanity. If there is a call-to-action here, it is to be more mindful of the spaces around you and the stories they tell. Where is power imbued? What arguments are being made, and what are the consequences? What bodies are involved? Answering these questions can lead to greater awareness about disparities and work toward building solutions to improve person-centered outcomes.

17.6 RECENTERING EMPATHY

Hospitals have made a dramatic shift over the last several centuries and are more patient-centered today than ever before. Today, healthcare leaders are actively engaging in a dialog to deliver care not in clinical practice alone, but through broad esthetic and spatial narratives designed to engage with patients and families by creating a desirable experience that makes them feel less "other." An emphasis on patient experience and empathy is at the center of this cultural shift. Theoretically, this culture can, and should, operate outside of hospitals. Central to this analysis is the question, "Can space heal?" It would be magical thinking to imagine that stained glass windows in a patient's room could improve outcomes or that whimsical art or flecks on a building exterior could cure pediatric cancer. A light-filled atrium and oversized iceberg will not make a bad prognosis any more bearable. But it does reflect a system that cares.

On your way out of Cleveland Clinic, you run into a sculpture of the human body, made up of letters of the alphabet and perched on a large river stone. The body appears to be meditative, to be reflecting. Usually, you can find a few people lingering around it. It is supposed to represent the universality of the human experience. The artist Jaume Plensa says, "Every second, every moment, our experiences are tattooed on our skin." Modern trends in hospital design seem to recognize patients are more than sick bodies shuffling in and out of hospitals. They have identities, they have stories to tell, and sometimes just feeling *seen* can be efficacious (Figure 17.6).

FIGURE 17.6 Cleveland Clinic (2016). Soul Man, a sculpture from artist Jaume Plensa is perched near the atrium of the main entrance at Cleveland Clinic (Cleveland, Ohio).

Source: With permission via the Cleveland Clinic Newsroom. *https*://newsroom.clevelandclinic.org/wpcontent/uploads/sites/4/2016/12/Soul_Man_008808_ST_01986CB.jpg.

KEYWORDS

- **design efficiency**
- **esthetic and patient experience**
- **hospital aesthetic**
- **hospital design**
- **medicine and social institution**
- **neuroaesthetics**
- **religion, medicine, and design**

REFERENCES

Barco, M. D., (2017). *At Design Miami, the Future Isn't so Bright.* Retrieved from: http://www.npr.org/2017/12/06/568920342/at-design-miami-the-future-isnt-so-bright (accessed on 14 July 2021).

Boissy, A., (2014). From the editor. *Journal of Patient Experience, 1*(2), pp. 2–2. https://doi.org/10.1177/237437431400100201.

Boissy, A., (2017). Labels matter: Challenging conversations or challenging people? *Cleveland Clinic Journal of Medicine, 84*(7), 543–544. doi: 10.3949/ccjm.84a.16128.

Buklijaš, T., (2008). Medicine and society in the medieval hospital. *Croatian Medical Journal, 49*(2), 151–154. doi: 10.3325/cmj.2008.2.151.

Crampton, J. W., & Elden, S., (2014). *Space, Knowledge and Power Foucault and Geography.* Farnham, Surrey: Ashgate.

Ellerbe, T. F., & Bonnie, J. H., (1980). *The Ellerbe Tradition: Seventy Years of Architecture & Engineering: From the Papers of Thomas Farr Ellerbe*. Ellerbe, Inc.

Hughes, S. (2010). *Beauty and the Brain: Understanding Our Responses to Art.* Science Friday, www.sciencefriday.com/articles/beauty-and-the-brain-understanding-our-responses-to-art (accessed on 14 July 2021).

Hutson, M., (2013). *The 7 Laws of Magical Thinking: How Irrationality Makes Us Happy, Healthy, and Sane.* One world.

Kaiser, L. F., (2012). *The Art of Healing: Johns Hopkins Hospital by Perkins Will.* Retrieved from: https://www.interiordesign.net/projects/9735-the-art-of-healing-johns-hopkins-hospital-by-perkins-will/ (accessed on 14 July 2021).

Kisacky, J., (2017). *When Fresh Air Went Out of Fashion at Hospitals.* Smithsonian.com, Smithsonian Institution. www.smithsonianmag.com/history/when-fresh-air-went-out-fashion-hospitals-180963710/ (accessed on 14 July 2021).

Kisacky, J., (2017). *Rise of the Modern Hospital: An Architectural History of Health and Healing, 1870–1940.* University of Pittsburgh Press.

Matambanadzo, S., (2017). *Natural Remedy: Biophilic Design Supports Behavioral Health.* Retrieved from: http://www.healthcaredesignmagazine.com/projects/natural-remedy-biophilic-design-supports-behavioral-health/ (accessed on 14 July 2021).

Plummer, H., (2016). *The Experience of Architecture*. Thames & Hudson.

Smith, T. W. (2017). *The History of Human Emotions.* TED: Ideas Worth Spreading, www.ted.com/talks/tiffany_watt_smith_the_history_of_human_emotions (accessed on 14 July 2021).

Theriot, N., (1989). Diagnosing unnatural motherhood: Nineteenth-century physicians and 'puerperal insanity. *American Studies, 30*(2), 69–88. JSTOR, JSTOR, www.jstor.org/stable/40642344 (accessed on 14 July 2021).

Waddington, K., (2011). *An Introduction to the Social History of Medicine: Europe Since 1500.* Palgrave Macmillan.

Wheeler, T., (2017). *Cleveland Clinic Appoints Chief Design Officer.* Cleveland Clinic Newsroom. newsroom.clevelandclinic.org/2017/05/31/cleveland-clinic-appoints-chief-design-officer/ (accessed on 14 July 2021).

Ziff, K., (2012). *Asylum on the Hill.* Athens: Ohio University Press. muse.jhu.edu/book/13519 (accessed on 14 July 2021).

CHAPTER 18

HOSPITALITY MEETS HEALTHCARE IN ONCOLOGY AT CANCER TREATMENT CENTERS OF AMERICA®

PETER C. YESAWICH,[1] ANANTH MOHAN,[2]
CAROLYN LAMMERSFELD,[3] and KELLY MURRAY[4]

[1]*Former Chief Growth Officer of Cancer Treatment Centers of America Global, Inc., USA*

[2]*Senior Vice President, Innovation, and Network Operations at Cancer Treatment Centers of America Global, Inc., USA*

[3]*Vice President of Integrative Oncology at Cancer Treatment Centers of America Global, Inc., USA*

[4]*Manager, Enterprise Innovation at Cancer Treatment Centers of America Global, Inc., USA*

ABSTRACT

Specialty hospitals typically attract a large percentage of patients from distant areas because of the unique appeal of their clinical expertise. The complexion of this demand poses numerous challenges for these "destination" care providers. This is particularly true in oncology because of the clinical condition of patients, especially those with an advanced-stage diagnosis, as well as the possible side effects of treatment. One national cancer care provider has evolved a unique service strategy that anticipates and accommodates the special needs of these patients (Cancer Treatment Centers of America (CTCA)).

18.1 INTRODUCTION

Cancer Treatment Centers of America (CTCA) is a comprehensive network of cancer care hospitals and outpatient care centers in Atlanta, Chicago, Philadelphia, Phoenix, and Tulsa. Founded in 1988, CTCA® specializes in the treatment of adult cancer patients through an integrative approach to care that combines surgery, radiation, chemotherapy, and immunotherapy with advancements in precision cancer treatment and supportive therapies designed to manage side effects and enhance quality of life both during and after treatment. CTCA also offers a range of clinical trials for patients with the objective of revealing new treatment options supported by scientific and investigational research. The company has treated over 90,000 patients since its founding 30 years ago.

Today, CTCA represents the manifestation of the founder Richard J. Stephenson's dream about the way cancer patients should be treated, borne of the adversity through which his mother fought and lost her valiant battle with cancer because she did not have access to the most promising treatment options available at the time. The experience led him to acquire his first hospital in suburban Chicago in the late 1980s and subsequent focus on cancer treatment, drawing upon the most promising treatments as the science of cancer care evolved. The rapid growth of the enterprise was guided by his resolute belief that all cancer patients should receive the same level of clinical expertise and attention each of us would insist upon for our mothers in the event they received a cancer diagnosis, something he christened the Mother Standard® of care. This prevails as the manifestation of the company's ethos today.

Given its singular focus on the treatment of cancer, CTCA attracts many patients with a complex or advanced-stage diagnosis. Many also come to CTCA in pursuit of its integrative approach to care, offering therapies that complement surgery, radiation, and chemotherapy, including naturopathic medicine, psychosocial support, chiropractic care, acupuncture, and spiritual support, all guided by the latest evidence-informed science. The specialized nature of the care CTCA provides is the primary reason the average patient travels more than 300 miles to receive treatment. This is a remarkable observation given the fact that proprietary research conducted by CTCA has revealed that 90% of adults who are currently in treatment for cancer receive care from a provider located within 100 miles of their primary residence.

CTCA also believes an informed patient is an empowered patient. Consistent with this belief, CTCA was the first national cancer care provider to make its treatment results readily available to patients and caregivers who wanted to understand the efficacy of care delivered, defined by both length and quality of life. The company remains in the vanguard of providing patients with this type of transparency today, publishing results for 11 different cancer types on its website (www.cancercenter.com), including results for the most common adult cancers. This remains an important point of brand distinction for CTCA, given that fewer than 7% of the over 1,500 providers of cancer care in the U.S. who report clinical outcomes to the National Cancer Data Base, jointly sponsored by the American College of Surgeons and the American Cancer Society, make their results available to the public.

Consistent with its philosophy of respecting patients' desire to exercise informed choices about their care, yet unlike most other prominent cancer care providers, CTCA has pursued a direct-to-patient communications strategy designed to inform and educate individuals seeking to learn more about its unique approach to care. The company uses sophisticated, multi-channel communications to reach and engage prospective patients, the majority of whom travel a considerable distance to seek care at one of CTCA's five destination hospitals. Many of these patients decide to come to CTCA after consulting with their primary care physician and/or local medical specialist. Many also explore treatment options at CTCA because of concern about, or dissatisfaction with, the outcome of treatment they have received elsewhere. The company has evolved a sophisticated support team to accommodate the logistic needs of these patients, including a skilled travel services unit that handles all required air transportation arrangements, 24/7 concierge-style reception and transfers from airports and train stations, and either in-house or on-property lodging in accommodations that are owned and/or operated to the high-quality standard imposed by CTCA. Today, hundreds of thousands of individuals contact CTCA annually to inquire about its unique approach to the delivery of compassionate care. Literally millions more seek information from the company's website, www.cancercenter.com, the most frequently visited website of any cancer care provider in the world, hosting an average of over 3.5 million unique visitors monthly.

CTCA's commitment to the delivery of compassionate cancer care is evident in its enviable patient satisfaction scores. According to the 2018 in-patient satisfaction metrics published by Press Ganey® in the HCAHPS (hospital consumer assessment of healthcare providers and systems) national survey conducted for the U.S. Department of Health and Human Services, CTCA's scores rank among the highest for all cancer care providers who participated in this highly-regarded survey: the top 2% of

2,371 hospitals ranked nationally, with 93% of patients stating they would "definitely" recommend CTCA to family/friends. Patient satisfaction scores for outpatient surgical procedures during the same year were equally impressive as revealed in the OAS CAHPS survey conducted for the U.S. Department of Health and Human Services: top 3% of 2,566 hospitals ranked nationally, with 94% of patients stating they would "definitely" recommend CTCA to family/friends. And the CTCA brand currently ranks third in the national Brand Index rankings of 42 prominent hospitals and hospital systems authored by YouGov, following the Mayo Clinic and Johns Hopkins Medicine, ahead of Cedars-Sinai and Cleveland Clinic as revealed in Table 18.1.

TABLE 18.1 YouGov 2018 Brand Index Scores*

Rank	Brand	Score
1	Mayo Clinic	40.4
2	Johns Hopkins Medicine	34.0
3	Cancer Treatment Centers of America	22.6
4	Cedars-Sinai	18.4
5	Cleveland Clinic	16.7

**1 July 2017 to 30 June 2018.*

18.2 WHAT CANCER PATIENTS ARE SEEKING?

It is axiomatic to state that all cancer patients seek one thing: to conquer their disease and get well. Yet, most patients understand the insidious nature of cancer and the uncertainty of achieving a specific clinical outcome. This acknowledgment explains why many patients place an equal and often greater emphasis on the quality of life they experience while undergoing treatment.

All healthcare providers attempt to be "patient-centric" yet; realistically, few have more than a general sense of how to operationalize this desire because they lack specific insights into the nuanced expectations of the patients they serve; not so at CTCA.

Each year CTCA fields a national probability survey of non-CTCA cancer patients (*The Cancer Experience)* to discern the relative importance of a comprehensive array of specific aspects of care and the extent to which patients experience them. Respondents (n = 1,000) are quota sampled across all four stages of cancer (I to IV) to reveal any meaningful differences in the perspectives maintained by individuals with an early versus advanced-stage diagnosis. The results are evaluated with respect to a companion survey of 1,000 non-CTCA cancer patient caregivers to compare and contrast their perceptions along the same dimensions of care. The highlights provide an interesting perspective on the aspects of care that are both most important to patients and closely aligned with their expectations, versus those that are not:

- Indicators of clinical excellence top the list of attributes rated most important by patients and include “has the best qualified doctors” (96%), “good treatment results for my type of cancer” (96%), “uses the most advanced and sophisticated treatments and technology” (93%) and “has a reputation for cutting edge cancer research” (86%);
- The attributes rated lowest in importance by patients include “has clinical trials available” (58%), “uses mobile devices to remotely monitor vital signs” (43%) and “uses telecommunications technology” (38%);
- Almost half (48%) of all patients seek a second opinion (one out of 10 seeks a third opinion), with adults who have a more advanced-stage diagnosis and those <44 years of age more likely to do so;
- One out of six (16%) cancer patients have switched cancer care providers. “Changing insurance coverage” is cited as the primary reason why (23%), yet “dissatisfaction with their treatment plan” (21%), “declining confidence in their treatment team” (20%) and “a desire to explore other options” (20%) follow closely.

Respondents were asked to rate the relative importance of 25 aspects of care classified in one of three ways: (i) clinical offerings; (ii) care logistics; and (iii) care management. They were then asked to rate the extent to which they experienced each aspect of care to reveal any gaps in their “experiences” versus “expectations.” The results were illuminating.

One “logistics” consideration trumps all others in importance: “getting timely information” on the status of their care. This is followed by several “care management” considerations, including “doctor spends as much time as patient needs,” “comforting and stress-free environment,” “involves patient and caregivers fully,” and “integrated care team.” It is interesting and perhaps perplexing to note that all but two clinical considerations (“pain management” and “survivorship programs”) are cited by *less than* 50% of patients (Figure 18.1).

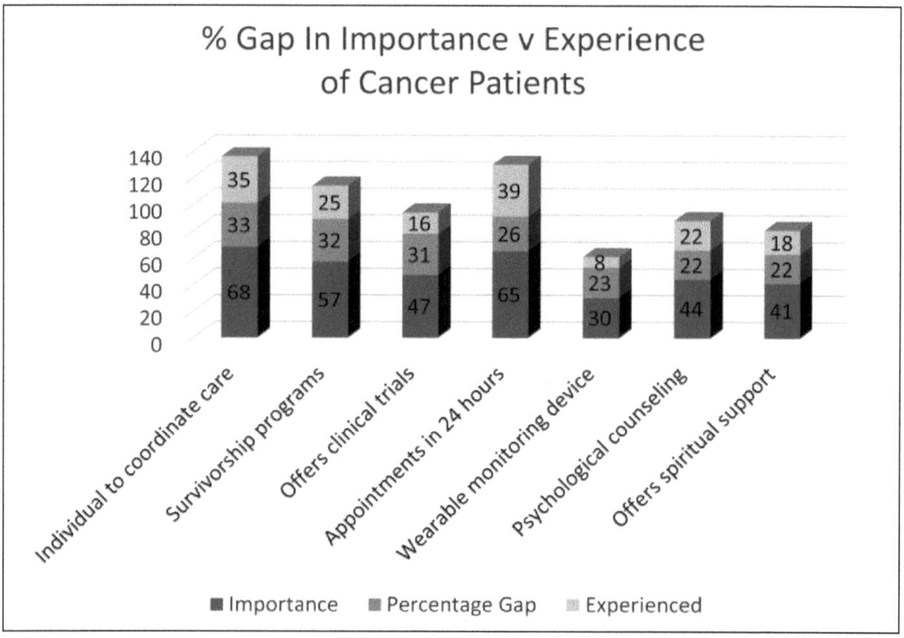

Figure 18.1 Percentage gap in importance vs. experience of cancer patients.
Source: National Cancer Experience (2017).

The insights revealed through this research have guided the development and/or refinement of several operational priorities at CTCA. They have also been used to guide the composition of brand communications highlighting how CTCA's existing clinical and/or patient service strategies address specific experience "gaps."

The Chairman of CTCA, Richard J Stephenson, is fond of reciting a refrain with which all company stakeholders are very familiar: "It's always and only about the patient." The resolute pursuit of this conviction is evident in the considerable time and expense the company invests in primary research to illuminate the needs, desires, hopes, and fears of cancer patients.

18.3 MEDICAL TRAVEL AT CANCER TREATMENT CENTERS OF AMERICA (CTCA)

The average CTCA patient travels more than 300 miles to receive the care that CTCA provides, even though 90% of adults in treatment for cancer receive care from a provider located within 100 miles of their primary residence.

This remarkable observation begins with a patient getting in touch with a CTCA Oncology Information Specialist. This is a role dedicated to serving the unique support needs of traveling cancer patients. Oncology Information Specialists compassionately and rapidly provide information on services and treatment options available, inquire about and address any barriers that might restrict care or travel to CTCA, and act as an overall guide for the patient and his/her caregiver(s) from the point of initial inquiry to arrival at a CTCA hospital. This provides the patient with one point of contact throughout the intake process, ensuring the process is convenient and stress-free. Behind the scenes, a number of CTCA stakeholders work to gather prior medical records, schedule clinical and integrative care appointments, obtain approval from health insurance companies, and arrange travel and lodging. From the moment a patient picks up the phone, they experience the CTCA commitment to patient-empowered care.

Each year, with over 35 million unique visitors to its website, hundreds of thousands of calls to its Oncology Information Specialists in Chicago and more than two-thirds of patients traveling to its hospitals from out of state, CTCA's geographically diverse demand requires an operating model tailored to address the many barriers that may hinder patient travel. Regardless of distance, traveling for medical care adds stress to an already enormously difficult situation. As a result, CTCA processes, operations, services, and, indeed the very culture, is dedicated to seamless coordination of medical travel and destination oncology care. Possible patient travel barriers on which Oncology Information Specialists are particularly focused include:

1. **Financial Burden:** Offering several financial programs and discount partnerships to assist with travel. For certain patients, CTCA will cover the cost of the flight/train ride or reimburse for miles driven. This may include caregivers when medically necessary. In addition, CTCA offers on and off-site guest rooms at reduced daily rates and has also partnered with nearby hotels to offer rooms at reduced rates. While at the hospital, CTCA offers healthy and affordable meal options for patients and their caregivers.
2. **Coordination and Logistics:** Providing patients with access to a dedicated team of travel and accommodations specialists who provide concierge services. The staff is well trained to ensure each patient's needs are understood and met, and all arrangements are communicated in a timely and thorough manner to the patient and family members. In addition, CTCA helps arrange ground transportation

services, including trips to and from the hospital, local lodging sites, and shopping areas.

3. **Lifestyle Disruption:** Minimizing quality of life disruptions with mindful scheduling that reflects personal situations (including employment and family care needs), coordinates provider calendars around the patient, and sequences consultations to reduce the number of visits required. A patient may choose to move into treatment immediately after an evaluative second opinion, often the same day, instead of scheduling a return visit. CTCA also pays particular attention to helping manage treatment side effects such as pain, nausea, or malnutrition, allowing patients to return to everyday life more quickly.
4. **Clinical Conditions:** Ensuring travel plans are designed with individual patient clinical situations and health acuity in mind. CTCA screens each patient to ensure travel to the hospital does not create additional risk. Some patients may require the assistance of a caregiver to help ease the burden of travel. Patients are scored on functional status, and physicians ensure travel teams are aware when a caregiver is medically necessary. Issuance of a transportation security administration (TSA) notification card is another aid in the process, allowing inconspicuous alerting of TSA personnel to the presence of a traveler with a medical condition. Finally, updates on the patient's functional performance score, clinical appointments, or personal commitments are monitored to determine any appropriate changes in travel plans.

On-site, CTCA ensures each patient's time away from home is manageable. This includes providing concierge services, healthy meal options, shuttle services, fitness facilities, activities, support groups, on-site pharmacies, and free Wi-Fi throughout each of the hospitals. Accommodations are provided in comfortable rooms with a variety of amenities designed to make patients feel at home. Patients stay at CTCA for varying lengths of time based on their treatment plans, and these amenities create a more enjoyable visit.

Globally, the future of cancer care presents a bit of a paradox. On the one hand, experts foresee cancer diagnoses and fatalities are increasing sharply in the coming decade. At the same time, technology and treatment breakthroughs (such as immunotherapy, genomic sequencing, and precision medicine) offer the hope of more accurate prognoses and higher survival rates. At present, however, the latter remain concentrated in mature markets and capital cities. Hence patients who reside in less well-developed geographic areas must travel to access the best care. For international patients, the CTCA operating model provides a seamless solution. Market-specific services such as providing language translation and interpretation, documentation support for U.S. medical visa applications, creating tailored food menu offerings, and engaging international health insurance account and case managers help augment the CTCA Oncology Information Specialist, travel support, accommodations, and care management processes provided to U.S. patients.

Further opportunities to improve the medical travel experience will continue to emerge. For example, with the advent of telehealth technology, CTCA has introduced services such as remote surgical follow-ups that allow patients to visit with their surgical care team from the comfort of their home, and time and plan their next treatment visit more accurately. With additional treatment locations in free-standing, community outpatient care centers, CTCA has partnered with several sites to offer patients the choice of receiving certain aspects of care that do not require a higher acuity setting at a location closer to home.

18.4 INTEGRATIVE ONCOLOGY

By definition, integrative oncology is a patient-centered, evidence-informed field of cancer care that utilizes mind and body practices, natural products, and/or lifestyle modifications from different traditions alongside conventional cancer treatments. It aims to optimize health, quality of life, and clinical outcomes across the cancer continuum and to empower people to prevent cancer and become active participants before, during, and beyond cancer treatment (Witt et al., 2017, p. 3).

People being treated for cancer may seek an integrative approach to their care in order to help manage symptoms, have a sense of control and to improve quality of life. Published surveys demonstrate high utilization of complementary treatments utilized by people with cancer. Some of these approaches, for example, herbal and dietary supplements, come with the risk of adverse interaction with conventional cancer treatments, thereby underscoring the need for coordination among all of the patients' providers. An integrative approach may help achieve several important objectives that address patient safety and quality of life concerns:

- Care is provided by clinicians and providers experienced in working with people with cancer;
- Care is documented in each patient's medical record;
- Patients are educated and counseled on practices that may have benefit and those unlikely to have benefit or cause harm. (Cassileth, 2013, p. 114).

In order to meet patient demand and deliver a seamless experience all under one roof, CTCA was an early adopter of an integrative approach to cancer care over 30 years ago. Patients are introduced to the supportive services available to them throughout their care. New patients are screened and assessed for their possible need for these services by nurse navigators who can schedule the necessary appointments during initial visits. New patients also complete a symptom inventory tool at baseline, then again no less than 21 days later, to isolate and monitor specific indications of integrative therapies. Depending on the severity of symptoms, referrals are then made to appropriate support service providers. Similar processes are in place to assess and manage needs on an ongoing basis, including team rounds, symptom assessment, and the incorporation of these modalities into treatment workflows.

Over the past decade or so, the presence of integrative medicine programs in both community hospitals and academic medical centers has grown; an increasing number of medical schools offer courses in integrative care in their curriculum, and clinical research has expanded in this area. Cancer is a common diagnosis for referral into these programs and inclusion in these studies. Access for patients varies by program and degree of integration. A well-designed integrative model is collaborative and multi-disciplinary, with highly trained, educated, and experienced providers who possess expertise in oncology care. Modalities are personalized based on the patient, their cancer type, treatment plan, reported, and anticipated symptoms and quality of life concerns. Modalities may be integrated into clinical pathways and other workflows that address patients' needs. And clinicians evaluate several factors when considering integrative options for their patients:

- The level of scientific evidence supporting the potential efficacy in the context of other available options;
- The cost of the therapy;
- Patient safety, contraindications, and possible interactions.

Many resources are available to assess these factors when evaluating options for patients. The National Comprehensive Cancer Network (NCCN), Multinational Association of Supportive Care in Cancer (MASCC), the society for integrative oncology (SIO), and many other professional organizations have outlined clinical practice guidelines for integrative care. Recently, the SIO Clinical Practice Guidelines on the Evidence-based Use of Integrative Therapies During and After Breast Cancer Treatment were endorsed by the American Society of Clinical Oncology (ASCO) with a few added discussion points.

Most of these guidelines recommend the use of mind-body therapies such as yoga, meditation, and relaxation training to help manage symptoms such as anxiety, depression, and insomnia. Guidelines also support acupressure and acupuncture as considerations for the management of chronic pain and chemotherapy-induced nausea and vomiting that is not completely relieved by anti-emetics.

Nutrition counseling and physical activity may also help with fatigue, physical functioning, and improved quality of life. Much research still needs to be done on the clinical benefit of natural products. However, open communication in an integrative setting regarding their use, along with guidance by qualified providers, may help patients identify those which should be considered and, more importantly, those that are contraindicated for symptom management during cancer treatment. The integrative approach to care may also be beneficial in steering patients away from "alternative" treatments and delays in seeking conventional cancer care.

Diet recommendations and nutrition counseling at CTCA are aimed at moving individuals towards a plant-focused diet. Most, if not all, population-based dietary guidelines advise consumption of vegetables, fruits, whole grains, and legumes or lean sources of animal proteins as the foundation for a healthy diet. The World Cancer Research Fund/American Institute for Cancer Research's *Diet, Nutrition, Physical Activity and Cancer: A global perspective Continuous Update Project Expert Report-2018* recommends this dietary approach for cancer risk reduction. These types of foods contain fiber, vitamins, minerals, and phytochemicals, which may protect cells from damage that may lead to cancer. In addition, a plant-based diet may help individuals achieve and maintain a healthy body weight, which is a critical risk factor that must be addressed in cancer survivors.

Given CTCA's personalized approach, diet recommendations vary based on the unique needs of each patient. Registered dietitians spend considerable time reviewing each patient's medical history in the context of their lifestyle and value system then design a plan to support them during and after treatment. Approaches may range from a diet that includes smaller, less frequent portions of animal foods, along with fish a couple of times per week for omega-3 fatty acids, to the full adoption of a well-planned, nutritionally adequate vegan diet. The meals served to in-patients and available in CTCA's dining rooms reflect this personalized approach. Plant proteins and dairy alternatives are readily available, as are a number of whole grains, vegetable, and fruit options. Variety and choices are very important for patients who may be trying to balance a change in their appetite and other symptoms with healthy eating.

There are several key differentiators in CTCA's delivery of integrative care that may otherwise not be available to patients close to home, which, in turn, may motivate them to travel a considerable distance for their care. CTCA's truly integrative approach relies on a collaborative clinical environment and multi-disciplinary approach, all managed under one roof. Patients have a complete team of providers who can address multiple needs during a visit. These teams have extensive experience working with the oncology population and are either certified or in the process of becoming certified in their area of oncology specialization. This compares with other medical generalists who may not see many patients with cancer and therefore not be conversant with the clinical benefits of integrative care. Finally, the level of integration, communication, and availability of CTCA's symptom inventory tool allows for a high degree of personalization.

An integrative approach to care facilitates coordination of the use of these modalities, so the care provided is safe, appropriate, and, guided by the best available evidence, designed to enhance the patients qualify of life while undergoing treatment.

18.5 THE MOTHER STANDARD OF CARE

How would you want your mother to be treated if she had cancer? The answer to that simple question has evolved into the CTCA Mother Standard of care. The Mother Standard of care represents a unique, vibrant approach to care that ensures a complete, patient-focused treatment experience.

CTCA was founded in 1988 by Richard J Stephenson, following the death of his mother, Mary Brown Stephenson, from cancer. After his mother's diagnosis, Richard, and his family searched for the most advanced and effective cancer treatments available. They were disappointed to discover that the world-renowned cancer hospitals at which she was receiving treatment were unable to provide the advanced treatment options that were otherwise available at the time. Sadly, Mrs. Stephenson succumbed to her disease before she could watch her grandchildren grow and mature.

To keep his mother's spirit alive, Richard vowed to change the face of cancer care. He created a new model of care that changed the way oncologists, surgeons, and other clinical professionals treat patients with cancer. This whole-person approach to cancer treatment combined the latest evidence-informed treatment options with a compassionate, nurturing environment, and is now known as the Mother Standard of care.

The CTCA Comprehensive Care Network of hospitals and outpatient care centers offers a wide spectrum of state-of-the-art cancer treatments-both conventional and integrative-as part of this model of care, which treats every person differently based on their unique clinical needs. For this reason, patients, physicians, employers, and insurers can depend on CTCA to offer comprehensive, compassionate, and truly personalized cancer care. At CTCA, patients are served by a dedicated, multidisciplinary team of physicians, nurses, registered dieticians, and other clinicians, all working collaboratively. These teams consist of individuals with extensive experience treating cancer. Together they develop and implement a personalized treatment plan for each patient that honors the individual's health and life goals. Some treatment services, such as chemotherapy, are available seven days a week for the convenience of patients and their caregivers. In fact, CTCA was among the first U.S. cancer care systems to offer weekend appointments.

In keeping with this spirit, as a demonstration of its positive effects on patients and their families, and to help patients and their families make more informed decisions about their cancer treatment, CTCA publishes its treatment results, including survival rates and comprehensive data on the patient experience, quality of life and patient safety program results[3]. These include:

1. **Survival Statistics and Results:** These comparing the length of life results for CTCA patients with national SEER (surveillance, epidemiology, and end results program) survival rates. CTCA treatment results are available for 11 cancer types, including the most common.
2. **Quality of Life Results:** These with CTCA among the first U.S. cancer hospitals to use quality-of-life metrics as part of its routine assessment of patient well-being and quality of care. Patient self-reported outcomes (PRSO) by cancer type are used to evaluate and demonstrate success in improving 13 core symptoms such as pain, fatigue, nausea, and distress, and six additional conditions that interfere with patients' everyday functioning.
3. **Patient Experience Results:** These including in-patient ratings via the third party administered HCAHPS survey; ambulatory surgical outpatient ratings via the outpatient and ambulatory surgery consumer assessment of healthcare providers and systems (OAS CAHPS) survey; outpatient experience ratings using a Press Ganey administered outpatient survey to all eligible patients within one week of the completion of any CTCA appointment for service; and

a Physician Transparency Star Rating program providing greater insight into the quality of patients' experiences with CTCA medical oncologists, radiation oncologists, and gynecologic oncologists.

4. **Quality and Patient Safety Results:** These demonstrating the outcome of a set of very thoughtfully designed, reliable procedures and systems that connect people, processes, knowledge, and technology in the delivery of high-quality, safe care. Metrics reported include infection prevention, complications, and length of stay, medication safety, nursing-sensitive indicators, and ensuring patient discussions around key topics such as chemotherapy intent and infertility risk. All five CTCA hospitals have achieved and maintain the quality oncology practice initiative (QOPI) certification conferred by the ASCO.

18.6 HOSPITALITY MEETS HEALTHCARE

Consistent with the company's commitment to the Mother Standard of care, awareness of the unique needs of cancer patients and appeal to patients who travel a considerable distance for their care, CTCA acknowledged the need to adopt key principles of hospitality as part of its service culture from its inception. This is evident in practically every touchpoint with patients, from responding to their initial inquiry to celebrating their survivorship. Some examples are provided below:

- Every inquiry about treatment options is handled by a full-time Patient Empowerment representative. Whether by phone or online chat, at 11:00 AM or 11:00 PM, on a typical Tuesday in May or Christmas Eve, the Patient Empowerment team is the first line of live engagement with prospective patients. Calls are typically answered in less than three rings, and these highly-trained representatives are skilled at responding to questions about treatment options, care logistics, travel assistance, insurance, and related financial matters;
- Once a patient has been scheduled for an appointment, the CTCA Travel Management Team assists with booking all travel arrangements and coordinates airport or rail station reception by one of the CTCA transportation professionals (waiting in baggage claim with a discrete, personal greeting displayed on an iPad);
- All patients are greeted upon arrival at the hospital by a lobby steward whose sole job is to provide a warm, personal welcome (by name) and assist with directions on how to proceed with registration;

- Hospital public areas have been designed and furnished to reflect the ambiance of a four-star hotel, thereby making the statement "it's different, less stressful and more welcoming here;"
- Patients receive a detailed schedule of appointments upon registration and those who are planning to spend one or more nights are escorted to their on-site accommodations which have also been designed to welcome caregivers, and from which they can watch the latest movies, access complimentary Wi-Fi, even order room service;
- The main dining room provides a relaxed atmosphere (often accompanied by the soothing sounds of a pianist) in which to enjoy a creative array of healthy, organic cuisine prepared by professional chefs in consultation with staff nutritional counselors;
- The well-equipped fitness center is available for use 24/7 and offers professionally-supervised fitness classes, massage therapy and chiropractic care, all complimentary;
- The unique Patient Empowered Care® model puts the patient in charge. So, instead of the traditional "doctor will see you now" expression recited numerous times throughout the day as patients wait for appointments in other clinics, at CTCA it's "the patient will see you now," after which the team of clinicians who will be serving the patient arrives for a group consultation in an examination room that has been thoughtfully designed to reduce anxiety, promote dialog, and enhance comfort;
- And patients who have completed 5 years of therapy are recognized in an inspirational event that is one of the annual highlights at each hospital: Celebrate Life®. This day-long event honors their courage, celebrates their victory over their disease and concludes with the placement of a personal "leaf" on an interactive Tree of Life that is displayed prominently on a large-screen television monitor at each hospital. These leaves may be "clicked" by visitors from a station appearing in front of each tree, opening to reveal the story of each survivor's journey told through a combination of personal narrative and video.

However, it is the resolute commitment by stakeholders to deliver truly compassionate care that reflects the spirit of hospitality that represents the foundation of the CTCA culture. Compelling examples occur every day: the Empowerment representative who took personal time away from her family to meet an incoming patient at the airport on a snowy Chicago evening to help alleviate the anxiety associated with her first visit; the nurse navigator who volunteered to take a patient's nervous children out for a movie when it was apparent they needed a break from their father's long day in the clinic; the pastor who responded to the late call from a deeply-concerned daughter yearning for spiritual support; the physical therapist who pulled strings to get a star linebacker from the local NFL team to pay a surprise visit on a fan struggling through surgical recovery; the nursing team on the 5th floor who helped a bedridden patient stage the wedding she thought would never happen because of her illness…in her room, with the requisite flowers, and cake. To CTCA stakeholders, there is no request too frivolous, or too difficult, if its fulfillment will enhance the patient experience. That is how hospitality meets healthcare at CTCA.

KEYWORDS

- **American Society of Clinical Oncology**
- **Cancer Treatment Centers of America®**
- **Multinational Association of Supportive Care in Cancer**
- **National Comprehensive Cancer Network**
- **patient self-reported outcomes**
- **quality oncology practice initiative**

REFERENCES

Cancer Treatment Centers of America, (2018/19). Global, Inc. Patient Treatment Results, www.cancercenter.com/ctca-results (accessed on 14 July 2021).

Cassileth, B. R., (2014). Psychiatric benefits of integrative therapies in patients with cancer. *Int. Rev. Psychiatry, 26*(1), 114–127.

Witt, et al., (2017). A comprehensive definition for integrative oncology. *JNCI Monographs, 2017*(52), 3–8.

CHAPTER 19

STRATEGIC MEDICAL TOURISM DESIGN: A CASE OF SWITZERLAND, HOSPITALITY BRIDGING HEALTHCARE (H2H©)

FREDERICK J. DEMICCO,[1] PETER P. TSCHIRKY,[2] MARTIN JEFFREY,[3] MENGYU LI,[4] and MARDELLE M. SHEPLEY[5]

[1]*School of Hotel and Restaurnat Management, the W.A. Franke College of Management, Northern Arizona University Flagstaff, Arizona, USA*

[2]*Former Managing Director Grand Resort Bad Ragaz, Switzerland*

[3]*Professor, The Swiss School of Tourism and Hospitality (SSTH), Switzerland*

[4]*University of Delaware, USA*

[5]*Cornell University, Ithaca, New York, USA*

ABSTRACT

The purpose of this chapter is to describe the desirable design elements that are essential in the rapidly evolving dynamics of Hospitality Bridging Healthcare (H2H©). This design thinking in H2H provides an effective means for integrating the leisure hotel guest with the medical tourist guest in the same physical structure (in this case, a medical resort or medical campus) where the hotel guests are located nearby with healthcare/medical guest patients in the same facilities. A case study of a prominent resort in Switzerland is described in this chapter.

19.1 INTRODUCTION

The design of the physical building (e.g., medical resort) should not disrupt the key reasons for each guest or patient staying there. The design must be universally accepted by each group, and allow for each to function independently of one another in a seamless environment, yet share many of the same resort/medical resort services. For example, the spa can be used for pleasure for the medical resorts' leisure guests, and simultaneously for rehabilitation guests (following medical treatment for the Patient/Guest or the "Mediguest"). Through the innovative design elements used at the Grand Resort Bad Ragaz, surrounded by the snowcapped mountains in picturesque Bad Ragaz, Switzerland, each group will be independent of one another, and not know the main reason that other groups of guests are there (e.g., either for leisure/pleasure vs. for medical reasons or both as in medical tourism (MT)) (Figure 19.1).

FIGURE 19.1 The Grand Resort in Bad Ragaz, Switzerland. The resort and medical campus are in the lower left quadrant.

19.2 THE STRUCTURE OF THIS CHAPTER

This chapter will use a systems model approach to describe the Grand Resort. The Systems Model has "Inputs" (the guests and or guest/patients) arriving into the system (the Grand Resort for Leisure and/or MT). Many Processes take place once the guests and/or guest/patients are inside the system. When they depart (check out of the resort), they are technically referred to as the Output in the Systems Model. A Feedback Loop links back from Output (where guests and/or Guest/Patient depart the system) to provide feedback to make improvements and obtain guest comments on the satisfaction level of their visit and constructive feedback for continuous improvement (Figure 19.2).

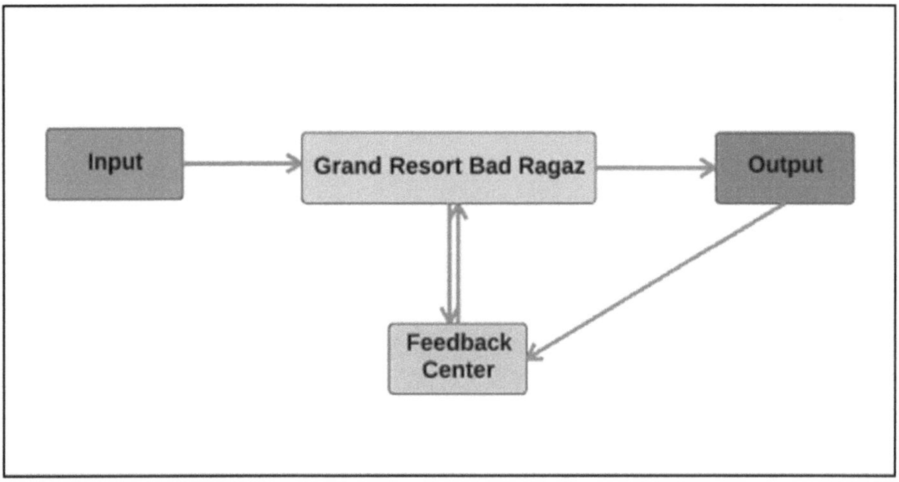

FIGURE 19.2 Systems model for the grand resort hospitality bridging healthcare.

Figure 19.3 illustrates the user case view within Grand Resort Bad Ragaz that fulfills primary functions and services that its resort guests seek. For example, resort guests check-in and go to a designated desk. Specifically, leisure guests check-in with resort associates at the Leisure Front Desk, while medical guests arrive at the resort and reports to medical and wellness associates at the Admissions Front Desk at the other resort entrance. Also, clinic guests utilize a rehabilitation system with various specialties that cater to their specific wellness needs. In general, all resort guests are encouraged to use resort amenities and facilities to enrich their stay experience and invigorate their energy and health. Furthermore, all guests, medical or leisure, share their feedback about the resort experience via an established feedback center. This communication channel is critical for the success of Grand Resort Bad Ragaz. Continuous improvement and refinement are based on the collected feedback from all participating guests (refer to Figures 19.3–19.5).

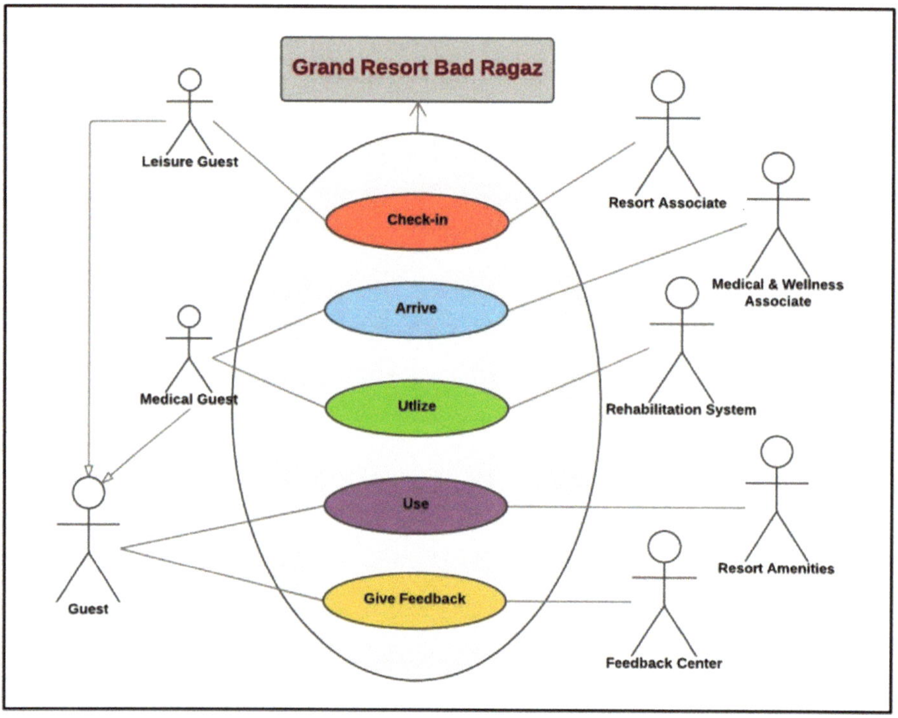

FIGURE 19.3 User case view within grand resort: "H2H."

19.3 THE DESIGN ELEMENTS IN PLACE AT THE GRAND RESORT

We will now describe each design element for check in, and all the processes/procedures for keeping the leisure guests separated from the medical patient/guest. These represent all the distinct processes in the system at the Grand Resort and include check-in, transportation (to medical procedures, dining, transportation, staff training, communications, and rehabilitation in the spa) and, using separate hallways and paths, for the Tamina Therme (the thermal water pool and spa), the restaurants, the restaurants, the entertainment, etc.

19.4 METHODOLOGY FOR THIS STUDY

A questionnaire was developed to ascertain the design elements at the Grand Resort in Bad Ragaz, Switzerland (about 25 Kilometers from Davos and 50

KM from Zurich). The leadership at the Grand Resort was asked a series of questions, and the responses appear below. We had a 100% response rate (N=12) from the Hotel and Medical Center at the Grand Resort. The questions were as follows:

1. **Check-In:** Do the leisure guests and the medical guests check into the same front desk/reception area- or are they separate? Can we get a photo of the front desk and the patient "admissions" desk here?
2. **Website:** Is there a different website for leisure guests and for medical guests- or are they the same website? Can we get a screenshot here of the website pages pertaining both to "leisure guests" and for "medical guests"?
3. **Guest Room Design:** Discuss the design of each guest room set up for the leisure guest and the medical guest? Are there separate walkways, corridors, elevators, etc.?
4. **Separating Guests:** Discuss how guests there for medical procedures are kept out of view of the leisure guests. Are there special physical designs in the hotel to maintain their privacy?
5. **Spa:** Discuss the separate areas in the spa for the leisure guests and the medical guest? Discuss the design features and provide photos and visuals for the design.
6. **Dining:** Any differences in the dining options for the restaurants in the hotel spaces for the leisure guests versus the medical guest?
7. **Differing Amenities:** Can you describe the different room design and amenities for the leisure guest as compared to the medical guest? Any photos would be good here to include.
8. **Training:** Can you discuss any unique training for the staff on the medical guest side of the resort?
9. **Unique Design:** Any other unique design features at the Grand Resort to separate the leisure guests from the *Mediguests*?
10. **Other:** Anything else?

19.5 SURVEY RESPONSES

Below are the responses to the questions listed in the previous section.

After the renovation of the Grand Hotel Quellenhof in 2019, we will also have a check-in for medical guests at the Hotel Front Office area (Figures 19.4 and 19.5).

FIGURE 19.4 Front desk for leisure guests.

FIGURE 19.5 Admission front desk for medical guests.

At the reception for the clinic, guests' medical-related issues such as doctor's appointments, therapies, etc., can also be booked, but no bookings for hotel rooms.

In the process of our re-branding, we have created four different websites: www.resortragaz.ch; www.healthragaz.ch; www.clinic-badragaz.ch; www.resortragaz-gruppe.ch, our corporate website.

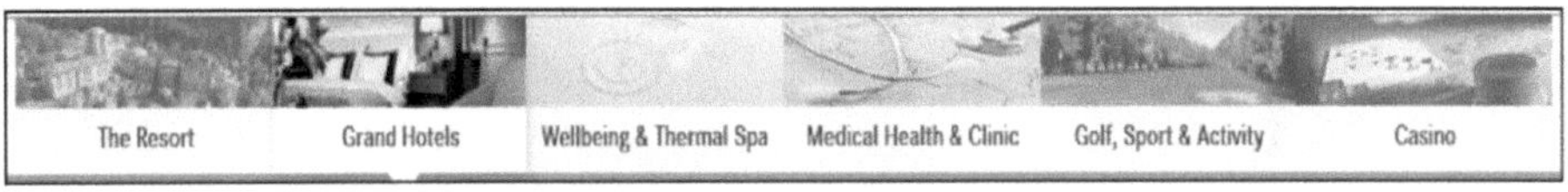

1. **Guest Room Design:** It is a misconception to think that you can simply use hotel rooms for clinic patients. That was a lesson for us when we modified our hotel rooms into clinic rooms. Initially, we did a soft-redesign which led to our patients not having a good sense of well-being because the level of comfort was simply not enough. Specifically, the beds, although height-adjustable were never adequate due to an incorrect matrass. Secondly, it was not possible for these beds to work with a hydraulic lift, which is an absolute must for bedridden patients. The entire bathroom concept had to be totally redesigned, as in Switzerland, there is a differentiation between normal, disabled, and wheelchair compatible rooms. Subsequently, we decided to make half the clinic rooms compatible for guests with walking disabilities and the other half wheelchair compatible. The biggest difference between these two categories is that the wheelchair patient cannot walk, in other words, everything must be done in a sitting position.

   Although in the same building, the top 5 floors are part of the hotel complex while the lower 4 floors are clinic. In total, Grand Resort Bad Ragaz offers 263 hotel rooms and 17 Clinic rooms. There are separate entrances, separate signage in the building as well as separate bed elevator and guest elevator for the clinical wing (Figure 19.6).
2. **Separating Guests:** The Clinic guests are, unless they are self-mobile, not visible on the hotel side. But as soon as the clinic guests feel fit enough, the hotel amenities such as swimming pool, restaurant, etc., can be used. We have experienced that people must have a good sense of well-being after an operation in order to show themselves in public. In terms of structure, the clinic wing is in fact outside the hotel, except as written before, the rooms above the clinic. Of course, we have the enormous advantage that in the clinic area, all further

FIGURE 19.6 Separate entrances.

medical installations are to be found. The number of leisure guests is certainly higher, given that there are substantially more hotel rooms than clinic rooms. Should a clinic guest cross paths with a hotel guest, then only because both require medical assistance, and then the difference between clinic- and hotel guest is less apparent. Clinic patients have a list of individual options to preserve their privacy. It is worth noting that patients that are administered to the clinic have to prove a certain level of health. The clinic is strictly meant for musculoskeletal and internal-oncological rehabilitation. If a patient's laboratory results fall below a certain requested threshold, they are transferred to a nearby partner hospital until they are well enough to continue their rehabilitation. Surgeries for Clinic guests are not performed in-house.

3. **Spa:** The clinic set-up has been discussed previously, with its separate entrance and separate elevators. Concerning the medical and the therapy centers, they are situated in their own buildings. Since we are a rehabilitation clinic, the therapy department has its own spa area with a specially designed swimming pool for water therapies (see photo attached; Hotel spa vs. therapy pool). This area also has special changing rooms and showers as well as a dedicated hydraulic lift to lower disabled patients into the water. All the therapies are done either in single cubicles or in group cubicles outside the normal leisure guest area. The advantage of being able to have the clinic on our premises, is that we have been dealing with medical care and therapy for 60 years for our spa guests and the clinic wing (now with 17 rooms, 25 beds) does not get in the way of the hotel (Figures 19.7 and 19.8).
4. **Dining:** The clinic guests eat mainly in their rooms, and as already explained, once they are able to walk and feel themselves again, they have the opportunity to eat in all of our restaurants (Figure 19.9).
5. **Differing Amenities:** We have already discussed our room design, in which the rooms were modified according to impaired mobility or wheelchair accessibility. The size of the amenities, the style, quality, etc., are identical. Our slogan is "the finest art of rehabilitation" and we are currently the most expensive rehabilitation clinic in Switzerland, classified as private under Swiss law. An exception to the private care is the area of internist-oncology, where we have an official order of the canton. This order of the canton is accessible to patients with general insurance, all other rehabilitation patients are either semi- or privately insured patients who expect a certain added value (Figure 19.10).

FIGURE 19.7 Therapy pool.

FIGURE 19.8 Hotel spa.

FIGURE 19.9 In-room dining in the clinic.

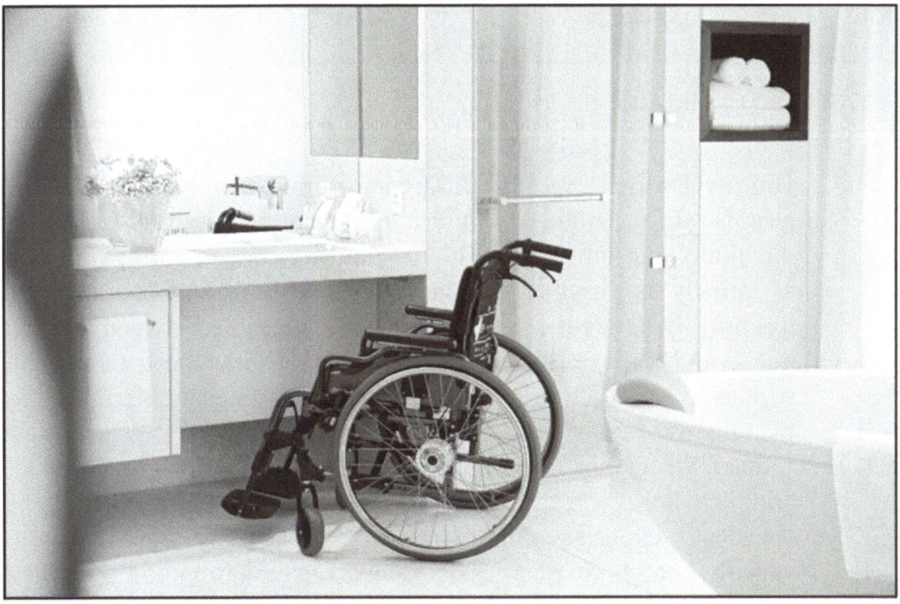

FIGURE 19.10 Wheelchair-accessible rooms.

6. **Training:** The clinic is completely run by trained medical staff, except for cleaning, food and beverage, and maintenance. These three departments are governed by contracts between the clinic and the hotel. Even the wellness and spa program operated by our hotel is operated only by qualified professionals in medicine. This is our so-called "Care Team" (Figure 19.11).

FIGURE 19.11 Medical staff.

7. **Unique Design:** No, the difference is much more in our service promise. In the hotel, our aim is to "make people happier," and in the clinic, our aim is that patients should return to normal daily life as quickly as possible. In other words, the healing process is the top priority hence the main difference noticeable between the hotel guests and clinic guests is the daily routine. Clinic guests have a solid stipulated plan, while hotel guests or resort guests determine their plan themselves.
8. **Other:** We would like to state that such an expansion of services clinic vs. hotel was only possible due to our circumstances, we had decades of experience in medical care and therapy. I think there are very few hotels that can easily convert excess rooms into medical- or clinic rooms. In fact, the reconstruction is not the issue, rather it is a question of culture and process. What I could imagine would

be new buildings, part hotel, part rehabilitation, but it would be a necessity for them to be connected to a hospital or themselves have the size to provide all medical requirements. Moreover, in such a hybrid environment, it is of utmost importance to have a perceptual balance between healthy and sick because you want to actually be a hotel but with extended medical business divisions. This balancing act has to be constantly monitored, redefined, sustained, and above all implemented.

19.6 CONCLUSIONS

This chapter describes a unique design for the hotel, with hospitality taking a central role. The design illustrates the blending between hotel hospitality and healthcare and wellness service quality delivery. The Grand Resort Bad Ragaz is a model for others looking to find a stable balance between excellent hotel and hospitality services and quality wellness and medical care on one physical medical, healthcare, and wellness campus. Even though the resort caters to an exclusive core clientele above the age of 55, the very efficient structures of the clinical and leisure divisions that operate side by side provide an excellent basis for other institutions that aim to build or implement similar business models.

KEYWORDS

- **hospitality**
- **medical resort**
- **methodology**
- **physical building**
- **rehabilitation**
- **seamless environment**
- **tourism**

PART VII

Emerging Trends in Health, Wellness, and Medical Travel

CHAPTER 20

ADVANCES IN TELEHEALTH AND PRIMARY CARE

ALI A. POORANI, AMANDA ADKINS, TERRI HOLLEN, and KATHRYN RABORN

Alfred Lerner College of Business and Economics, Department of Hospitality and Sport Business Management, University of Delaware, 14 W., Main Street, Newark, DE–19716, USA, E-mail: poorani@udel.edu (A. Poorani)

ABSTRACT

Telehealth has the potential to improve patient access to healthcare, enhance the overall delivery of care, and the success of patient outcomes. The majority of the U.S. hospitals and healthcare organizations in us use some form of telehealth. In this chapter we address we address the historical development of telehealth in primary care; and how patients receive telehealth in primary care. We also report issues related to telehealth in practice and provide three examples Northern Light Health, MedStar Washington Hospital Center, and the U.S. Department of Veteran's Affairs. Then we discuss how telehealth can improve patient experience and engagement, using the case study of *Presbyterian Healthcare Services* and their findings. Based on the literature research, we discuss issues related to *patient adoption of telehealth, consumers of telehealth, obstacle of the growth of telehealth, and the future.*

Finally, we discuss important issues relating to the use of telehealth in medical travel both within the U.S. and outside of the U.S. Our findings support that there are many challenges that healthcare organizations need to address before engaging in telehealth beyond their state borders.

20.1 INTRODUCTION

Finding a common definition for telehealth is not as easy as expected. There are many definitions, ranging from very narrow to very broad. It is also common for the term telemedicine, to be used interchangeably with telehealth. In its most basic form, the American Telemedicine Association defines it as the use of medical information that is exchanged from one site to another through electronic communication to improve a patient's health.

The first recorded use of hospital-based telehealth was in the late 1950s and early 1960s when a closed-circuit television link was established between the Nebraska Psychiatric Institute and Norfolk State Hospital for psychiatric consultations. Since that time, it has evolved to take on many forms and an ever-expanding range of use cases, throughout the world.

The global implications support the World Health Organization's definition of telehealth (Telehealth, 2016), which is the most universal and comprehensive. In it, they clearly articulate the ways in which telehealth has the potential to improve patient access to healthcare, enhance the overall delivery of care, and the success of patient outcomes. Interestingly, it also addresses the opportunity for improved sustainability by minimizing the physical footprint of healthcare.

20.2 LITERATURE REVIEW OF TELEHEALTH AND PRIMARY CARE

Telehealth can be delivered clinician to clinician (Asynchronous), clinician to patient (Synchronous), remote patient monitoring, and patient to mobile health technology. Telehealth services include clinician to clinician for dermatology, radiology, and emergency trauma not used in real-time. Clinician to patient services with real-time interactive technologies under telehealth included care for chronic conditions such as diabetes, wound care, and mental health. Remote patient monitoring and patient-to-mobile health technology use telehealth for health education, monitoring of diet and physical activity, and cognitive fitness (American Telemedicine Association, n.d.). The Department of Health and Human Services estimates that more than 60% of all health institutions and 40%–50% of all hospitals in the U.S. currently use some form of telehealth (Department of Health and Human Services, 2016). Healthcare systems reports (Tuckson, Edmunds, and Hodgkins, 2017) using telehealth to:

- Fill gaps in care that result from provider shortages;
- Provide services after normal working hours;

- Reduce the travel burden for patients;
- Facilitate services such as appointment scheduling and refilling prescriptions; and
- Respond to business needs and meet consumer expectations.

Primary care providers (PCP) play a critical role in our health care system. With the predicted shortage of the PCPs and other providers in the years to come, telehealth can be an alternative to onsite care (Doarn, 2016). Via telehealth, PCPs can provide ongoing care for patients suffering from chronic conditions such as diabetes, hypertension, or heart disease. PCPs can monitor symptom progress, adjust medications, and modify treatment plans without asking the patients to come into the office. They can also offer counseling services for managing depression, weight loss, and smoking cessation. When receiving lab results or reviewing the results from wellness exams, PCPs can discuss the results with the patients remotely without losing money on follow-up phone calls. Finally, PCPs can manage medication requests remotely using telehealth. It is more efficient for both the patient and the provider without sacrificing PCP reimbursement (Chiron Health, 2019).

According to Daniel and Sulmasy (2015), they offered policy recommendations for the practice and use of telehealth in primary care put forth by the American College of Physicians. These position statements and recommendations create a framework for the practice, use, and reimbursement of telehealth in primary care. Some of the recommendations by the American College of Physicians were to support the expanded role of telehealth to enhance the patient-physician relationship, to improve regular access to relevant medical expertise in their area, to address the needs of all patients without leaving behind financially disadvantage, low literacy, and low technologic literacy populations, to provide telehealth in a secure, compliant, and private environment, and to reimburse for appropriately structured telehealth communications (Daniel and Sulmasy, 2015).

20.2.1 HOW DO PATIENTS PERCEIVE TELEHEALTH IN PRIMARY CARE?

In a study done by Powell et al. (2017), most participants reported a positive experience with video visits. Patients identified convenience, efficiency, communication, privacy, and comfort are important areas to consider when deciding on using video visits compared to on site office visits. Cost savings were also noted by the participants in the study. This was due to cost savings from copays in the office, reduced transportation costs, and were absent from work for less time (Powell et al., 2017). In another study performed by Orozoco-Beltran et al. (2017), they evaluated the impact of telemonitoring on patients with chronic conditions with a high risk of rehospitalization or an emergency room visit. The chronic conditions used in the study were diabetes, hypertension, heart failure, and chronic obstructive pulmonary disease. Results showed that out of 521 participants, there was a significant reduction in weight loss, and high blood pressure as well as a reduction in emergency room visits for primary care services and in emergency situations or disease exacerbation.

20.3 TELEHEALTH IN PRACTICE

Based on a 2017 survey conducted by the American Hospital Association (American Hospital Association Fact Sheet: Telehealth, 2019), the use of telehealth in U.S. hospitals, increased from 35% in 2010 to 76% in 2017. A 2019 research report by Global Market Insights projects the telemedicine market size to exceed $130.5 billion by 2025, an 18.9% compound annual growth rate (CAGR) from its 2018 size of $38.3 billion.

Telehealth adoption varies by provider type and use case, ranging from 25% adoption for rural hospitals to 61% for large academic medical centers. Continued growth is evident. A 2018 Forrester research survey found that 17% of healthcare organizations are making a new investment in virtual care, with 38% planning to increase their existing investment (Forrester Analytics, 2018). The numbers indicate that telehealth is an established market disruptor, with larger institutions clearly establishing themselves as industry leaders.

In late 2016, Kaiser Permanente of Northern California reported that its virtual communications (e-mail, telephone, and video) exceeded in-person visits. Other large health systems, including Geisinger Health System, Partners HealthCare, the University of Virginia Health System, and the Veterans Health Administration report plans to continue to expand their telehealth services based on proven success (Tuckson, Edmunds, and Hodgkins, 2017).

As the insurance industry and U.S. government provide expanded coverage for telehealth and consumers apply continued pressure for service improvements, providers that do not engage are likely to be challenged by market entrants and other disruptors. Faced with a significant reduction in their share of the consumer healthcare market, we may see more mergers and acquisitions in this space.

Understanding the challenge before them, providers of various sizes, are engaged in the exploration of opportunities to embrace telehealth to enhance their services. Here are some examples:

- **Northern Light Health:** This is a local hospital in Portland, Maine that provides service to many older patients, in rural locales, with multiple, chronic conditions. To improve the convenience and effectiveness of the ongoing monitoring that they require, they partnered with their visiting nurse association to use telemonitoring. A pilot that began in 2006 with 32 patients, grew to 400 patients by 2017. It has effectively reduced heart-failure re-hospitalizations and also provides an ongoing connection with these patients that enables pro-active intervention (Telehealth: A Path to Virtual Integrated Care, 2019).
- **MedStar Washington Hospital Center:** In Washington, D.C., established a TeleTriage Program in their emergency department (ED). As a large urban hospital, they are combining live triage assessment by a nurse with remote physician evaluation to effectively reduce ED waiting times. Following nurse assessment, through the use of 2-way audio and video communications and electronic health records, the physician is able to provide a preliminary diagnosis, a treatment plan and patient orders. This streamlines care and shortens the overall duration of the patient's ED visit. This program also contributes to better alignment of ED staffing with predictably, unpredictable patient volume (Telehealth: A Path to Virtual Integrated Care, 2019).
- **The U.S. Department of Veteran's Affairs:** This supports 3 different telehealth programs that are associated with VA hospitals. One of these programs, Home TeleHealth is available at 900 VA sites nationwide. The program has improved care coordination for non-institutional care needs and chronic care management through health informatics, disease management, and in-home and mobile monitoring. By enrolling veterans in Home Telehealth, in 2016, the VA realized a 59% decrease in VA bed days of care and a 31% decrease in VA hospital admissions (VA Telehealth Services Fact Sheet, n.d.).

20.4 HOW TELEHEALTH CAN IMPROVE PATIENT EXPERIENCE AND PRACTICE EFFICIENCY

According to research published by Chiron Health (Smith, 2016), a provider of telehealth technology services, medical practices can realize as much as a 46% reduction in per visit patient costs, through the delivery of virtual vs. in-person care. Based on qualitative feedback there is evidence that, even at a lower cost, the quality of telehealth consultations exceeds the clinical effectiveness of in-person care for a number of reasons:

- Physician has less distractions;
- Patient is less stressed because they are in a more familiar, comfortable environment;
- Majority of the visit is focused on the patient;
- Self-monitoring can provide immediate information for diagnosis and treatment; and
- There is typically better compliance with the recommended treatment plan. With 83% of common conditions deemed suitable for virtual treatment, this is encouraging news.

20.4.1 BUILDING LOYALTY

Improving patient access to healthcare is also believed to build loyalty. Awareness of provider accessibility is proven to lessen patient dependence on the ED and urgent care clinics, which can impact primary care billables. Alternatively, telehealth provides the physician the convenience of meeting with patients outside of typical office hours, from any location. This can contribute to quality of life for on-call physicians and, with the expansion of health insurance coverage, can also provide additional revenue opportunities.

20.4.2 A CASE IN POINT: PRESBYTERIAN HEALTHCARE SERVICES: MOVING FROM INNOVATION TO OPERATIONS WITH SMARTEXAM

Patient Engagement HIT, Xtelligent Healthcare Media, just recently published the results of a case study conducted within the New Mexico-based Presbyterian Healthcare Services. The main challenge was how to expand patient access while maintaining patient and provider satisfaction. They employed a pilot study using a *non-video telehealth platform*, namely SmartExam, defined as follows: (Sponsored by: BrightMD, 2019).

SmartExam is a new breed telehealth solution, enabling online healthcare—from anywhere without video—to patients eligible under New Mexico laws and regulations for more than 400 common primary care conditions. Unlike video visits, with SmartExam, patients can log in anytime from their smartphone or computer and complete a clinical interview about their symptoms and health history in about 10 minutes. They then receive a diagnosis and treatment plan from a Presbyterian clinician within an hour.

- **Phase 1: Pilot Program and Results:** In the first phase of the program, they initiated a pilot study using their own employees. As soon as the number of cases reached to 100, the initial goal, the results were analyzed. The findings were astonishing and exceeded their expectations:
  - Prior to SmartExam, provider time to deliver care was 20 minutes;
  - Target time with implementation of SmartExam was 5 minutes;
  - Actual provider time to deliver care with SmartExa was reduced to under 2 minutes.

  As a result, providers could review interviews and provide diagnosis and treatment plans in 1–2 minutes. Meanwhile, patient satisfaction was high, and staff were comfortable using the platform.
- **Phase 2:** Based on the success of the initial program, the Innovation Hub, Operations, and Bright.md teams decided to expand the program for a greater audience. Soon after they realized that a majority of episodic care cases were funneled through their urgent care, where the waiting time to see a provider was nearly 4-hours. Utilizing the SmartExam, a typical patient was in the care queue for under three minutes. When the SmartExam was not the right avenue for care, the system moved the patient along the care continuum, saving time for both providers and patients to the next stage of care, be it via video, in person, or in the ED. According to the case study, "This quick response time was the result of clinicians' growing comfort with the application. Because the patient's health history, interview responses, and other relevant information are presented in a repetitive, consistent manner, providers know what to look for and where to find it, creating a streamlined process and incredible efficiency." Some of the important criteria for the success of the program were:
  - Those clinicians who were comfortable using virtual care were selected;
  - A minimum of one-year clinical experience;
  - Interest in SmartExam;
  - Confidence in diagnosing patients without actually seeing them in person.

Another added advantage of the program was that providers now could deliver care via their smart phones anywhere anytime.

20.4.3 RESULTS

- A year into the system-wide implementation, more than 20,000 patients have registered to use SmartExam.
- Patient wait times have been reduced to less than six minutes, while providers are now able to deliver care in a minute or less.
- According to the case, SmartExam has achieved a net promoter score (NPS, measures customer loyalty) of 72 among patients who completed a visit; which is substantially higher than the healthcare industry average (NPS 12); the Humana's scores 32 points; and the Amazon's scores (61).

20.5 PATIENT ADOPTION OF TELEHEALTH

Statistics from the Government Accountability Office (GAO) show that 90% of U.S. providers offer access, but only one-third of patients are regular users of a portal (Heath, 2017). Results like this, lend themselves to the question, "if you build it, will they come?" The answer probably lies somewhere between "yes" and "no," depending on your age and gender, and where you live.

A June 2019 consumer survey, by Power (Truex, 2019) found that 9.6% of American consumers have used telehealth services. Despite the assumption that telehealth is most effective in rural locations, it is actually more popular in west coast cities and suburbs. Women, the primary U.S. healthcare consumer demographic, have been the strongest adopters. Younger patients have also gravitated to telehealth with more confidence than seniors. The age bias is somewhat aligned with tech-savvy, but there may be a shift as the generational tech gap is shrinking.

The most interesting aspect of the survey findings is the lack of awareness about access to telehealth services, and it is the biggest impediment to patient usage. According to the survey, 74.3% of respondents said they either do not have access or are unaware of telehealth options. If it seems these numbers are inconsistent with the information previously provided on the availability of telehealth, you are correct. According to J. D. Power, major initiatives in both patient education and consumer experience are critical steps in making telehealth a staple for healthcare delivery in the United States (U.S.).

20.6 WHY SOME PATIENTS ARE CONSUMERS OF TELEHEALTH?

Although the growth of telehealth services in the healthcare industry, appears to be outpacing patient usage, there are many reasons some consumers are excited about it. These proof points are perfectly suited to more global "marketing" of telehealth to dis-engaged consumers to raise awareness and utilization.

Patients seeking value-based care are prime candidates for telehealth. The data proves that patients can obtain high quality telehealthcare at a lower cost than in-person physician consultations. With a growing number of private healthcare plans providing coverage for virtual care and the U.S. Department of Health and Human Services expanding Medicare and Medicaid reimbursement, beginning in 2020, patients will no longer need to worry about out of network expenses.

Additionally, various sources estimate that the average in-person consultation with a physician involves about 2 hours of a patient's time. Of those 121 minutes, only 20 minutes involve interaction with a clinician. In addition to the co-pays, travel costs, lost wages, the potential cost of childcare, and the inconvenience of rearranging one's life around the appointment, it is estimated that a consumer absorbs approximately $43 in incremental costs per visit. Telehealth has the potential to transform the patient experience by delivering quality medical care, at the patient's convenience and at a potentially lower cost.

20.7 OBSTACLES TO CONTINUED GROWTH OF TELEHEALTH

Despite the proven success of telehealth to date, its ongoing evolution, and the optimism of multi-disciplinary analysts and researchers, universal adoption will require resolution of existing obstacles. Center for Innovation (Telehealth: A Path to Virtual Integrated Care, 2019) cites the following barriers:

- Inconsistent payer reimbursement policies;
- Licensure laws and regulations limiting telehealth services across state lines;
- Access to and strength of broadband services required to support telehealth;
- Lack of organizational commitment to embrace and integrate into care delivery;
- Funding for required technology and infrastructure;

- Inadequate clinical engagement and preparedness;
- Consumer awareness of telehealth services.

An additional barrier is privacy concerns. Paula Manley, Executive Director Behavioral Health with Prevea Health, addressed this topic at the American College of Healthcare Executives-Wisconsin Chapter, the 5th Annual Conference on September 17, 2019. Ms. Manley was part of a panel discussion on the need to increase value for an organization and meet community needs when it comes to behavioral health. Telehealth is an option her organization is exploring; however, she has concerns regarding privacy if a patient utilizes behavioral health e-visits from their home. She specifically commented on the inability to control the environment in a patient's home and was concerned about friends or family members being in close proximity and hearing the confidential discussion. One would hope that the patient would take steps to have the conversation in private, but there is no guarantee.

20.8 THE FUTURE OF TELEHEALTH FOR THE INDUSTRY

Despite some of the challenges identified, if the projections are accurate, it is likely that telehealth will follow the success of online banking. What seems uncomfortable or even impossible today, will become as natural as depositing a check or transferring money to a friend through your cell phone.

Hospitals and healthcare systems have unlimited opportunities to extend the use of telehealth to address disparities in medical access and improve health outcomes for all. Some of the programs and services currently being explored and piloted include (Telehealth: A Path to Virtual Integrated Care, 2019):

1. **Pharmacy Services:** Remote patient access to professionals for post hospitalization consultation, periodic medication reviews, ongoing patient counseling and confirmation of patient adherence.
2. **Opioid Crisis:** Remotely provide consistent treatment in rural areas where resources are insufficient to support needs. Access to clinicians and experienced counselors who can administer medication-assisted treatment and provide more frequent oversight.
3. **Sleep Disorders:** Ongoing monitoring of patients with sleep apnea through remote transmission of patient breathing, oxygen levels, sleep patterns, body positions, etc. Provides more oversight and convenience than sleep clinics at less cost.

4. **Stroke Services:** Remote access to specialized stroke related resources, including stroke neurologists and targeted therapies to preserve brain function and save lives of chronic stroke victims.
5. **Provider-to-Provider Networking:** This can enable medical training, collaboration on patient diagnosis and care, as well as real-time peer-to-peer coaching on procedures and surgeries. This also has the potential to address workforce shortages in rural locales and 3rd world countries and enables more efficient use of health professionals.

Some industry experts believe that telehealth will change healthcare as we know it, with hospitals existing only in larger urban communities and essentially providing only intensive care. Moderately ill patients will be cared for outside of a hospital setting (Wachter, 2016).

Those patients who wish to use telehealth will feel that they have more ways to access care; they will no longer need to schedule an in-person visit when they need to see their provider. Having the flexibility to communicate with their provider via text or messaging in the patient portal may enhance their experience. Adding e-visits for primary care will also provide immediate benefits. Patients may be more willing to seek care if it can be done easily and conveniently from their home or office.

In the long-term, implementing telehealth may be necessary to survive in the healthcare environment. Generation Xers and Millennials are not willing to wait weeks to see a primary care provider. If we are not able to utilize options such as telehealth to minimize that wait time and provide care when patients want it, our organizations will cease to exist (Jones, 2018). Compounding the issue of wait times is that there is a shortage of healthcare providers. Currently, the top U.S. patient/physician ratio is 1,050:1 (County Health Rankings and Roadmaps, n.d.). Telehealth can offset part of this shortage by improving the efficiency of visits enabling providers to see more patients. Telehealth may also improve patient engagement. Remote monitoring options will enable providers to work together with their patients to monitor their health, leading to increased patient engagement.

20.9 TELEHEALTH AND MEDICAL TRAVEL

Obviously, the majority of patients traveling abroad are in pursuit of some form of surgery or procedures that requires the patients' visit to the destination and an in-person treatment. However, medical travel can greatly benefit from telehealth during pre-and-post visits. Some of the general benefits include pre-departure preparations. For example, a secured communication between the home-based provider and treating physician (attending physician) can help mitigate incorrect diagnoses or incomplete lab work. Or, upon return, a home-based physician (HP) may be unfamiliar with the procedures the patient has received abroad or may be unfamiliar with devices or equipment prescribed. In such cases, telehealth can help by coordinating communication and activities before travel or upon return, which all can optimize patient and family costs and experience. This is particularly important involving inbound and outbound travel.

20.9.1 MEDICAL TRAVEL WITHIN THE UNITED STATES (U.S.): TELEHEALTH CHALLENGES

There are a number of challenges for telehealth within the U.S., which were explained earlier. Two of them demand detailed attention. Those are reimbursement and license issues:

1. **Reimbursement:** While congress is trying to consolidate telehealth laws at the Federal level, at the moment, there are 37 states that have telehealth parity laws. Parity laws require private insurance companies to reimburse providers for care delivered remotely via telemedicine. However, states have many guidelines and restrictions regarding parity laws, conditions that vary state by state. Table 20.1 highlights major variations in state coverage.

   License Issues: In the U.S., physicians are licensed by state-by-state. That is, if a physician is licensed in one state cannot practice in another state, unless the practitioner maintains licenses in both states. Table 20.2 highlights some important licensing issues to consider when practicing telehealth.

20.9.2 INBOUND AND OUTBOUND MEDICAL TRAVEL: TELEHEALTH BENEFITS AND CHALLENGES

Without a secure telehealth, a patient will travel without a proper orientation to the practices of the host institution; carrying a bunch of health documents, CDs, etc., which sometimes introduces translation issues. As stated before, without a coordination between the home-based provider and the foreign practitioner, there are possibilities of misdiagnosis, or inadequate preparations, such as lab works, etc. Consultation between patient, patient family, HP and host-practitioner can help to mitigate some of these problems, and save time and money, as well as risks for patient and the HP. The following steps using telehealth can be very effective and provides a great experience to all parties involved. Let us look at a hypothetical patient traveling abroad, and the steps involved, and how telehealth can assist in each of the steps:

TABLE 20.1 Major Variations in State Coverage

Variations	Brief Descriptions
Patient location	In most states, these parity laws prevent health plans from withholding reimbursement for telehealth services based on a patient's location, such as home or office vis vi. travel to a qualified originating site. This might inconvenience certain groups.
Reimbursable services	Certain legislation will define "telehealth" or "telemedicine" as services provided in "real-time," thereby excluding store-and-forward.
Eligible providers	Each state determines which licensed professionals may practice telemedicine-this is often determined by the state medical board. For reimbursement purposes, the general rule is any provider that can bill for an in-office visit can also bill for remote telemedicine encounters.
Reimbursement levels	The amount providers are reimbursed for telemedicine will vary depending on a state's legislation. Some states specifically mandate those private payers reimburse the same amount for telemedicine as if the service was provided in-person. However, most states with reimbursement mandates leave this determination up to the payers.
Exceptions	Many states reimbursement mandates do have exceptions for certain types of insurance plans. For example, small group plans and worker's comp plans sometimes have the option to opt-out of telemedicine coverage.

Source: Adapted from: CHION a Medici Company; https://chironhealth.com/telemedicine/reimbursement/private-payer/.

1. **Pre-Departure:**
    i. The home-based physician (HP) makes an online consultation with the practicing physician (PP) and confirms the need for such a treatment;
    ii. The patient then is oriented to the host practice and procedures;
    iii. HP and PP share patient information using a secured platform;
    iv. HP and PP complete the treatment plan and steps to follow.

TABLE 20.2 Licensing Issues in Telehealth

Requirements	Brief Descriptions
Location	It is the location of the client that determines the state in which the practitioner must be licensed. Operating otherwise, such as obtaining waivers or consents from patients, is considered operating without license.
Licensing authority	The state licensure board within the state where a practitioner practices is regarded as the leading authority on the use of telehealth as a service delivery model within the state. Information about licensure boards is publicly available.
Regulatory requirements	When a practitioner holds a license in his/her home state and the state where the client resides, the practitioner is required to comply with the laws, regulations, rules, and policies where the licenses are held.
Continuing education credits (CE)	States vary considerably in their CE requirements. It is essential that such differences be reviewed and complied with when practicing telehealth across the states.
Malpractice policies	Since malpractice policies for services delivered through telehealth vary by carrier, practitioners must consult with their malpractice insurance carrier. In addition, the insurer's licensed coverage area must be considered when practicing in multiple states via telehealth.
Most stringent laws	When states have differing requirements for privacy, security, and informed consent, practitioners are encouraged to follow the most stringent laws and regulations. The regulations of the state where the client resides carry more weight.
Exceptions	Department of Defense and the Veterans Health Administration, which have their own licensing, are exempt from such requirements. Some states may allow a provider to practice in certain fields, such as speech-language pathology and audiology; such exemptions are temporary.

Source: Adapted from: International Journal of Rehabilitation, by Jana Cason, DHS, OTR/L and Janice A. Brannon, MA.

2. **Encounter: While Abroad:**
    i. All parties can involve in continued conversation and consultation as needed;
    ii. HP and PP can consult about necessary changes in treatment plans and diagnosis;
    iii. HP will have a chance to monitor the progress using telehealth platform.
3. **Pre-Discharge:**
    i. HP and PP or the foreign operations can review the discharge process and plans;

ii. HP and PP confirm that the medications and their availability at the home-country;
iii. HP and PP review the equipment if necessary.

4. **Arrival at Home Country:**
   i. Follow up telecommunication between HP, PP, and if necessary, involving patient and family;
   ii. Electronic files are updated;
   iii. Follow-up steps are continued as needed.

20.10 CONCLUSION AND RECOMMENDATIONS

Telehealth is here to stay and is changing the face of healthcare. If a healthcare system wants to remain a vibrant competitor in the healthcare marketplace, they need to develop a plan and implement telehealth initiatives that will meet patient expectations:

1. Senior leadership team should meet to determine their strategy regarding telehealth. Do they want to simply meet what the rest of the market offers, or do they want to become the market leader?
2. After determining their overall strategy, evaluate initiatives that are available within their current EMR and infrastructure followed by initiatives that may require outside resources in the form of purchased services or perhaps collaboration with other larger healthcare organizations in the region.
3. Once opportunities are evaluated, prioritize initiatives based on the needs and expectations of their patient population. Priority should be given to initiatives that would be more highly utilized in the short term. Long-term strategic plans should be made to budget for and implement additional technologies that will become necessities in the future as healthcare evolves.
4. Implementation of each and every initiative should include a comprehensive marketing and education plan. Patients need to be introduced to the initiative in a positive manner and have the opportunity to learn how to best utilize it. Providers also need to be part of the education process. They need to be not only aware of the technology but also be able to promote it to their patients and be advocates for usage.

By keeping the needs and expectations of patients at the forefront, healthcare organizations can successfully implement additional telehealth options to better serve their patient population now and in the future.

KEYWORDS

- **compound annual growth rate**
- **emergency department**
- **U.S. Government Accountability Office**
- **home-based physician**
- **net promotor score**
- **practicing physician**
- **primary care providers**

REFERENCES

About Telemedicine, (n.d.). Retrieved from: American Telemedicine Association; https://www.ortelehealth.org/content/telemedicine-or-telehealth-definitions (accessed on 14 July 2021).

American Hospital Association Fact Sheet: Telehealth, (2019). Retrieved from: https://www.aha.org/system/files/2019-02/fact-sheet-telehealth-2-4-19.pdf (accessed on 14 July 2021).

American Hospital Association, (2019). *Telehealth: A Path to Virtual Integrated Care.* Retrieved from: https://www.aha.org/system/files/media/file/2019/02/MarketInsights_TeleHealthReport.pdf (accessed on 14 July 2021).

BrightMD, (2019). *Scaling Telehealth at Presbyterian: Moving from Innovation to Patient Satisfaction with Smart Exam.* Retrieved from: Patient Engagement HIT; https://patientengagementhit.com/resources/white-papers/presbyterian-health-services-moving-from-innovation-to-operations-with-smartexam (accessed on 14 July 2021).

Chiron Health, (2019). Retrieved from: Telemedicine for Primary Care; https://chironhealth.com/telemedicine/providers/primary-care/ (accessed on 14 July 2021).

County Health Rankings and Roadmaps, (n.d.). Retrieved from: https://www.countyhealthrankings.org/app/wisconsin/2019/rankings/manitowoc/county/outcomes/overall/snapshot (accessed on 14 July 2021).

Daniel, H., & Sulmasy, L., (2015). Policy recommendations to guide the use of telemedicine in primary care settings: An American college of physicians positions paper. *Annals of Internal Medicine,* 787–789.

Doarn, C., (2016). Telehealth: A very useful tool that enables and improves patient access. *JABFM,* 430, 431.

Forrester Analytics, (2018). *Forrester Analytics Global Business Technographics Priorities and Journey Survey.* Retrieved from: https://www.forrester.com/GlobalBusinessTechnographics PrioritiesAnd JourneySurvey2018/-/E-sus4251 (accessed on 14 July 2021).

Heath, S., (2017). *Patient Portal Use Lagging Despite Strong Provider Support.* Retrieved from: https://patientengagementhit.com/news/patient-portal-use-lagging-despite-strong-provider-support (accessed on 14 July 2021).

Manitowoc County Demographics, (n.d.). Retrieved from: https://datausa.io/profile/geo/manitowoc-county-wi (accessed on 14 July 2021).

National Policy: Telehealth and Medicare, (n.d.). Retrieved from: https://www.cchpca.org/telehealth-policy/telehealth-and-medicare (accessed on 14 July 2021).

Office of Health Policy, Office of the Assistant Secretary for Planning and Evaluation, (2016). *United States Department of Health and Human Services*. Retrieved from: Report to Congress: E-health and telemedicine; https://aspe.hhs.gov/system/files/pdf/206751/TelemedicineE-HealthReport.pdf (accessed on 14 July 2021).

Orozco-Beltran, D., Sanchez-Molla, M., Sanchez, J., Mira, J., & Group, V. R., (2017). Telemedicine in primary care for patients with chronic conditions: The ValCronic quasi-experimental study. *Journal of Medical Internet Research.*

Powell, R., Henstenburg, J., Cooper, G., Hollander, J., & Rising, K., (2017). Patient perceptions of telehealth primary care video visits. *Annals of Family Medicine*, 225–229.

Smith, A., (2016). *2016 Telehealth Quote Roundup*. Retrieved from: https://chironhealth.com/blog/2016-telehealth-quote-roundup/ (accessed on 14 July 2021).

Telehealth, (2016). Retrieved from: https://www.who.int/gho/publications/world_health_statistics/2016/EN_WHS2016_TOC.pdf (accessed on 14 July 2021).

Truex, G., (2019). *2019 JD Powers Telehealth Satisfaction Study, As Telehealth Technology and Methodologies Mature, Consumer Adoption Emerges as Key Challenge for Providers.* JD Powers.

Tuckson, R., Edmunds, M., & Hodgkins, M., (2017). Telehealth. *The New England Journal of Medicine,* 1585–1592.

Veteran's Administration, (n.d.). *VA Telehealth Services Fact Sheet. VA Telehealth Services Fact Sheet*. Retrieved from: https://www.va.gov/COMMUNITYCARE/docs/news/VA_Telehealth_Services.pdf (accessed on 14 July 2021).

Wachter, R. M., (2016). *The Hospital Visit of the Future*. Retrieved from: https://www.youtube.com/watch?time_continue=2&v=pO3hXlA86D8 (accessed on 14 July 2021).

ADDITIONAL RESOURCES

AHA Annual Survey Information Technology Supplement Database. Chicago: American Hospital Association. 2017 Report. www.ahadata.com (accessed on 14 July 2021).

Board on Health Care Services, (2012). Institute of medicine. *The Role of Telehealth in an Evolving Health Care Environment: Workshop Summary*. Washington (DC): National Academies Press (US).

Chamberlain, G. (2018, Spring). *Deconstructing the Telehealth Industry: Part II: Improving the Access Points of Healthcare Delivery*. Ziegler. https://www.ziegler.com/newsroom/2018/04/deconstructing-the-telehealth-industry-part-ii-improving-the-access-points-of-healthcare-delivery/ (accessed on 14 July 2021).

FY 2017 AHA Annual Survey Database. Chicago: American Hospital Association. www.ahadata.com (accessed on 14 July 2021).

https://news.careinnovations.com/blog/report-va-telehealth-services-revolutionizing-healthcare-for-veterans (accessed on 14 July 2021).

https://www.gminsights.com/pressrelease/telemedicine-market (accessed on 14 July 2021).

https://www.groupmgmt.com/about-us/testimonials/telemedicine-testimonials/(accessed on 14 July 2021).

https://www.who.int/gho/publications/world_health_statistics/2016/EN_WHS2016_TOC.pdf (accessed on 14 July 2021).

Nesbitt, T., S. (2012). *The Evolution of Telehealth: Where Have We Been and Where Are We Going?* Available from: https://www.ncbi.nlm.nih.gov/books/NBK207141/ (accessed on 14 July 2021).

CHAPTER 21

DELIVERING AN EXCEPTIONAL HOLISTIC PATIENT EXPERIENCE WITH EMERGING TECH-ENABLED SERVICES

TAMMY RICHMOND

CEO/President, Go2Care, Inc., USA, Occupational Therapy Registered/ Licensed, Fellow of the American Occupational Therapy Association, E-mail: tammy@go2care.com

ABSTRACT

In this chapter, the authors discuss the emerging use of tech-enabled services to create enhanced patient health and wellness experiences, or a "holistic" approach that can lead to better outcomes and protect your medical tourism (MT) business.

21.1 INTRODUCTION

As the medical tourism (MT) industry continues to be impacted by competitive marketplaces and the empowerment of patient choices, MT operatives and advocates should become aware of emerging innovative tools in the traditional medical business models that can enhance and benefit the outcomes of destination medical services.

The global MT paradigm is changing from the mobility of consumers and the advancements of tech-enabled systems such as electronic medical records and digital health and the use of daily mobile devices. In fact, over half of the world's population has access to the internet or mobile devices. In addition, the use of information and communication technologies (ICT) in health and wellness has increased dramatically over the last several years. A majority of all adults go online first to seek out health information and education before contacting a medical professional. Virtual visits for medical care or "telemedicine" have reached over 77 million, and as younger generations are entirely raised on mobile devices, the use and convenience of technology to engage in all types of services will be expected. Consumers want access, follow up and follow-through from their health care providers. Technology provides just that.

Potentially 11 million people a year will travel for medical treatments and the expansion of MT will increase to over 25%, the use of technology as a delivery model for specialty and enhanced services will allow for increased access to patients and providers, improved quality of care and a more coordinated and valued care process making it the number one emerging model for growth and sustainability in destination medical travel.

In this chapter, I will discuss how to use technology to bridge the gaps in existing medical travel services to create a "holistic" patient engagement experiences to benefit the patient and destination facility and promote value-based services, revenue, and business sustainability.

21.2 CURRENT ENCOUNTER RISKS AND BENEFITS

There are some known and inherent risks with traveling to another country or facility for medical services. These include quality control, lack of social support or preparation, cultural differences and sensitivities, lack of international standards, native language barriers, general lack of follow-up care, and general lack of medical record transfer. Several solutions are available to decrease these risks including site certification, accreditation, and training. However, let me bring your attention and intention to the concept that these risks could be mitigated by integrating the use of technology or tech-enabled services.

The benefits of destination or medical travel for medical services are reduced costs, reduced waiting times, and most importantly, access to specialty services. In addition, just the humanistic and ethical efforts to provide necessary health care services in countries where hospitals or medical services or specialties do not exist are vital.

What do these risks and benefits have in common? They can both benefit from integrating touchpoints with patients through the care process by implementing technologies. The use of technology-driven service solutions can ensure valuable end-to-end continuum of care for patients receiving destination medical services. Patients can gain virtual access to a dedicated medical, rehab, and wellness specialists trained in their unique conditions and procedures, through the use of everyday technology. More specifically, I am talking about online, virtual visits by HIPAA compliant video conferencing platforms or software or what is known as telemedicine or telehealth.

21.3 HOLISTIC PATIENT ENGAGEMENT DEFINED

Patient engagement is officially defined as "the active participation of patients and their families in the process of making medical decisions." Basically, this means to educate, inform, and encourage patients while allowing them to participate and manage their own care. The approach involves collaboration and coordination between patients, facilitators, and practitioners to provide a continuum of care from arrival to departure and beyond. Strategically, you need to improve the patients' care experiences and their understanding of their medical choices, including preparation, recovery, and ongoing monitoring through tools and ways to assess or communicate their needs and outcomes.

There is mounting evidence about the benefits of patient engagement. Involving patients through technology can lead to greater patient adherence to pre- and post-surgical guidelines, fewer patient complaints and grievances and fewer and fewer malpractice claims. As a result, much better outcomes. This generates valuable and measurable repeat business and revenue sustainability.

What is the value of the patient experience? Something called the "Holistic" experience. There are three elements that have to be included in the holistic experience; access, engagement, and value.

The four goals of a holistic patient experience are improved outcomes, touchpoint data, loyalty, and retention and continuous engagement. Let us take a look at what that means as a destination medical location.

21.4 MARKETPLACE OPPORTUNITIES

Virtual visits are the only non-traditional setting where expected future usage exceeds expectation, according to industry experts. In a recent survey of patients (Health Technology Forum, Phoenix, 2017) 62% want to communicate to their health care providers by e-mail, 9/10 want to share their decision making with their provider, 61% of patients say that digital services are important when selecting a physician and 76% of patients say technology has the potential to improve their life. Patients want access, engagement, and best possible outcomes by being prepared, becoming a part of the process, and engaging in the recovery and wellness. That equals "value" and value means better health and better care.

The bridge to "holistic" care is using technology to not only have an online website, but the integration of service information, robust digital resources, patient-specific tools to drive decision making including access to experts beyond just the physician but anyone who will become part of the care team; nurses, therapists, facilitators, family members, governments, payers, and so on. Your current business online model should include patient care profiles, health risk assessments, service options, cost management, outcome surveys, and the opportunity for new emerging tech-enabled services. In other words, a destination "portal" where the patient has enough information and tools to make an informed decision about your destination location and specialty services based on their individual need, personalized medical reporting, and access to specialty providers for the full team approach and follow up.

Let us explore some services packages that can be delivered by current technologies and currently emerging in destination or global MT. These are just a few:

1. **Complimentary Care Services (CCS):** In this enhanced service package, the facilitator collaborates with the destination medical hospital or clinic for both pre and post-treatment solutions to ensure end-to-end continuum care. That can include an online assessment that identifies pre-trip symptoms, physical, and emotional health indicators to triage rehab and recovery and wellness needs before and after destination health care. Patients gain on-demand, virtual access to the specialist before and after their procedures to ensure outcomes are achieved and personalized care is given. To achieve this, I recommend standardized wellness assessments such as the SF Rand 36 Item Short Survey, Modified Perceived Stress Test, and other behavioral and emotional assessments that can be performed online, within the destination's patient profile account. These assessments should also be performed post-treatment to identify potential risks in outcomes and success. Add virtual visits via HIPAA compliant videoconferencing software such as VSEE, ZOOM, and DOXYME where facilitators or destination care coordinators can discuss pre-assessment results and prepare for a personalized approach to the entire care package from travel to treatment and back home again. Meeting online should be part of every treatment package to ensure all care questions are answers, expectations are known and shared,

any health risks are identified, and solutions are put into place before the patient leaves home. In my experience working with destination services, having these early online touchpoints establishes part of the "holistic" engagement that patients want and expect and are willing to pay for.

2. **Early Mobility and Patient Monitoring Packages (EMPM):** EMPM online service packages enhance outcomes, prevent avoidable complications, and maximize early function and well-being following specialty care medical services with continued CCS during destination health stays. Patients meet with a physical rehabilitation or wellness provider through the use of videoconferencing technologies. Patients receive customized services that can include symptom monitoring, health coaching, wellness services (stretching/mobility), ADL training, and environmental modifications pre- and post-treatment. Facilitators and caretakers have access to safe patient handling consulting/instruction. Early Mobility Solutions and Patient Monitoring tools ensure increased patient satisfaction with services.
3. **Travel Accommodation Consulting:** Many patients have pre-existing conditions or special challenges prior to destination medical services. Outstanding destination packages should include the ability to identify and remedy potential challenges that can occur with extended air travel and out-of-country hoteling ahead of time to eliminate avoidable problems. Prior to leaving, patients can meet online with dedicated travel accommodations consultants who can work with families, medical facilitators, and patients to directly address air travel/airport transition concerns such as wheelchair accessibility, vehicle transport challenges such as getting in/out of cars or taxis, bathing/toileting challenges, time zone challenges, and assistive device needs or ADL equipment (canes, walkers, wheelchairs, raised toilet seats, shower chairs, etc.). Using a pre-travel online assessment and survey allows for early identification of travel accommodation needs. Post-treatment guidelines such as icing, hydration, and other basic first aid and prevention strategies can easily be handled online by remote monitoring and access to care coordinators. The patient's vitals such as blood pressure, heart rate, operation incisions, and more can be performed by mobile devices or wearables and synchronously sent to the physician or care coordinator as post-surgical, or treatment conditions need to be monitored and surveillance of symptoms important to outcomes.
4. **Transitional Care Consulting:** An often-overlooked part of destination health service packages is providing what I call transitional

care consulting. Patients continue to need follow-up care when they return home to create long-term wellness and longevity plans to prevent avoidable declines, maintain gains, and improve functionality. There needs to be a reassessment of symptoms and health indicators to establish post level of well-being and function. Online or virtual service package can include rehab or wellness specialists who can consult with the local physician or rehab providers to assist in the transfer of medical information and care plans to facilitate continued care with local and community partnerships. In addition, patients can continue to receive remote rehabilitation and wellness recommendations for the first few days to weeks within the comfort of their homes, before being referred to the local provider if needed. Specialists such as physical or occupational therapists, wellness coaches, or prevention specialists can assist the destination facility with follow-up care. Partnerships with specialty providers can continue to perform virtual concierge patient monitoring and health coaching per recommendations with Destination Health facilitator, surgeon or practitioners and assist in transition of post-recovery, rehab, and wellness when the patient returns home and/or to their local providers. This can be accomplished by the use of telehealth or online videoconferencing services.

21.5 CONCLUSION

Utilizing tech-enabled services enhances collaborative care through virtual concierge-style physical rehabilitation and wellness solutions to ensure a valuable end-to-end continuum of care for patients receiving destination medical services. Patients gain virtual access to dedicated specialists trained in their unique conditions and procedures, through the use of everyday technology. This "Holistic" patient engagement can provide improved health care outcomes by access to better health through timely access to pre and post-service package enhancements, providing patient triaging and monitoring, and potentially avoiding health issue pre/post-care escalation or outcomes health risks. In addition, implementing emerging technology-driven patient education, tools of engagement, real-time, virtual access to all types of providers that can benefit the patient outcomes is vital to standing out from other medical destination choices. Patient engagement is about engaging a patient in their own care from start to a healthy and successful finish. Technology bridges the gap in patient-provider relationships. Technology touchpoints allow for data gathering, risk management and proliferate successful marketing and revenue value for repeat business. The importance of an exceptional patient experience has been shown to not only improve outcomes but is a key priority in medical destination growth.

I recommend taking the time to build out a patient engagement strategy that goes beyond just the primary care site but evolves into the full care cycle of a "holistic" experience. This will build brand loyalty and retention and value.

KEYWORDS

- **complimentary care services**
- **early mobility and patient monitoring packages**
- **empowerment**
- **global medical tourism**
- **information and communication technologies**
- **medical tourism industry**

REFERENCES

Conversa WBR Insights, (2018). *Healthcare 2020; Automated Patient Experiences Will Transform the Communication and Delivery of Care.* [White Paper] Retrieved from: Conversa Health WBR Insights. https://conversahealth.com/research/ (accessed on 14 July 2021).

Lumeris, (2014). *Risk Adjustment and Revenue; Enhancing the Provider/Patient Encounter to Optimize Payer Revenue.* [White Paper] Retrieved from: Lumeris. https://www.lumeris.com/wp-content/uploads/2015/11/LUMERIS_POV.EE_.05-14.v11.pdf (accessed on 14 July 2021).

Richmond, T., (2015). Administration and management: Paradigm shift in patient engagement: Key actions for administrators. American occupational therapy association. *Special Interest Section Quarterly, 31*(1).

CHAPTER 22

OVERVIEW AND TRENDS OF MEDICAL MARIJUANA TOURISM: A CASE OF COLORADO

SOO KANG

Colorado State University, Fort Collins, Colorado, USA

ABSTRACT

This chapter investigates medical marijuana (MMJ) tourism, focusing on its history and overview of the market in the wake of the 2014 recreational cannabis legalization in Colorado, USA. Emerged as an important SIT market, medical cannabis has been a hot topic in a broad medical tourism (MT) spectrum, ranging from a wellness offering at a high-end spa to an effective treatment for chronic children's illnesses. This chapter also discusses current trends and issues surrounding MMJ tourism and concludes with pressing challenges facing the MMJ tourism market in Colorado.

22.1 INTRODUCTION

Medical marijuana (MMJ), also known as medicinal cannabis, refers to the use of marijuana to treat specific health-related conditions (Nussbaum, Boyer, and Kondrad, 2011). While recreational marijuana market has received much of the attention due to the recent legalization, the MMJ market is also important in two ways. First, MMJ legalization has been considered as a pathway to full legalization for recreational marijuana. All states that have approved recreational marijuana in the United States (U.S.) have already allowed MMJ legally in the past decades (Caulkins, Hawken, Kilmer, and Kleiman, 2012). This gradual approach has been favored because it can alleviate the public's negative attitude toward the drug, thereby making the full and formal legalization less threatening. Also, the amicable approach also has been well received by the general public: an overwhelming 94% of the American adults supported the MMJ (Quinnipiac University National Poll, 2017) and 62% of the American adults approved the recreational marijuana legalization in 2018 (Hartig and Geiger, 2018).

Second, the MMJ legalization serves as a pilot test to examine the important legal and operational procedures, including retail policy, production, transportation, tax structure, labor laws, etc. This pilot trial would give state policymakers and industry practitioners with operational advantages when the full legalization goes into effect. This means an easier transition for state and local governments and industry enterprises in laying out the landscape of the business environment, rather than creating brand new policies and procedures in overseeing and managing the marijuana industry. From a medical tourism (MT) perspective, the current state of MMJ does not explicitly address the needs and interests of tourists as part of its ecosystem. The purpose of this chapter is to present the overview of MMJ tourism as a special interest tourism (SIT) segment since the 2014 recreational cannabis legalization in Colorado and to discuss current tourism trends and challenges facing the MMJ tourism market.

22.2 HISTORY OF MEDICAL MARIJUANA LEGALIZATION

In 1996, MMJ was first legalized in California. As the treatment of ADIS, muscular spasticity, cancer, migraines, and other disorders, patients, and their primary caregivers can cultivate and purchase marijuana with a physician's written recommendation. The legalization also prohibits physicians from being criminalized if they issue marijuana prescriptions. As of November 2019, 33 states and Washington D.C. have fully legalized MMJ (DISA, 2020). Another 13 states have rather limited legalization such as allowing only some form of MMJ or a lesser MMJ extract (Historical Timeline, 2019). Only four states in U.S., Idaho, Kansas, Nebraska, and South Dakota, still ban marijuana entirely (Horn, n.d.). Figure 22.1 shows the map of U.S. states that have legalized medical and recreational marijuana as of 2020 (Legality of Cannabis by U.S. Jurisdiction, 2020).

Colorado approved Amendment 20 on November 7, 2000 and became the 7th state to pass the MMJ legalization law, which took effect on June 1, 2001. The law decriminalized patients from cultivating, possessing, and using marijuana once they obtain written prescription from their medical providers. Qualifying patients are also required to register with a state-operated patient registry that issues identification cards (Historical Timeline,

2019). As of April 2019, there were 467 MMJ retail dispensaries operating in Colorado, which has one of the largest numbers of dispensaries in the country (MED Licensed Facilities, 2019).

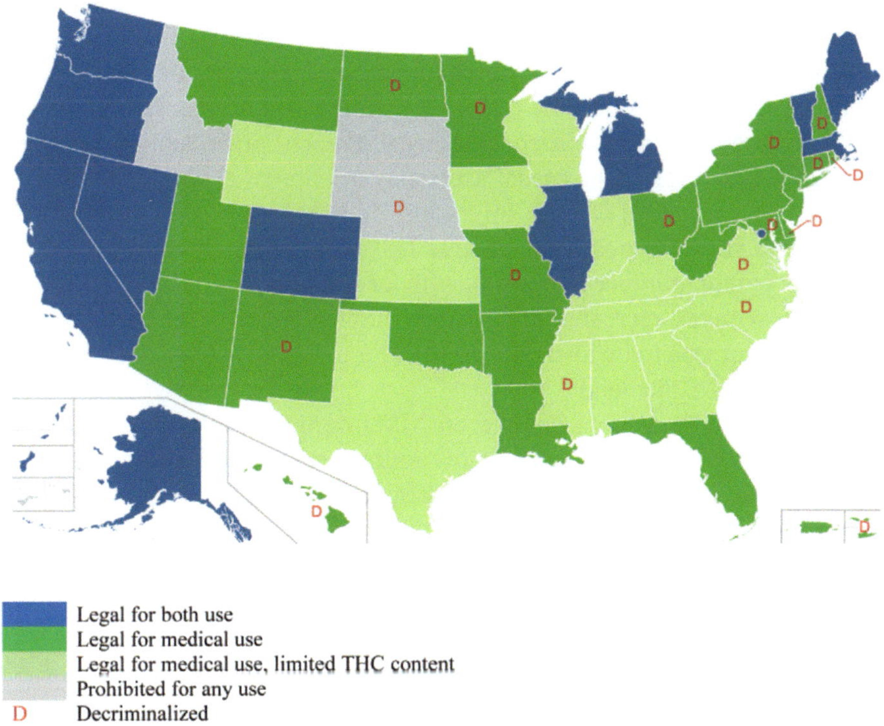

FIGURE 22.1 Current legalization status of medical and recreational marijuana in the U.S.
Source: Legality of Cannabis by U.S. Jurisdiction (2020).

22.3 DECLINING MMJ PATIENTS AND MARKET SHARE

According to data from MMJ programs in Nevada, Oregon, and Colorado, which maintain a pre-existing MMJ patient database, its patient counts are steadily declining after the launch of each state's recreational market (McVey, 2018). Colorado witnessed that the patients have decreased by 22%, from 114,774 in 2014 to 88,417 in 2017, since the state's recreational market launched in January 2014 (Medical Marijuana Registry, 2019). Compared to Oregon and Nevada, the Colorado's relatively smaller decline can be ascribed to low-cost MMJ cards and reduced tax on medical purchases

(McVey, 2018). Still, it appears that the passage of recreational laws had shifted some MMJ users out of the healthcare system.

Convenience is the main reason of the conversion from medical to recreational side. Obtaining a medical card is relatively burdensome; from visiting and paying for a physician, obtaining a recommendation, filing paperwork, finding a dispensary, designating a caregiver, to paying fees. (Associated Press, 2018). Alternatively, a patient can obtain the same or similar product on the recreational side. The patient must pay a higher tax rate to purchase recreational marijuana, but considering the recent price decline and expansion of product selection, casual users may not notice a significant difference in cost (McVey, 2018). The migration of MMJ users to the recreational marijuana side also poses a challenge for some dispensaries as they need to match a certain number of patients with a certain number of plants. Tracking down how many recreational users are purchasing for medicinal reasons is an important area to be investigated in order to evaluate the accurate size and volume of MMJ demands.

Even with the recent decline on MMJ demand, however, it is hard to conclude that MMJ market will be defunct soon. High-potency cannabis products are available only to MMJ patients for chronic illnesses and severe conditions. Thus, some MMJ patients are willing to renew their medical cards despite the availability of recreational marijuana in their states by enduring regulatory hurdles and paying the fees (McVey, 2018). In addition, an MMJ cardholder can purchase more marijuana per visit (usually 4 to 8 times) than recreational customers (1 ounce per visit) (Rough, 2017).

Figure 22.2 shows total sales comparison between medicinal and recreational marijuana markets in Colorado over the past 5 years. In the 1st year of legalization, MMJ sales were higher than those of recreational marijuana. However, the recreational marijuana caught up with its counterpart promptly. Recreational marijuana segment generated more than three-quarters (78%) of the total sales in 2018. The total sales of MMJ continued to shrink with a 10% drop since 2014. This decline might be strongly related to the decrease in patient numbers as presented earlier. On the other hand, the total sales of recreational marijuana significantly increased by 300% from $303 million in 2014 to more than $1 billion in 2018 (Marijuana Sales, 2019).

Marijuana tourism in Colorado was reported to continuously increase since 2014 (Kovacevich, 2018). According to a survey conducted by the Colorado Tourism Office (CTO), the recreational marijuana had attracted an estimated 6.5 million out-of-state visitors to Colorado in 2016 (Beth, 2018). Furthermore, it was projected to grow by 6% in 2017, mainly due to its geographic proximity to strict marijuana prohibition states, Utah,

Texas, and Kansas (Giannotti, 2017, p. 268). Despite the estimate that the economic impact of legal marijuana will increase 223% from 2017 to 2022 (Kovacevich, 2018), more reliable research data is urgently needed to determine how much medical MMJ related purchase contributes to the total tourism visitation.

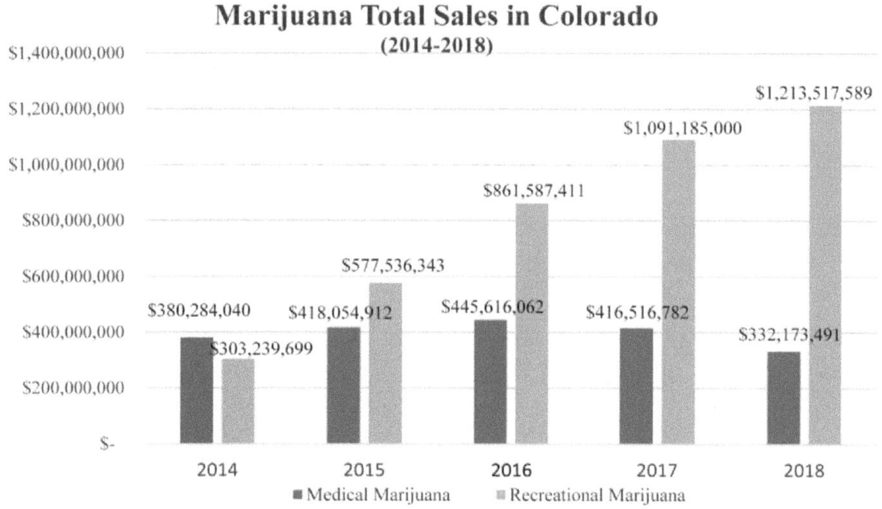

FIGURE 22.2 Total revenue comparison between medical marijuana and recreational marijuana. *Source:* Marijuana Tax Data (2019).

22.4 SCARCITY OF MEDICAL MARIJUANA (MMJ) TOURISM RESEARCH

Little research on marijuana has been conducted within the context of tourism perspective as it was not widely accepted due to the prohibition at the federal level (Belhassen, Santos, and Uriely, 2007; Kang, O'Leary, and Miller, 2016). Once dubbed as drug tourism and drug tourists, the recent development of legalized marijuana for recreational use in the U.S. has called for scholarly attention from various academic disciplines, including tourism (Kang et al., 2016).

Kang et al. (2016) identified four existing themes of the current tourism and leisure literature regarding marijuana tourism, which need to be challenged and expanded owing to rapidly changing legal environment: (i) focus only on selected topics (e.g., addiction or deviance); (ii) skewed data collection methods; (iii) limited data collection locations; and (iv) viewing the consumption of the drug mainly as a leisure or an extension of leisure activity. Especially for MMJ tourism, most current MMJ research is primarily concerning either medical/health or social challenges. The lack of academic research in both quantity and quality postulates a huge gap in the body of knowledge within the context of MMJ tourism (Kang and Lee, 2018; Tham, 2018).

22.5 CURRENT TRENDS

22.5.1 HEALTH AND WELLNESS MARKET

Among the many emerging trends around marijuana legalization, the health and wellness market has rapidly evolved to accommodate marijuana and its potential benefits (MPG, 2018). Even though marijuana has not always been received as an optimal offering in the health and wellness market, these perceptions are quickly altered owing to its emerging medicinal benefits. CBD-based wellness brands across all major categories, including food, beverage, beauty, and more have begun to accept marijuana as consumers increasingly seek out its benefits (Allway, 2018). Furthermore, Canada's nationwide legalization of recreational marijuana has allowed some health and wellness companies to open legal businesses offering marijuana-based products to consumers (MPG, 2018).

In hospitality and tourism setting, some resorts have embraced legalization as part of their therapeutic offerings by applying for marijuana production licenses or even developing their own craft strain for sale to their guests. Some lodging properties may consider participating in the creation of boutique marijuana edibles, either for sale at various onsite food and beverage outlets or for distribution elsewhere. Originally, the trend was only popular with small independent medical spas. Recently, however, some national luxury hotel chains such as Hiltons have jumped on the bandwagon, primarily targeting affluent consumers (Heckel, 2018).

In Colorado, Ketanga Fitness Retreats offer a new type of trip: a 4-day, 3-night cannamaste (cannabis + namaste) package. The main purposes of this retreat are to educate consumers highlighting the benefits of marijuana in the leisure activities like yoga practice. Also, it is designed to debunk the negative myths surrounding the marijuana culture. The participants will enjoy yoga sessions twice daily, join a mountain hiking excursion, visit a local marijuana shop, learn how to cook with cannabis, including a CBD workshop, and dine at a top Denver restaurant with a hemp farm tour as an option (Homan, 2019).

Luxury beauty retailers are also aggressively responding to increasing demands from their target markets with upscale marijuana products (Ackeret, 2019). Recently, Barneys New York, the luxury specialty retailer, announced that it will launch "The High End" shop, selling CBD products, vaporizer pens, and pastilles. It will be the first major retailer capitalizing on a luxury marijuana lifestyle and wellness concept (Tyko, 2019).

The notable shift in marijuana product mix occurred in both markets between 2014 and 2017 has also contributed to wellness prosperity. As Figure 22.3 illustrates, the proportion of flower sales in the medical market decreased from 74.5% to 61.2%, while concentrate sales grew steadily from 14.0% to 28.3% in the medical market. These numbers demonstrate the popularity and continuous growth of concentrates. Considering that concentrates are more potent and cover a variety of marijuana products, MMJ patients can save money by taking less dosage to achieve desired health effects. Concentrates are also consumed in unique forms, such as vaporized pens, creams, and ointments; thus, they do not need to be smoked and avoid the issues associated with that method of administration, which may also appeal to certain demographics including female users (Evergreen Organix, 2018).

Many consumers report that they can achieve desired outcomes by using significantly less concentrates compared to traditional flowers. This provides more convenience with less time spent on actual consumption but with more relief. Additionally, concentrates have longer shelf lives than flower, offering more convenience and economical value. Even though the price per gram for concentrates is higher than flower, its potency and effectiveness due to high cannabinoid content make up for the differential (Ward, 2017).

22.5.2 TAPPING FEMALE CONSUMER BASE

In conjunction with a growing health and wellness market, marijuana market has also been tapping a new demographic segment by offering more diverse products and services. Traditionally, either a broad audience or male consumers were mainly targeted by most marijuana brands as they are either seeking relaxation in the recreational market or pain relief in the medical market. While many companies have been successful with these strategies, the cannabis consumer coalition (CCC) recently reported that women may consume more marijuana than men (Allway, 2018). In addition to selling dried marijuana, some marijuana companies in Canada have repositioned its marketing focus on the female demographic by developing its new brands and products, such as nutraceuticals, cosmetics, and wellbeing products using a white-labeling facility (MPG, 2018).

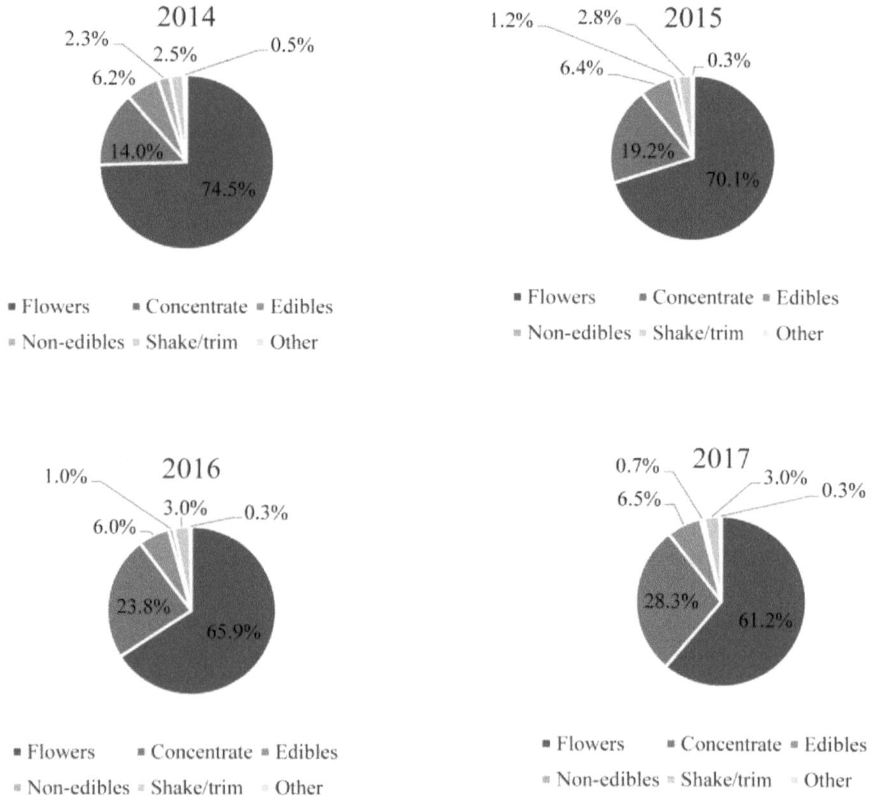

FIGURE 22.3 Medical marijuana: Product type market share (2014–2017).
Source: MPG (2018).

In the MMJ market, a small but highly engaged client base with unique needs is actively sought after as the patient base is also evolving demographically (McVey, 2018). Moreover, there was a steady increase in female patients in both Colorado and Oregon after the launch of recreational sales, accounting for 38% of all marijuana patients in Colorado and 43% in Oregon. That represents a 5% increase in Colorado compared to the launch of recreational sales and a 3% uptick in Oregon. (McVey, 2018).

22.5.3 TREATMENT FOR MORE ILLNESS AND MEDICAL REFUGEES

Current law allows MMJ use for common ailments like arthritis, cancer, chronic pain, Cronhn's disease, epilepsy, glaucoma, HIV/AIDS, multiple sclerosis, and nausea to alleviate patient conditions (Marijuana Doctors, 2019). Nearly all Colorado MMJ patients (94%) were prescribed the drug to relieve severe pain (Belenko and Spohn, 2014). However, further investigation on more treatments with marijuana has been demanded as a pressing issue. Recently, in February 2019, Colorado's House passed a bill allowing MMJ use to treat autism disorders, which permits minors to be registered at Colorado's MMJ registry. The law will also encourage more medical research on the benefits of marijuana for more illnesses, including dementia, ovarian cancer, and other medical conditions (Colorado House OKs, 2019). However, unlike Colorado, some conservative states continue to keep its strict stance on MMJ and are reluctant to expand its usage beyond what is already spelled out in its laws or to encourage more research experiments (DePhillis, 2018).

Due to different legal environment, some patients have moved to another state with a more liberal regulation. There was a term coined referring to those people who migrate to another state because of the availability of the legal drug or medical procedures that are prohibited in their original residence, medical refugees (Medical Marijuana Refugee, 2018). Especially some families with ill children may consider moving to one of the seven states where MMJ are allowed for children patients (DePhillis, 2018). It is still unknown how many families have migrated to Colorado because of MMJ (Garcia, 2016).

22.6 CHALLENGES

22.6.1 STIGMA

Despite overwhelming public support, the stigma surrounding marijuana remains. It is mainly due to the fact that the federal government still prohibits simple possession of marijuana. For example, CBS declined a commercial advertising the benefits of MMJ during the 2019 Super Bowl, where a veteran with combat injuries and a child with seizures were featured as an advocacy campaign (Broker, 2019). The stigma that survives will be one of the important factors that determine whether marijuana becomes an accepted habit of mainstream life in the U.S.

Recently, the stigma is slowly changing across the U.S., and this could be due in part to the role of pharmacist inclusion in dispensing marijuana. Connecticut was the first state to mandate that a registered pharmacist be onsite at all MMJ dispensaries. Other states have followed suit, including Arkansas, Minnesota, New York, and Pennsylvania. The professional engagement of a pharmacist in the process of MMJ could be a huge step in reducing stigma; they can answer specific questions, oversee drug interactions, and counsel patients (Medical Marijuana and Cancer, 2019).

22.6.2 EDUCATING CONSUMERS

Much of the marijuana being produced in Colorado is more potent than marijuana from only a few years ago. Compared to the 1990s, THC levels of some commercial products, marijuana's psychoactive component, have increased by more than 200 times (Jones, 2018). Thus, the public education targeting tourists on potential harms of marijuana products has been one of the biggest challenges since legalization in Colorado (Kang et al., 2016). Other challenges involve inculcating the perception that marijuana is not completely harmless, and careless usage could result in adverse health effects (Noel, 2016). Five years after Colorado's legalized marijuana for adult use, a study at one large Denver hospital reported inhaled marijuana to be a culprit for the most serious problems along with edibles, marijuana-infused foods, and candies. Furthermore, since the sales of recreational marijuana in 2014, emergency room visits related to marijuana cases have been increased by 300% (Johnson, 2019). Future studies, therefore, need to investigate how marijuana providers and suppliers adapt themselves to develop sustainable relationships with its stakeholders, including visitors, residents, law enforcement, and state and local entities (Kang et al., 2016).

22.6.3 TRAVELING PATIENTS

From the tourism perspective, one important issue that MMJ legal states or nations should address is whether they would accept out-of-state/out-of-country authorizations so that visitors can purchase MMJ in destinations. MMJ laws and regulations in the U.S. have been enforced differently over their scope and implementation, including the regulation of dispensaries. Some states only allow terminally ill patients to legally use marijuana, for example, while others are much liberal. This is a list of states that have recognized authorizations issued by other states with a reciprocity program for out-of-state MMJ patients with valid cards in 2018: Alaska, Connecticut, Delaware, Hawaii, Nevada, Rhode Island, and Puerto Rico (State Medical Marijuana Law, 2019). Once more states legalize marijuana for medical purposes, it is expected to have more standardized policies and procedures, crafting their own rules and policies with respect to their own constituents. Although some states are allowed to sell MMJ to out-of-state patients, this is still entirely up to the dispensary owners' discretion to accept it or not, especially in stricter states such as Rhode Island and Michigan. Therefore, it should be acknowledged that there is significant variation in accepting MMJ prescription in different states (Cerda, Wall, Keyes, Galea, and Hasin, 2012).

22.6.4 DECLINING PRICES

According to the MPG 2017 Market Update Report in 2018, in the past 4 years, prices for MMJ products have dropped at a similar pace with the recreational use market in terms of percentage decline. Between 2014 and 2017, average medical flower prices declined 40.9%, from $5.55 per gram to $3.28. Over the same period, medical concentrate prices changed from $25.83 per gram to $16.89, showing a 34.6% drop, and the average price for a 100-mg medical infused edible was assessed approximately $9.00 (MPG, 2018).

The rapidly declining product prices in both markets have several implications for consumers, producers, and governments. Lower prices will generate more overall demand and total sales from a consumer's perspective but may also increase addiction and dependency rates. For marijuana producers and retailers, their profit margins will be negatively affected as prices fall, especially for the less-efficient and often small operators. Consequently, the governments will experience a decline in per unit tax revenues as sales tax revenues are based on retail prices. Consequently, with slower market growth and price decline, tax revenues will eventually plateau (MPG, 2018).

In Colorado, the MMJ market still makes up a major component of the overall marijuana market in certain parts of Colorado, such as the Front Range and Denver Metro areas. However, this market may encounter serious operational challenges since the recreational use market provides much easier access except proof of age, while medical patients must qualify and apply for a medical card. Still, the lower pricing trends and numerous MMJ centers can be attractive to the patient base that lives closely or travel to the more developed Front Range and Denver medical dispensaries.

22.7 CONCLUSION

This chapter offers an overview of MMJ tourism in Colorado, the first state that legalized recreational marijuana in the U.S. Because the legalized part of the industry is unprecedented, few states have addressed tourism issues. Along with residual stigma against marijuana use, the legal uncertainty means that the promotion of marijuana tourism is not a top priority for many destination marketers (Kang et al., 2018).

MMJ tourism market is considered as a SIT that hospitality and tourism operators can capitalize on as a niche segment. With recent trends in the MMJ market, there are several opportunities and challenges facing the MMJ tourism segment. Regardless, when it comes to policy and regulations surrounding MMJ tourism, a more systematic approach guided by objective and valid data will be helpful for policymakers and industry professionals to understand this continuously evolving industry. There is no doubt as legalization continues, MMJ tourism will become a more studied field (Giannotti, 2017). Even though it is premature to conclude that marijuana tourism will be a mainstream theme in travel and tourism soon, but it may serve as an additive offering for some destinations. Therefore, targeting specialized or niche market and offering unique services as a SIT sector, instead of competing with recreational counterpart, would be a better positioning strategy for a sustainable growth in the era of "Green Rush."

KEYWORDS

- **Cannabis Consumers Coalition**
- **Colorado Tourism Office**
- **medical marijuana**
- **medical marijuana (cannabis)**
- **medical refugees**
- **MMJ trends**

REFERENCES

Ackeret, A., (2019). *Beauty High: Cannabis is the Next Big Trend in Luxury Beauty*. Robb Report. Retrieved from: https://robbreport.com/muse/beauty-wellness/cannabis-luxury-beauty-2839770/ (accessed on 14 July 2021).

Allway, R., (2018). *Trends to Watch as Cannabis Moves into the Health and Wellness Market*. CFN media group. Retrieved from: https://www.cannabisfn.com/trends-watch-cannabis-moves-health-wellness-market/ (accessed on 14 July 2021).

Associated Press, (2019). *Registered Medical Marijuana Patients Dropping Fast in Oregon*. The Oregonian. Retrieved from: https://www.oregonlive.com/marijuana/2018/11/oregpm.html (accessed on 14 July 2021).

Belenko, S., & Spohn, C., (2014). *Drugs, Crime, and Justice*. Sage Publications: Thousand Oaks, CA.

Belhassen, Y., Santos, C. A., & Uriely, N., (2017). Cannabis usage in tourism: A sociological perspective. *Leis. Stud., 26*(3), 303–319.

Beth, M., (2018). *Colorado Sprinkles its Tourism Market with Cannabis: Is There a Lesson for Michigan?* South bend tribune. Retrieved from: https://www.southbendtribune.com/news/local/colorado-sprinkles-its-tourism-market-with-cannabis-is-there-a/article_d7e3c164-add8-5613-9485-16625e9ab1fc.html (accessed on 14 July 2021).

Broker, T., (2019). *The Stigma on Cannabis*. Above the law. Retrieved from: https://abovethelaw.com/2019/01/the-stigma-on-cannabis/ (accessed on 14 July 2021).

Caulkins, J. P., Hawken, A., Kilmer, B., & Kleiman, M., (2012). *Marijuana Legalization: What Everyone Needs to Know*. Oxford University Press: New York.

Cerda, M., Wall, M., Keyes, K. M., Galea, S., & Hasin, D., (2012). Medical marijuana laws in 50 States: Investigating the relationship between state legalization of medical marijuana and marijuana use, abuse, and dependence. *Drug Alcohol Depen., 120*(1–3), 22–27.

Colorado House OKs., (2019). *Colorado House OKs Bill on Medical Marijuana Use for Autism*. The Aspen Times. Retrieved from: https://www.aspentimes.com/news/colorado-house-oks-bill-on-medical-marijuana-use-for-autism/ (accessed on 14 July 2021).

DePhillis, L., (2018). *Louisiana is Trying to Keep Medical Marijuana Medical. It's Harder Than It Sounds*. CNN Health. Retrieved from: https://www.cnn.com/2018/11/27/health/louisiana-medical-marijuana/index.html (accessed on 14 July 2021).

Disa Global Solution, (2020). *Map of Marijuana Legality by State*. Disa Global Solution. Retrieved from: https://disa.com/map-of-marijuana-legality-by-state (accessed on 14 July 2021).

Evergreen Organix, (2018). *Flower vs. Concentrates: Is One Better Than the Other*? Evergreen Organix. Retrieved from: https://www.evergreenorganix.com/cannabis-blog/2018/8/16/flower-vs-concentrates-is-one-better-than-the-other (accessed on 14 July 2021).

Garcia, A., (2016). *Do People Move to Colorado for Legal Weed? Probably Not, But No One Knows for Sure*. Denvrite. Retrieved from: https://denverite.com/2016/07/07/people-move-colorado-for-pot/ (accessed on 14 July 2021).

Giannotti, S., (2017). A cottage industry emerging in medical tourism: Medical cannabis. In: DeMicco, F., (ed.), *Medical Tourism and Wellness: Hospitality Bridging Healthcare (H2H)* (pp. 257–276). Apple Academic Press: Waretown, NJ.

Hartig, H., & Geiger, A., (2018). *About Six-in-Ten Americans Support Marijuana Legalization*. FACTANK. Retrieved from: https://www.pewresearch.org/fact-tank/2018/10/08/americans-support-marijuana-legalization/ (accessed on 14 July 2021).

Heckel, A., (2018). *The Coolest Cannabis-Inspired Spa Treatment in Colorado.* Tripsavvy. Retrieved from: https://www.tripsavvy.com/cannabis-inspired-spa-treatments-colorado-4155862 (accessed on 14 July 2021).

Homan, J., (2019). *Would You Try a Luxury Yoga Cannabis Retreat?* Jetsetter. Retrieved from: https://www.jetsetter.com/magazine/cannabis-yoga-retreat/ (accessed on 14 July 2021).

Horn, J., (n.d.). *The Brave New World of Cannabis Tourism*. Travel channel. Retrieved from: https://www.travelchannel.com/interests/food-and-drink/articles/the-brave-new-world-of-cannabis-tourism (accessed on 14 July 2021).

Johnson, C. K., (2019). *Marijuana ER Visits Climb in Denver Hospital Study.* Fox25. Retrieved from: https://okcfox.com/news/local/marijuana-er-visits-climb-in-denver-hospital-study (accessed on 14 July 2021).

Jones, R., (2018). *This is How Legal Marijuana Changed Colorado and Could Change Michigan.* 7abc WXYZ Detroit. Retrieved from: https://www.wxyz.com/news/this-is-how-legal-marijuana-changed-colorado-and-could-change-michigan (accessed on 14 July 2021).

Kang, S. K., & Lee, J. S., (2018). Support of marijuana tourism in Colorado: A residents' perspective using social exchange theory. *J. Destin Mark Manage, 9*, 310–319.

Kang, S. K., O'Leary, J., & Miller, J., (2016). From forbidden fruit to the goose that lays golden eggs: Marijuana tourism in Colorado. *SAGE Open, 6*(4). doi: 10.1177/2158244016679213.

Kovacevich, N., (2018). *The Next Big Thing in Cannabis: Tourism*. Forbes. Retrieved from: https://www.forbes.com/sites/nickkovacevich/2018/08/16/the-next-big-thing-in-cannabis-tourism/#618fd6855d9b (accessed on 14 July 2021).

Legality of Cannabis by U.S. Jurisdiction. Wikipedia, (2020), Retrieved from: https://en.wikipedia.org/wiki/Legality_of_cannabis_by_U.S._jurisdiction (accessed on 14 July 2021).

Marijuana Doctors, (2019). *Medical Conditions Treatable with Marijuana*. Marijuana Doctors. Retrieved from: https://www.marijuanadoctors.com/conditions/ (accessed on 14 July 2021).

Marijuana Sales, (2019). *Colorado Department of Revenue*. Retrieved from: https://www.colorado.gov/pacific/sites/default/files/0219_MJSalesCalendarReport_PUBLISH.pdf (accessed on 14 July 2021).

Marijuana Tax Data, (2019). *Colorado Department of Revenue*. Retrieved from: https://www.colorado.gov/pacific/revenue/colorado-marijuana-tax-data (accessed on 14 July 2021).

McVey, E., (2018). *Chart: Can Medical Marijuana Programs Survive in States with Recreational Markets?* Marijuana Business Daily. Retrieved from: https://mjbizdaily.com/chart-can-medical-marijuana-programs-survive-in-states-with-recreational-markets/ (accessed on 14 July 2021).

MED Licensed Facilities, (2019). *Colorado Department of Revenue*. Retrieved from: https://www.colorado.gov/pacific/enforcement/med-licensed-facilities (accessed on 14 July 2021).

Medical Marijuana and Cancer, (2019). *Fight Colorectal Cancer*. Retrieved from: https://fightcolorectalcancer.org/blog/medical-marijuana-and-cancer/ (accessed on 14 July 2021).

Medical Marijuana Refugee, (2018). *What is a Medical Marijuana Refugee?* Presto doctor blog. Retrieved from: https://prestodoctor.com/content/cannabis-industry/medical-marijuana-refugee (accessed on 14 July 2021).

Medical Marijuana Registry, (2018). Patient numbers. *Colorado Department of Public Health and Environment*. Retrieved from: https://www.colorado.gov/pacific/sites/default/files/CHED_MMR_2001-2017-average-count-year.pdf (accessed on 14 July 2021).

MPG, (2018). *Market Size and Demand for Marijuana in Colorado 2017 Market Update*. Marijuana Policy Group. Retrieved from: https://www.colorado.gov/pacific/sites/default/

files/MED%20Demand%20and%20Market%20%20Study%20%20082018.pdf (accessed on 14 July 2021).

Noel, J., (2016). *Colorado Struggles to Educate Marijuana Tourists*. Chicago Tribune. Retrieved from: http://www.chicagotribune.com/lifestyles/travel/ct-pot-tourism-travel0224-20160224-story.html (accessed on 14 July 2021).

Nussbaum, A. M., Boyer, J. A., & Kondrad, E. C., (2011). But my doctor recommended pot: Medical marijuana and the patient-physician relationship. *J. Gen. Intern. Med., 26*(11), 1364–1367.

Quinnipiac University National Poll, (2017). *U.S. Voters Say 68–27% Let Transgender People Serve, Quinnipiac University National Poll Finds; Voters Disapprove 5-1 of GOP Handling of Health Care* [Press release]. Retrieved from: https://poll.qu.edu/images/polling/us/us08032017_Ujm62prt.pdf/ (accessed on 14 July 2021).

Rough, L., (2017). *Which U.S. States Accept Out-of-State Medical Marijuana Authorizations?* Leafly. Retrieved from: https://www.leafly.com/news/cannabis-101/which-us-states-accept-out-of-state-medical-marijuana-authorizati (accessed on 14 July 2021).

State Medical Marijuana Laws, (2019). *National Conferences of State Legislatures.* Retrieved from: http://www.ncsl.org/research/health/state-medical-marijuana-laws.aspx (accessed on 14 July 2021).

Tham, A., (2018). Sand, surgery, and stakeholders: A multi-stakeholder involvement model of domestic medical tourism for Australia's sunshine coast. *Tour. Manag. Perspect., 25*, 29–40.

Tyko, K., (2019). T*he High End: Barneys to Open 'Cannabis Lifestyle and Wellness' Shop in Beverly Hills*. Herald-mail media. Retrieved from: https://www.heraldmailmedia.com/news/usa_today/the-high-end-barneys-to-open-cannabis-lifestyle-and-wellness/article_fb155917-71b6-555f-9f35-042384f74723.html (accessed on 14 July 2021).

Ward, A., (2017). *Are Cannabis Concentrates Becoming More Popular than Flower?* Potguide. Retrieved from: https://potguide.com/pot-guide-marijuana-news/article/are-cannabis-concentrates-becoming-more-popular-than-flower/ (accessed on 14 July 2021).

CHAPTER 23

MEASURING QUALITY OF HEALTHCARE ACROSS BORDERS: THE ROLE OF INTERNATIONAL HEALTHCARE ACCREDITATION IN MEDICAL TRAVEL

ELIZABETH ZIEMBA and CLAUDIA MIKA

Temos International Healthcare Accreditation, Cologne, Germany

ABSTRACT

In this chapter, we address what is accreditation is healthcare; the importance of quality and accreditation and working together to create a better healthcare; creating a comprehensive international patient services; and the role of innovation, transparency, and leadership during a crisis.

23.1 INTRODUCTION

23.1.1 QUALITY AND ACCREDITATION: WORKING TOGETHER FOR BETTER HEALTHCARE

When asked, "How important is the quality of healthcare to you?" the majority of people answer, "Very important!" When asked, "How do you know if the healthcare services you receive are of high quality?" most people have difficulty answering that question. One mark of excellent quality services is to look for the seal from the best international accreditation organizations.

When medical tourists are asked, "Why do you travel abroad for medical treatment?" typical answers include, "I do not want to travel abroad but the waiting time is too long" or "I cannot afford the treatment here in my country" or "This special cancer treatment is not available in my country." Perhaps situations where dental or medical treatment abroad can be combined with vacation or family visits, the motivations may be different such as "I have the time to take care of myself," "The local prices are great!" or "My family or friends have gotten treatment here and it was excellent." For most people, leaving their own familiar environment and their usual healthcare system is a challenge.

Individuals who choose treatment abroad have different needs than local patients. Finding the right healthcare provider can be a challenge. Many people search the Internet for information about the providers and the services they offer. Often testimonials and "before and after" photos are used to help people assess "quality." While this information may be helpful, it takes experience, knowledge, and qualifications to accurately measure and assess the level of services offered by healthcare providers. That is the role of accreditation organizations.

23.2 THE PATIENT JOURNEY WHEN TRAVELING ABROAD FOR TREATMENT

Patients have unique needs when selecting treatment abroad. Pre-treatment services like sharing of medical information and virtual consultations, onsite healthcare treatment and procedures, and post-treatment care, including follow-up examinations, take place in different countries. Coordinating care across borders requires special knowledge that many providers do not possess if they do not usually treat international patients. These "extra" services must be managed professionally to assure high quality and cost-effective personalized services as well as medical care from beginning to end of the treatment process. It is challenging to know if the healthcare provider can deliver a high-quality level of service for international patients.

Healthcare systems around the world are unique, and some are better than others. Within each country, the quality of services may differ from provider to provider, depending on how well each country's government regulates its healthcare services. The medical service structure depends on a variety of factors, including the overall economy, including the standard of living as well regulatory, financial, and organizational conditions. Healthcare services are influenced by the scientific or professional standards that are implemented within countries. Culture, including language and other societal norms also have an impact on healthcare.

Difficulties in international patient management arise as a result of cultural differences and language barriers as well as legal issues, different understanding of medical treatments, or an interruption of cross-border processing. For example, the medical records and documentation used to assess the recommended treatment for a medical condition in one country may be different in another country. Two different approaches may both be acceptable or not. While individual patients are unable to make sense of this situation and others like it, a reputable accreditation organization can.

Quality in nursing and medical treatments as well as patient safety are preconditions to attract patients and are to be fulfilled by all healthcare organizations worldwide. Whereas "word of mouth" of the family physician, relatives, friends, or colleagues or credentialing and awarding of physicians or hospitals on the national level are important factors for the decision-making process on the local level, this information is difficult to obtain on an international level.

The question for the medical traveler is how to find the best cost-benefit solution, high-quality treatment, and respective services abroad?

For healthcare organizations worldwide, the question is how to attract patients? It is not sufficient to claim to be "the best" when there is an internationally accepted scientific method to verify whether a particular healthcare provider really is as good as it claims. That scientific method is the basis of accreditation.

23.3 WHAT IS ACCREDITATION IN HEALTHCARE?

Accreditation for hospitals and clinics is defined as "a system of self-assessment and external peer process used by healthcare organizations to accurately assess their level of performance in relation to established standards and to implement ways to continuously improve." With the right accreditation organization, the process is a collaborative journey between the healthcare provider and the accreditation partner to help the provider maximize its potential and deliver the best clinical and non-clinical services that it can. It is a validating process which determines whether a healthcare organization meets a series of standards of quality based on internationally accepted best practices, knowledge, and experience.

Different accreditation companies offer different programs that cover various topics involved in the delivery of healthcare services to patients. For example, Temos offers an extremely comprehensive set of standards compared to other international accreditation organizations. It has standards that include the physical plant, administration, professional staffing, staff, and patient safety, risk management, outcome, patient experience, legal, and ethical aspects in patient management as well as sustainability. Healthcare providers that treat international patients may have accreditation from Temos or other international accreditation organizations or no accreditation at all. It is not a requirement to have international accreditation but is a business decision by a healthcare provider to pursue recognition of excellence from a neutral, independent third-party accreditation organization.

23.4 THE ACCREDITATION JOURNEY

Once, a healthcare provider has started its "accreditation journey," it is a never-ending story where all staff members are involved in the continuous quality and service improvement of the hospital or clinic. This journey takes place over time, generally six to 24 months, and evolves over time in a formal process designed to improve the quality and efficiency of the clinical and non-clinical services resulting in better outcomes that can be measured and continuously improved over time.

As an example, Temos accreditation is an 8-step process.

The Temos process starts with a simple registration process that includes the healthcare provider giving information about its services, an agreement for services is signed, and then the hard work begins. The provider receives the standards and goes about the process of demonstrating that it meets those standards or, with the help of the Temos team and its self-evaluation tools, determines what additional work the provider must accomplish to meet those standards. This collaborative process readies the hospital or clinic for the visit by an assessment team to visit and continue the evaluation process in person.

The Temos onsite visit is conducted by a team of experts who may include a clinician such as a doctor, nurse, or dentist as well as an expert in quality assurance or quality management. Temos has a team of highly qualified assessors who speak numerous languages and who have experience in more than 100 countries. The assessor team shares its knowledge and experience as well as its cultural sensitivity when evaluating the healthcare provider. This collaborative, culturally flexible approach is unique to Temos.

Following the onsite visit, the hospital or clinic may receive suggestions or recommendations to improve services or may demonstrate that it has fully complied with the standards and the accreditation is awarded. The accreditation is valid for a three-year period; however, once a year, the provider must deliver updates to confirm its continuing compliance with the standards as well as report any major changes to its services or governance.

There is even an accreditation process for accreditation organizations like Temos. There is an international renowned accreditation body, the International Society for Quality in Health Care External Evaluation Association (IEEA/ISQua), that evaluates the work of international accreditation organizations to determine whether their systems and processes are at the highest possible levels. ISQua/IEEA accreditation is only awarded to those accreditation organizations that complete the challenging process. Temos is proud to be ISQua/IEEA accredited.

In summary, accreditation is a neutral, independent process to objectively assess, evaluate, and validate the quality of clinical and non-clinical services. It strives to assure optimal care, operational efficiency, and fair prices for services as well as the highest possible patient safety. Overall, the goal of accreditation is to improve the overall quality of care offered to all patients.

23.5 INTERNATIONAL PATIENT SERVICES: HOW COMPREHENSIVE ARE THEY?

For international patients, the journey begins with a search on the Internet to find a provider that matches their clinical and non-clinical needs. "Non-clinical" services may include assistance coordinating travel to the country where treatment will be delivered, in-country transportation once the patient arrives, accommodations, and other related hospitality or logistical services. Some hospitals and clinics do not offer any of these non-clinical services. Some healthcare providers offer these services themselves while others out-source the non-clinical services. Depending on the needs of the prospective patients, they may need a little or a lot of help planning the non-clinical as well as the clinical components of the patient journey. The outcomes of the treatment abroad depend in part of the successful coordination of all of these components of the patient journey.

Different international accreditation programs may have no, some, or comprehensive standards that evaluate the patient journey, including non-clinical services from beginning to end. It is important to understand which part of the patient journey is associated with each different international accreditation program. For example, JCI, a popular international accreditation program for hospitals, does not have any standards that review the non-clinical services used by international patients. In contrast, Temos offered the first and most comprehensive set of clinical and non-clinical standards that includes the end-to-end patient journey.

Accreditation of the end-to-end patient journey is needed to assure the complete continuum of care from the first contact between the patient and the healthcare organization to the last element of post-discharge follow-up.

The international patient care cycle is divided into five stages: (i) pre-travel; (ii) travel; (iii) onsite treatment; (iv) discharge; and (v) follow-up segments. The focus in the pre-travel, travel, and follow-up phases is on non-clinical service provision, while the onsite treatment and discharge stages are mainly based on the quality of the medical services and the clinical outcomes.

At this time, there is only one accreditation body worldwide-Temos-that covers the continuum of care for patients from all over the world, including medical and non-clinical services for domestic and international patients at each step of the process in one accreditation program.

With its ISQua/IEEA accredited standards "excellence in medical tourism" and "International Quality in Medical Care," Temos International Healthcare Accreditation fills a gap in international accreditation and supports hospitals and clinics worldwide in their continuous quality improvement initiative from the first contact to the follow-up. Hospitals and clinics that are Temos accredited have successfully completed the accreditation process demonstrating their commitment to excellence and quality of the medical and patient experience services that they offer.

23.6 WHAT ACCREDITATION IS NOT

Consumers, hospitals, and clinics have common misconceptions about accreditation. This confusion results from lack of knowledge about the accreditation process itself, which may appear to be more complex than it really is, false hopes or expectations about what accreditation can achieve, or a reluctance to change the way of delivering services "because we have always done it this way." Here are the facts about what accreditation is not and what it can do.

First, international accreditation is not a guarantee that everything that the accredited hospital or clinic does will be perfect. Human beings make mistakes, and the delivery of healthcare services is complex. Accreditation is confirmation that the healthcare provider is capable of providing services according to the standards offered by the accrediting organization. That means that the provider has the capacity to reduce errors and maximize the likelihood of positive outcomes.

Second, accreditation is not completing a checklist of items. Accreditation is a holistic approach to ensuring that the systems and processes within a clinic or hospital are integrated and functioning optimally to maximize the benefits to patients, staff, and the organization itself. For example, a checklist approach wants to see that the hospital has fire extinguishers. An accreditation approach wants to know if the fire extinguishers work, if people know where they are located, who knows how to operate them, and so on.

Third, accreditation is not an automatic guarantee that a hospital or clinic will get more patients. Accreditation should increase the overall quality of clinical and non-clinical services offered by the hospital or clinic. The better the services, the more likely that patients will want to go to that particular hospital or clinic for treatment. The accreditation process is designed to help hospitals and clinics achieve their best possible services. With improved services, more patients should find their way to the best quality of care.

23.7 QUALITY, ACCREDITATION, AND THE PATIENT JOURNEY

Quality of healthcare services is difficult to measure, especially for people who are not experts. Relying on "word of mouth" or testimonial is one way to get information about the services offered at hospitals and clinics around the world. Finding healthcare providers who have achieved international accreditation is one measure of quality. Looking for hospitals and clinics that have achieved ISQua/IEEA accreditation is an indication that the medical services should be of high quality. Looking for Temos accreditation is an indication that the healthcare provider wants to offer excellent medical and non-clinical services for a seamless end-to-end patient journey.

23.7.1 COVID-19 AND ACCREDITATION STANDARDS

The emergence of the highly infectious and contagious coronavirus disease, COVID-19, has turned the world upside down, impacting the delivery of all healthcare services. Since the earliest cases in December 2019, mortality rates have dropped as detection and treatment have improved, but the number of cases has skyrocketed across the globe. While our understanding of this disease increases and potential vaccines emerge, certain lessons have already been learned.

One of the key lessons learned is the importance of hospitals and clinics adhering to international best practices, rigorous scientific research, and practical experience to create and share systems and protocols to reduce the risk of transmission of COVID-19. The scientific and medical communities are engaged in these collaborative endeavors. Temos is playing its part in making those best practices readily available. Here is a look at how Temos is responding to the COVID-19 crisis with expertise, innovation, focus, and heart.

23.7.1.1 REJECTING THE VIRTUAL OPTION FOR ACCREDITATION

When borders closed, causing the postponement of onsite accreditation assessments, the Temos team briefly discussed and quickly dismissed the idea of offering virtual accreditation assessments. With more than 400 standards to be evaluated and assessed, the onsite visit is an indispensable part of the process for shared learning, relationship building, and experiencing the hospital or clinic operations first-hand. No virtual experience can replace what assessors see, hear, smell, taste, and feel just as patients would. The decision was made to postpone onsite visits until travel can safely resume in order to maintain its high standards rather than compromise the process that all clients deserve.

23.7.1.2 CLOSING THE BUSINESS-NOT AN OPTION

Pausing onsite accreditation visits means lost revenue and significant financial consequences. QHA-Trent accreditation made its difficult business decision to close its doors in October 2020 after more than a decade in business. The leadership of QHA contacted Temos and recruited its team to help QHA client's transition to Temos accreditation. The two organizations are working together to ensure that QHA-Trent accredited providers and the patients they serve are well cared for by Temos.

23.7.1.3 INNOVATION: THE WAY FORWARD

The severity of the COVID-19 situation quickly became apparent to the Temos team. There was an immediate need for solutions and guidance among hospitals and clinics to assure patients and staff that the risk of transmission was being reduced. The Temos response is its "COVID-19 Safe" certification program.

Rigorous review of best international practices, available research, and the team's experience in other pandemic and emergency situations was rapidly distilled and standards created to form the foundation of Temos' first of its kind "COVID-19 Safe" program. Temos' unique online tools, used by clients to measure their compliance with each standard, were adapted to the "COVID-19 Safe" certification program. Then some business decisions were made.

23.7.1.4 TRANSPARENCY-LEADERSHIP DURING A CRISIS

Temos is the first accreditation organization to make the standards for its COVID-19 program free of charge to anyone, anywhere who wanted to download them. Doing the right thing and helping combat the pandemic are more important than profiting from the standards or denying access to providers that need the information.

Second, Temos' COVID-19 Safe certification program offers a paid option for clients who can engage the full array of Temos' services, including access to its online tools, international team of experts, and review of compliance with the program's 60+ standards. A virtual visit was developed exclusively for the COVID-19 Safe certification program to ensure compliance without compromising the integrity of the program. With its focus on minimizing the risk of COVID transmission, the team decided that the virtual visit, photos, documents, and a virtual tour assured an accurate assessment of compliance. By making the standards free and available to all, Temos took a calculated risk.

The COVID-19 certification program has been very successful-the standards have been downloaded by 170 organizations on six continents. Further, dozens of hospitals and clinics are engaged in the Temos certification process. New providers continue to sign up for the COVID-19 Safe program as it becomes apparent that the virus is going to be with us for a long time.

Third, the decision was made to continuously update the standards for free as more is learned about this disease. Temos recently issued version 3.0 of the COVID-19 Safe standards which are automatically forward to our clients. This process keeps our clients, assessors, and regional office affiliates informed of the latest research and developments.

Fourth, new standards—telehealth and telemedicine as well as crisis communications—were included in the COVID-19 Safe program. Temos clients were quickly creating or improving telehealth and telemedicine services and looking to Temos for guidance regarding international best practices. Our team of experts focused on the essential standards for these evolving service needs and launched them in this program. Based on the feedback from clients, these standards are vitally important now and for the future. While introduced as part of our COVID-19 Safe program, they are being incorporated into all of our accreditation programs.

23.7.1.5 POTENTIAL USE AND ABUSE

The risk of the program's standards being used without permission or copied by competitors was accepted as part of the company's commitment to corporate social responsibility. Both of these scenarios have happened. As the saying goes, "Imitation is the sincerest form of flattery." The Temos team is duly complemented and committed to leadership as well as innovation regardless of what others do. Patient safety comes first.

23.7.1.6 LONG-TERM CHANGES: TELEHEALTH AND TELEMEDICINE

The demand for telehealth and telemedicine services has skyrocketed as a response to COVID-19. Many hospitals and clinics canceled elective procedures to reduce the number of people who could be exposed to the coronavirus. Patients stopped going to hospitals and doctors' offices for fear of exposure. Instead, providers and patients rapidly adapted virtual visits to assess the patients' needs and discovered that many issues could be handled without the time and inconvenience of an in-person visit.

What started as a response to the pandemic has now been widely accepted by providers and patients as time-saving, efficient, and lower cost alternatives to in-person visits. There are issues that must be addressed to ensure that telehealth and telemedicine services protect the privacy and security of the patients' information. Rules, regulations, and statutes must be created, updated, and harmonized within countries and across borders to ensure that the services are being delivered by duly licensed healthcare professionals. Getting healthcare providers and systems to "talk to each other" requires compatibility of electronic health records and other technological challenges to be resolved.

Temos has contributed to this trend by developing basic standards for the implementation and use of telehealth and telemedicine services. These standards are based on universally accepted principles, allowing hospitals and clinics to implement them according to their healthcare and technological systems. As telehealth and telemedicine systems evolve, Temos will continue to update and expand its accreditation standards to ensure patient privacy and confidentiality.

23.7.1.7 IMPACT ON THE PATIENT EXPERIENCE

Healthcare providers are challenged to respond to the two new patient experiences that COVID-19 has created. The first new patient experience for many people in the virtual visit available through telehealth and telemedicine services. Most patients are not used to communicating virtually with doctors and most doctors are not used to communicating with patients virtually. A steep learning curve is being climbed by patients and providers alike.

The second new patient experience is the in-person visits with healthcare providers from receptionists to nurses to doctors and dentists who are dressed in personal protective equipment (PPE). With masks, shields, full sets of scrubs, and other gear, healthcare providers look more like astronauts dressed to land on Mars rather than deliver care. This new experience adds stress and unfamiliarity to an already stressful situation.

Temos was among the first international accreditation companies to include healthcare standards for delivering a superior patient experience. It continues to innovate and address the evolving patient experiences that are now being delivered.

COVID-19 has changed the way healthcare services are being delivered now to reduce the risk of transmission of the disease as well as accelerating changes to the patient experience with wider use of telehealth and telemedicine that will outlast the pandemic. What will remain of vital importance is the adherence to international best practices, rigorous scientific research, and the experience of healthcare professionals who create and deliver accreditation standards, systems, and protocols to assist hospitals and clinics to deliver their best possible care.

23.8 ABOUT TEMOS INTERNATIONAL HEALTHCARE ACCREDITATION

Founded in 2010 as a spin-off of a European scientific project by a governmental German research organization, Temos is the first international healthcare accreditation organization to incorporate standards specifically designed for hospitals and clinics that serve international patients. Since then, the company has evolved into an organization offering the highest standards of accreditation for hospitals and clinics globally, serving both domestic and international patients through its ISQua/IEEA accredited medical and dental standards.

The Temos accreditation process involves an independent assessment of the healthcare provider's policies and procedures evaluated against a backdrop of international best practices and evidence-based standards to help hospitals and clinics provide the best possible clinical and non-clinical outcomes.

Rather than take a one-size-fits-all approach, Temos offers a tailored approach to its accreditation. From single, stand-alone clinics to multi-site hospitals offering complex care, Temos accreditation programs meet the needs of:

- Hospitals offering primary to quaternary care;
- Clinics offering day surgery services;
- Clinics including primary care facilities and medical centers;
- Dental hospitals, clinics, and practices;
- Physical rehabilitation centers;
- Reproductive care clinics;
- Eye clinics.

These distinct accreditation programs are based on a core set of standards and principles that apply to every healthcare Temos accredited provider. Each program is tailored to the particular discipline to address the unique services provided to the patients they serve.

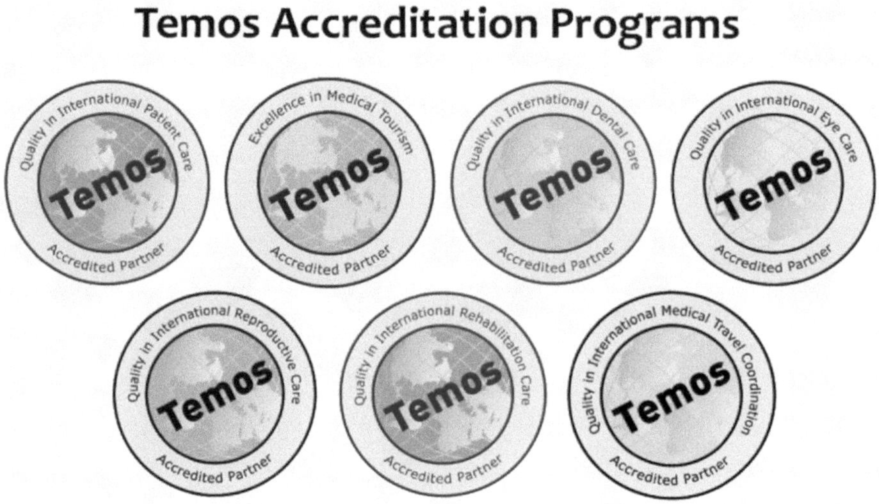

23.9 TEMOS FUTURE VISION

Accreditation is an important part of the process of improving healthcare services to patients around the world. It can raise the standard of care provided to patients in every corner of the world. The Temos team already serves accredited partners on five continents and is ready to do its part to improve access to affordable quality healthcare services for all patients worldwide.

KEYWORDS

- **accreditation journey**
- **health care external evaluation association**
- **healthcare provider**
- **healthcare system**
- **role of accreditation organizations**
- **self-assessment**

CHAPTER 24

HOSPITALITY BRIDGING HEALTHCARE: CAREER OPPORTUNITIES FOR THE FUTURE HOTEL SCHOOL GRADUATE

FREDERICK J. DEMICCO[1] and JOHN LA FORGIA[2]

[1]*The Franke College of Business, School of Hotel and Restaurant Management, Northern Arizona University, USA*

[2]*Mayo Clinic, USA*

ABSTRACT

Hospitality Bridging Healthcare or H2H© services can create a satisfying experience for the patient/guest, which can lead to long-lasting loyalty. Hospitality and healthcare services will play a critical role for success in the entirety of the process. University graduates of hotel and hospitality schools that focus on this intersection of H2H© will be rewarded with challenging and lucrative careers. The path forward includes developing a core curriculum and courses for this promising new H2H© career track. It will be an intersection of business management, healthcare administration, and hotel/hospitality business management theory and practice with an appropriate industry internship with hands-on learning in hospitals, healthcare, and medical centers.

24.1 PROLOG

Hospitality and healing have been intertwined from ancient times. In the parable of the Good Samaritan, a Samaritan came upon a severely beaten man in need of assistance along the roadway. He disinfected the man's wounds with wine and used oil to soothe his pain. After rendering medical assistance, the Samaritan transported the injured man to an Inn for a time of healing and rest.

The roots of the modern healthcare and hospitality industries can be traced back to the late 1800s. Pioneers in healthcare such as the Mayo Brothers traveled the world to learn about the latest advancements in medical care from colleagues in such far-flung places as China, India, and Europe. During their travels, they undoubtedly experienced the best- and worst-of hotel and hospitality offerings. Perhaps they stayed in a hotel run by Cesar Ritz.

These pioneers of healthcare and hospitality developed similar overarching beliefs. For the founders of the Mayo Clinic, it was that "The needs of the patient come first." In the case of Cesar Ritz of Ritz Hotel fame, it was "Le client n'a jamais tort" or "The customer is never wrong."

Today, many healthcare and hospitality organizations are looking to strengthen and broaden their brands. The wise leaders of these enterprises have come to understand that the future belongs to those who can combine the best of hospitality with healthcare delivery to deliver an outstanding experience. Indeed, the secret of success in the expanding world of medical tourism (MT) is the mastery of hospitality and healthcare.

In the book, *Hospitality Bridging Healthcare (H2H©) and Medical Tourism and Wellness* (DeMicco, 2016), we learn of insights about the past, present, and future of healthcare and hospitality. H2H© stresses that consumers expect quality and value from all products and services that they purchase. The concept of *Hospitality Bridging Healthcare (H2H©),* is a valuable cultural philosophy that makes excellence a habit as we strive to delight our patients/guests.

24.2 INTRODUCTION

Since the 1970s, with the advent of managed care, diagnostic Related Groups and other social and policy forces, healthcare providers have faced reimbursement constraints and an increasingly competitive marketplace. These developments and the later advent of digital communications hastened the emergence of the patient as a consumer. Arguably, once patients became consumers, it was inevitable that they would also be seen as customers, or guests. This paradigm shift fostered the now prevalent preponderance of hospitality style service enhancements throughout the healthcare industry.

Healthcare leaders now understand that medical and healthful patient outcomes are most successful when integrated with a hospitality driven process. The pre-planning of the medical journey leading up to the arrival at the medical/health/wellness destination begins a cascade of multiple Hospitality Bridging Healthcare or H2H© services to create an entire satisfying experience for the patient/guest which can lead to positive health and loyalty outcomes, if performance in the entirety of the process is at the highest level. Hospitality and healthcare services working in harmony and cadence will play a critical role for success.

24.3 THE H2H© MODEL

The H2H© model consists of different ways of merging the operations of hospitals with processes and operation from hospitality and hotel services in order to provide superior service quality. First, H2H proposes that hotels and hospitals can be merged operationally. Second, H2H can provide an entire service package for patients seeking treatment that spans the whole service spectrum from hospital to hotel. The third model merges both entities such that they represent one unit.

Some institutions have long recognized the health benefits of the non-medical aspects of caring for patients. At Mayo Clinic, the campus grounds abound with interesting architecture, colorful landscaping, fountains, even a sculpture garden. Likewise, the interior features artwork worthy of a museum. Perhaps more importantly, all staff understand the dictum that "the needs of the patient come first." Satisfaction surveys confirm that patients feel they have been treated as a person, not a "number."

Extending this recognition, our partner and co-authors from the Bad Ragaz in Switzerland administer a patient-centered medical resort equipped with various innovative facilities and services such as aquatic therapy and a spa that help patients experience better care and a faster recovery (Figures 24.1 and 24.2) (DeMicco, Cetron, and Davies, 2018; DeMicco, 2016).

In the future, patients will enjoy hotel-level hospitality while receiving medical services in their room via technological advancements offered by expert staff from the hospitals. The experience of being a patient will be improved considerably if hotels become partners of hospitals in delivering service experience for their customers (patients). For example, at Mayo Clinic, even those who have less than desirable outcomes (e.g., terminal cancer patients) express satisfaction with their overall care (author-observed focus group).

FIGURE 24.1 An example of a Switzerland H2H© model.

Source: https://www.clinic-badragaz.ch/fileadmin/user_upload/Clinic_Patienteninformation_EN.pdf

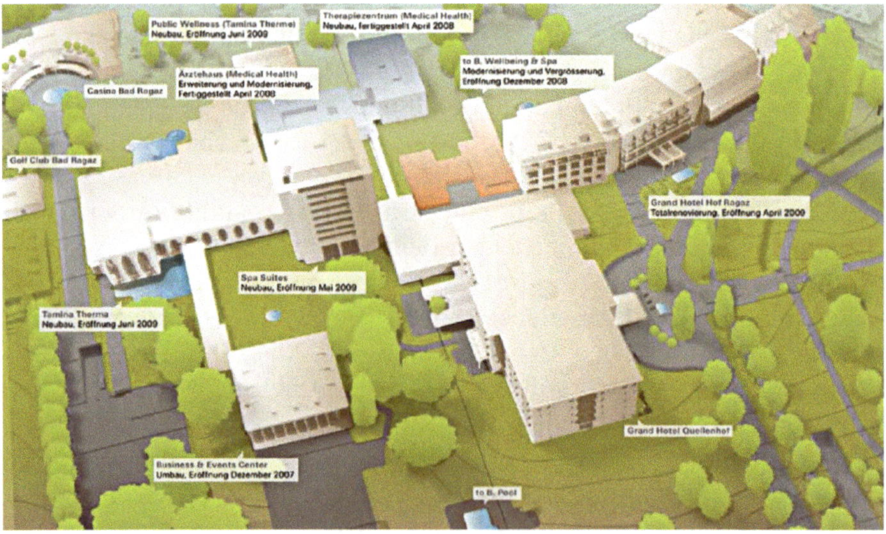

FIGURE 24.2 Swiss H2H©: The Grand Resort at Bad Ragaz Wellbeing and Medical Center in Switzerland.

Source: https://tuor-ag.ch/data/documents/04-2009_Bauten-im-Blickpunkt.pdf

24.4 HEALTHCARE HOSPITALITY MANAGEMENT AS A CAREER

By reinforcing the principles of service excellence and employing some of the tools and suggestions described below, every Hospitality/Hotel and Healthcare Manager ("*Hospitality*" Manager), can significantly strengthen Patient/Guest loyalty.

The key areas of needed focus are:

- Why patient or guest loyalty is very important;
- Why satisfied patients/guests do not equate to loyalty;
- Why each manager and employee have a role to play in the service setting;
- How attitude, courtesy, and compassion are the bedrock for success.

The term "healthcare" comprises hospitals, medical campuses (many with dedicated hotels), clinics, retirement communities, nursing homes, and life care and continuing care facilities. Approaching almost $4 trillion annually expenditure, healthcare is a part of the U.S. economy that shows no signs of slowing down. Given the aging population, the need for all services in the "healthcare" category will steadily increase. Furthermore, despite the abundance of data available, patients, and loved ones have a difficult time assessing the "technical" factors of healthcare quality. Service markers usually substitute as criteria for judging a hospital or provider. Consequently, with the new-found importance of service criteria, there will be many career openings for hotel and restaurant management graduates.

Generally, hospital-based hotel type services are administered by contract management companies and self-operated. For example, ARAMARK, which has contracted hospitals such as Hahnemann Hospital in Philadelphia and the Massachusetts General Hospital in Boston (where Fred DeMicco completed his dietetic internship for the RD certification), is self-operated (a non-contracted operator) medical center.

Graduates from hospitality, business, and healthcare management programs that join this segment of the healthcare/hospitality industry can look forward to solid growth for the future, stable work hours, good pay, and benefits (particularly medical insurance), oftentimes daycare for employee children, career growth, and the satisfaction of making a real difference by serving people in need. Usually, a graduate of a 4-year hospitality program begins as an assistant director, works up to a director of food services, and eventually can become a general manager or vice president (VP) for patient/guest services. The U.S. Labor force as a whole is expected to have slow if not anemic job growth as workers age. Indeed, much of the job growth in the coming decades will focus on services for seniors. Healthcare occupations and related health industries are expected to have the fastest employment growth and add the most jobs through 2024, according to the U.S. Department of Labor (The Wall Street Journal, 2015, PA2).

As more healthcare facilities grow to become "medical campuses" and *medical meccas* for care, they emulate hotels in their quality and delivery of health and wellness (H2H©) services. Many build hotels or have partner hotels for medical guests and tourists on their health campuses. An innovative approach for blending hospitality and healthcare can be found in Orlando, where the RosenCare® model could boost lodging revenue by using medical travel to the USA. In fact, in medical centers and hospitals, after the procedure (surgery or other treatment) approximately 75% of the services provided to patients are hospitality/hotel-related services (H2H©) including dining, environmental, transportation, front desk, educational, and entertainment services (Figure 24.3).

FIGURE 24.3 Hospitality bridging healthcare: Medical versus hospitality services in a health center.

And, for example, medical centers may add wellness classes, culinary education, and spa operations, which hospitality school graduates should be prepared to oversee. In addition, more foodservice directors are becoming general managers (G.M.'s) of the entire healthcare campus enterprise who leads not only the traditional hospital facilities but also, the hotel(s), spa, and wellness, environmental services (e.g., housekeeping), transportation services, parking, valet services, gift shops, employee markets, the grounds, snow removal, well-being education and all dining venues, including room services. This is usually a VP position on the health care campus and can provide challenge, excitement, personal fulfillment of caring, and very good pay and benefits in the six-figure salary range. In sum, hospitality schools prepare graduates to be GM's of hotels and now graduates should think of becoming GM's/VP's of the health care (non-medical) operations and H2H© patient/guest services in the future.

24.5 THE OPPORTUNITY FOR A NEW CAMPUS MAJOR: HOSPITALITY BRIDGING HEALTHCARE (H2H©)

This is clearly a field for hospitality, business, and healthcare graduates that provides challenge and growth opportunities well into the future. Therefore, a new major on campuses that concentrates on hospitality bridging healthcare prepares graduates for a growing field and challenging career holds outstanding opportunities for future graduates of Hotel and Restaurant Management schools and departments (Figure 24.4).

24.6 CONCLUSION

The pre-planning of the medical journey leading up to the arrival at the medical/health/wellness destination begins a cascade of multiple H2H© services to create a holistic and satisfying experience for the Patient/Guest. This can lead to long-lasting loyalty if performance in the entirety of the process is at the highest level. Hospitality and Healthcare services will play a critical role for success. Graduates of programs focusing on this intersection of H2H© will be rewarded with challenging, satisfying, and lucrative careers. The Path Forward includes developing a core curriculum and courses for this promising new career track. It will be an intersection of Business Management, Healthcare Administration and Hotel/Hospitality Business Management theory and practice with industry hands-on learning.

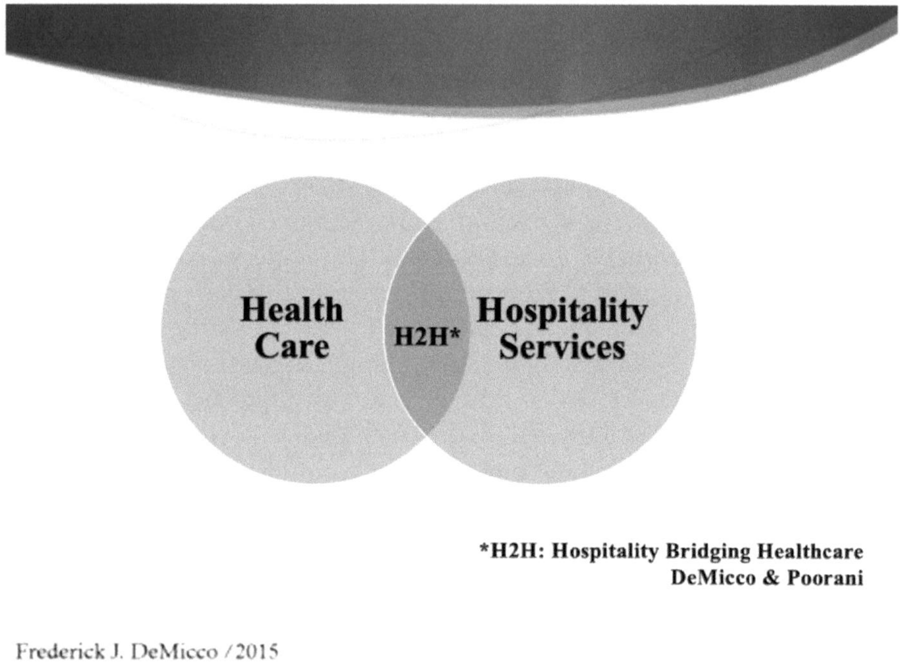

FIGURE 24.4 The intersection of hospitality with healthcare (H2H©).

KEYWORDS

- **careers**
- **digital communications**
- **general managers**
- **healthcare**
- **hospitality bridging healthcare**
- **leader**

REFERENCES

Cockerell, L., (2008). *Creating Magic: 10 Common Sense Leadership Strategies from a Life at Disney*. Doubleday Publisher. New York.

DeMicco, F. J., & Pizam, A., (2019). *RosenCare® Approach Could Boost Medical Travel to USA*. https://www.imtj.com/articles/rosencare-approach-could-boost-medical-travel-usa/ (accessed on 14 July 2021).

DeMicco, F. J., (2016). *Hospitality Bridging Healthcare (H2H ©) and Medical Tourism & Wellness*. Apple Academic Press.

DeMicco, F. J., (2018). How would Disney run a medical tourism business? *International Medical Travel Journal (IMTJ)*. https://www.imtj.com/articles/how-would-disney-run-medical-tourism-business/ (accessed on 14 July 2021).

CHAPTER 25

TELEHEALTH AND THE COVID-19 ERA

YAN CAO[1] and ALI A. POORANI[2]

[1]Spears School of Business, School of Hospitality and Tourism Management, Oklahoma State University, 365 Human Sciences, 122 N Monroe Street, Stillwater, OK–74078, USA, E-mail: yan.cao@okstate.edu

[2]Alfred Lerner College of Business and Economics, Department of Hospitality and Sport Business Management, University of Delaware, 14W, Main Street, Newark, DE–19716, USA, E-mail: poorani@udel.edu

ABSTRACT

Telehealth innovations made unprecedented expansion during the pandemic and will continue to become a permanent part of the health care ecosystem. In this chapter we provide information about the scope of telehealth, market size, and emerging telehealth providers, along with their major characteristics and service coverage.

To attract new patients and keep those that already exist, telehealth providers need to develop specific digital marketing plans to establish their brand. We will provide a list of well-established telehealth companies and their key characteristics for success.

Understanding customer acceptance and generational differences utilizing telehealth are significant for healthcare providers. We will share recent demographic data available telehealth usage before, during, and after the pandemic.

Finally, we will discuss factors affecting provider intention to continue practicing telehealth, regulatory and reimbursement issues, advances in telehealth technology, as well as telehealth etiquette, all are significant factors for implementing an effective telehealth strategy.

25.1 INTRODUCTION

Without question, COVID-19 has revolutionized telehealth and the medical travel industry can leverage the advances made in digital health and telemedicine. How did the revolution start?

According to the Center for Disease Control and Prevention (CDC, 2020), telehealth visits increased by 154% in the last week of March 2020 compared to the same time last year. This exceptional rise is due primarily to pandemic-related telehealth policy reforms and public health guidelines, which include social distancing. During the pandemic, telehealth provided many benefits, such as patient access, reducing disease exposure, and reserving personal protective equipment (PPE) for staff use. As such, telehealth policies may continue to post-pandemic periods, according to CDC.

Consequently, the COVID-19 health crisis propelled an unprecedented trend of blurring lines between hospital medical services and stay-at-home telehealth services. Such a trend has resulted in a fundamental change in customers' mindset toward telehealth ever; since then, virtual service has been integrated as an indispensable part of the overall care delivery model (Grossmann, 2020). When alternatives become necessities during the pandemic, discussions on trade-offs of telehealth utilization versus traditional in-person healthcare were replaced by the topic of how telehealth can efficiently be integrated and adapted to the care delivery model after the COVID-19 (Adams, 2021).

As expected, telehealth utilization declined throughout the first three months of 2021, because of CDC regulations being modified across the country, accounting for 5.6% of claims in March and 4.9% of claims in April, a drop of approximately 12.5% from March statistics (Drees, 2021). According to the same source, mental health conditions comprised the majority of telehealth visits (51.3%), followed by developmental disorders (3%), joint, and soft tissue diseases and issues (2.7%), acute respiratory diseases and infections (2.5%), and general symptoms (1.5%).

Despite the significant decreases, many experts believe that telehealth will continue to grow. As a convenient alternative to in-person appointments, telehealth will continue to play an important role. Telehealth innovations made unprecedented expansion during the pandemic (Michael, 2020), on the downside, those changes will not be sufficient to maintain the momentum gained throughout the outbreak, but on the upside the experience of millions of telehealth users (i.e., consumers, and providers) can help telehealth to become a permanent part of the health care ecosystem, which accounts for $4 trillion of the overall health care economy (Botta, 2021).

25.2 SCOPE OF TELEHEALTH PROVIDERS: MARKET SIZE OF TELEHEALTH BEFORE, DURING, AND AFTER THE PANDEMIC

Instigated by the COVID-19 pandemic, in March 2020, the US government expanded the use of telehealth services and reduced the Medicare program reimbursement requirements to help both patients and providers during the pandemic (Verma, 2020). Consequently, telehealth usage has increased from 7.8% in 2018 to 26.9% by January 2021 (NRC Health, 2021). In comparison to the circumstances prior to the current pandemic, providers raised their offers and treated 50 to 175 times the number of patients via telehealth service. As a result, the total revenues generated from telehealth service skyrocketed from approximately \$3 billion before the pandemic to \$250 billion (Bestsennyy et al., 2020).

As seen in Table 25.1, since 2019, the telehealth market is expanding and is expected to continue to grow at a rapid rate. According to the global telehealth service market report (The Business Research Company, 2020) in 2019, the global telehealth market size was approximately \$49.89 billion, rendering a compounded annual growth rate (CAGR) of around 22.5% since 2015. In 2019, COVID-19 acted as an impetus for the revolutionary expansion of telehealth. Consequently, it is reported that the market size will increase from \$49.89 billion in 2019 to \$194.05 billion in 2023, signifying a compound annual growth rate of 40.4%. Going forward, it is predicted that the telehealth market size will reach \$277.93 billion in 2025 and \$459.80 billion in 2030.

TABLE 25.1 Telehealth Market Size and Growth Rates

Year	Market Size (billion)	Compound Annual Growth Rate (CAGR)
2019	\$49.89	22.5% since 2015
2023	\$194.05	40.4% from 2019 to 2023
2025	\$277.93	19.7% from 2023 to 2025
2030	\$459.80	10.6% from 2025 to 2030

Adapted from The Business Research Company (2020).

From a geographic perspective, North America made up the largest telehealth market share of about 43.9%, followed by the Asia Pacific and Western Europe. It is also reported that the Middle East and the Asia Pacific are the two fastest-growing markets in terms of telehealth products and services, with CAGRs of 15.4% and 13.2%, respectively. Africa and South America are the next two regions on the list, with CAGRs of 12.4% and 11.4%, correspondingly (The Business Research Company, 2020).

Moreover, a host of countries have initiated governmental incentives to support telehealth implementations. For example, Australia launched a "Medical Support At Home" campaign, which supports the expansion of telehealth use across the country. For instance, the Australian government invested more than $1.1 billion to enable open access to telehealth services and further ease its penetration into the healthcare sector. Apart from Australia, China also has promoted the use of telehealth to relieve the strain on the healthcare system with Ali Health, JD Health, and WeDoctor providing COVID-19-specific online telehealth services in December 2019. Moreover, the Japanese Ministry of Economy, Trade, and Industry (METI) in partnership with Line and Medpeer, initiated free online health consultation services in March 2020 (The Market Analysis Report, 2021). These trends are projected to continue.

25.3 EMERGING TELEHEALTH PROVIDERS

According to Definitive Healthcare Data (2020 June), healthcare providers' telehealth utilization rate was relatively low in 2019. Telehealth services were only available in roughly 33% of inpatient facilities. On the contrary, in 2020, approximately 75% of hospitals in the United States provided telehealth services to their patients.

The growth of telehealth services has also led to an increase in the number of companies specialized in delivering telehealth. At the time of editing this chapter, there were around 260 telehealth companies in the United States, according to estimates (Dyrda, 2021).

Table 25.2 displays the top five telehealth providers, along with their major characteristics and service coverage depicted by Healthline Online Report (2020).

25.4 BRANDINGS TELEHEALTH

Does branding make a difference? The short answer is a "yes." In order to attract new patients and keep those that already exist, health care providers need to develop specific digital marketing plans to establish a niche in the market by providing exclusive medical support, patient experience, and engagement. To benefit from branding, and effective design is the one that is co-created by providers and their patients. With the increased demand during and after the pandemic, it is the right time for telehealth providers to rethink their branding strategy and position themselves in this emerging yet established market. In this segment, we draw on the 7Ps of marketing (Price, Place, Product, Positioning, Packaging, People, and Promotions) and illuminates branding strategies that healthcare providers can use to market their telehealth products and services during and after the COVID-19 (*see* Table 25.3).

TABLE 25.2 Sample Telehealth Providers

No.	Telehealth Companies	Key Characteristics and Service Scope
1.	Teladoc	• It is one of the earliest companies to offer telehealth services. • Both physicians and patients have favorable evaluations. • Its service scope includes pediatric services, dermatological conditions, mental health conditions, etc. • Physicians can send prescriptions to the patients' selected pharmacy and review lab test findings.
2.	MeMD	• Multiple symptoms or physical issues can be investigated without incurring additional charges during a single visit. • Lab results can be reviewed by physicians. Prescriptions can be sent to patients' preferred pharmacy by their healthcare providers.
3.	iCliniq	• A host of services are provided, including phone consultation and online video. • Service scope involves psychiatry, oncology, obstetrics, and dentistry.
4.	Amwell	• The services' price is more affordable. • Amwell's mobile application is more user-friendly. It is compatible with both iPhone and Android, which can direct the patients to the right doctor at the right moment. • Nutritionists are on hand to offer advice. It is well-known for its video treatment services.
5.	MDlive	• Both physicians and healthcare professionals are available by phone or video appointment 24 hours a day. • It deals with non-emergency health issues such as allergies, headaches, rashes, fevers, etc. • Mental health consultations are also available.

Adapted from Healthline (2020).

25.5 CUSTOMER PERCEPTIONS OF TELEHEALTH BEFORE AND AFTER THE PANDEMIC

As we described earlier, COVID-19 has altered the way how consumers search and approach healthcare services. As walk-ins become log-ins, understanding the viewpoints of patients on telehealth has become a must for telehealth deployment.

TABLE 25.3 Branding with 7Ps of Marketing: Descriptions and Examples

Seven Ps of Marketing Domains	Descriptions	Examples
Price	More affordable to marginalized populations and underserved communities.	Amwell repositioned its company with a more affordable rate. Specifically, before insurance, visits are $69 each (Healthline, 2020).
Product	• Patients can gain a taste of convenience. • Provider-reported higher office productivity, timeliness, and quality care. • Providers benefit from doctor-to-doctor easy connections. • Embedded website functionality, such as sign-in and follow-up contacts, as well as online scheduling appointments, can all be performed online (Gandolf, n.d.).	• Amwell's mobile app, which is available for iPhone and Android, connects patients with the right doctors at the right time while removing any spatial and temporal constraints (Healthline, 2020). • Amwell also has dietitians on hand who can provide advice, and it is well-known for its video treatment services (Healthline, 2020).
Place	Instead of the brick-and-mortar companies, telehealth products and services can be offered anywhere (Gandolf, n.d.).	Virtual healthcare facilities offer on-demand access from the comfort of one's own home, car, or office, allowing for less time away from work or school and eliminating the need to commute, stuck in traffic, and stay in waiting rooms.
Positioning	Convenience is one of the top-ranked benefits, followed by lower cost and higher or adequate service quality.	Value can be added via focusing on convenience, lower cost, and less spent time.
Packaging	Telehealth can be incorporated into a health system's or hospital's convenient variety of services (Gandolf, n.d.).	A mixed-use strategy can incorporate family care, urgent care, emergency department, and telehealth (Gandolf, n.d.).
People	Providers and Staff	Attention to detail and providing excellent customer service and experience to patients and family members.

TABLE 25.3 *(Continued)*

Seven Ps of Marketing Domains	Descriptions	Examples
Promotions	Products and services need to be advertised to targeted audiences with its offerings clearly distinguished and wining factors strongly identified (Gandolf, n.d.).	• To promote service and attract new patients, digital marketing tactics, such as Google Ads and social media can be used. • Adding a HIPAA-compliant chat function to the website can help existing and prospective patients to reach out to ask questions (Sauceman, n.d.). • Approaching social referral marketing to catch positive effects of word-of-mouth-WOM (Wicklund, 2020).

Consumers generally choose their providers based on online sources, family, and friends, healthcare professionals, insurance companies and other healthcare employees. Kyruus (2020 August) surveyed 1000 subjects from various generations and geographic locations and concluded that among all the sources internet is the top choice (57%), followed by family and friends (38%), healthcare professionals (37%), insurance company (27%) as well as healthcare staff (23%).

Physicians were the main drivers for the transition during the pandemic. According to (SYKES, 2021) study, the top factors for the adoptions of telehealth by customers were, "My physician's office moved my appointments to virtual visits: 61.76%," followed by convenience (28%), and patient choice asking for moving their appointment to virtual (24%). It is notable that the reported numbers were lower in communities with a high poverty rate.

A survey of 2000 American adults, conducted by SYKES (2021) examined if the participants had scheduled a telehealth appointment in March 2020, fall 2020, and March 2021. The findings revealed a considerable progression in their response from 2020 to 2021. In March 2021, only 19.5% of respondents had made at least one appointment via telehealth; that number increased to 29.1% in the fall of 2020; and jumped to an unprecedented 61.5% in March 2021.

COVID-19 pandemic has dramatically changed customer attitudes towards telehealth. Among the important factors for change include quality of care, ease of access, suitability for routine care, frequency of visits, and cost considerations. These factors have significantly contributed to the credibility and utility of telehealth as an alternative to in-person visits. Table 25.4 summarizes the changes in customer attitudes from March 2020 to March 2021.

SYKES's survey (2021) has also highlighted the shift in trusting telehealth before and after the COVID-19 signifying approximately 65% of Americans hesitated to use telehealth, and 56% believed that telehealth could not deliver them the same level of healthcare quality as compared to the traditional in-person visits. However, after COVID-19, a different picture has emerged. About 88% of respondents in the above study, said that they would continue using telehealth for nonurgent needs, while almost 80% thought that telehealth indeed can deliver quality care service. These high rates of adoption are also reflected in the forecasts for telehealth spending. Lagasse (2020) reported up to $106 billion of current U.S. healthcare spend could come from virtualized services by 2023.

Is telehealth here to stay? While unprecedented usage of telehealth emerged during the pandemic, mainly due to social distancing rules and payment procedures, those rates did drop as the pandemic impacts lessened. A 2021 trend report (Hrdy, 2021) posited that telehealth utilization rate increased rapidly in April 2020, and in the meantime, in-person visits plummeted to its lowest point in history. While an upward trend of telehealth utilization was prospected during the pandemics, for both privately insured patients and Medicare beneficiaries, its usage rate dropped as soon as the in-person visits resumed.

TABLE 25.4 Customer Attitudes Towards Telehealth Since the Pandemic

Factors	March 2020	March 2021
Quality of Telehealth	65.6% doubted the quality of telehealth.	87.82% are willing to continue using telehealth products and services after the pandemic.
Ease of Access	–	85.52% said telehealth has made it easier to get their care when they need it.
Routine Physical Exams	–	64.05% said they prefer part of their routine exam via telehealth
Frequency of Visits	–	51.64% said telehealth has made it possible to visit their doctors more frequently.
Number of Visits	19.5% had made at least one telehealth appointment.	61.05% had made at least one telehealth appointment.
Cost of Care	–	31.26% said that thanks to telehealth, their medical expenses decreased.

Adapted from Adams (2021).

Meanwhile, according to experts, telehealth will yet remain a viable component of healthcare (Adams, 2021). Despite the downward trend of telehealth utilization rate after COVID-19 compared to the during-pandemic situation, it is irrefutable that the current usage rate is higher compared to the almost negligible rate before the pandemic, which explains a sustainable and continued consumer demand after the COVID-19 (Hardy, 2021).

25.5.1 GENERATIONAL DIFFERENCE UTILIZING TELEHEALTH

During the pandemic, while 2/3 of patients expressed willingness to use telehealth, the actual usage rate was 46%, according to a 2020 study (Bestsennyy et al., 2020). Such a gap could be partially explained by:

- problems such as a lack of awareness of telehealth services;
- a lack of education on what type of care can be provided virtually; and
- a lack of understanding of insurance coverage.

Generational differences also explain some of the gap (Kyruus, 2020). As stated previously, Internet sources, such as Google, healthcare websites, insurance sites, etc., are the primary source of finding providers in telehealth, and Millennials and Gen Xers are more engaged in the internet sites. For example, In the Kyruus study, when asked if patients wanted to see a *virtual-visit function* in providers' and healthcare originations' websites, overall, one-third of the patients wished to see, and this was one of the reasons that influenced their decision to purchase. For both millennials and Gen Xers, this number was higher (40% or more).

Gen Xers and Millennials were also more open to use telehealth. Over 53% of Gen Xers and 49% of millennials in the survey were willing definitely or very likely to switch providers and consult via virtual care; only 17% of baby boomers were willing to do so (see Table 25.5).

TABLE 25.5 Generational Preferences on Consumer Telehealth Utilization

Age group	**Millennials 24–39**	**Gen Xers 40–55**	**Baby Boomers 56–74**
Whether consumers would switch providers to consult via virtual visits (definitely/very likely)	49%	53%	17%

Adapted from Kryuus (2020).

Likewise, a survey conducted in 2019 (SCP Health, 2020) consisting of 2000 adults, provided more detailed results for different age groups. Based on Table 25.6, younger generations were more willing to use telehealth compared to elder generations, and the actual usage rate was also higher for younger age groups.

TABLE 25.6 Patient Telehealth Adoption Difference by Age Group

	18–34	**35–44**	**45–54**	**55–64**	**65+**
Not at all/Not very willing	26%	28%	33%	39%	48%
Very/Somewhat willing	74%	72%	67%	61%	52%
Telehealth usage	16%	13%	6%	3%	1%

Adapted from SCP Health (2020).

The age-related *preferences* have also been noted in two fronts:

- users; and
- providers.

Among the users, there were notable differences in their preferences for services received through telehealth. Millennials and middle-aged consumers turn to telehealth for convenience and timesaving; while older population prefer telehealth for prescription services and renewal, and chronic disease management. By contrast, millennials use telehealth to approach mental health services (SCP Health, 2020). Baby boomers also highly valued patient reviews and provider insurances acceptance as a factor to use telehealth.

On the provider side, age also serves as a factor toward telehealth adoption. As shown in Table 25.7, overall, when compared to older physicians, younger physicians are more likely and less hesitant to utilize telehealth. One point that needs to be noted is that physicians age 25–34 expressed slightly less willingness (74%) compared to other younger physicians aged between 35 to 44 (77%). It can be inferred that these newer physicians were likely in the process of learning stage and thus less comfortable to provide diagnosis merely with the new technology (SCP Health, 2020).

TABLE 25.7 Physician Willingness to Have a Video Visit by Age

	25–34	**35–44**	**45–54**	**55+**
Not willing	11%	8%	9%	14%
Unsure	15%	15%	21%	27%
Willing	74%	77%	70%	70%

Adapted from SCP Health (2020).

In sum, there is a lot to be learned and researched about the future of telehealth, its usage in various practices, and catering to various age groups. Meanwhile, telehealth providers might consider such differences and market their capability to their patients in a more generation-tailored manner, especially for the younger generations. In the following section, we will address factors affecting provider perceptions of telehealth.

25.6 PROVIDER PERCEPTIONS OF TELEHEALTH BEFORE AND AFTER THE PANDEMIC

25.6.1 FACTORS AFFECTING PROVIDER UTILIZATION OF TELEHEALTH

Providers' main concerns include reimbursement, security issues, high costs of implementation, workflow integration, and effectiveness of virtual visits

(Schinasi et al., 2021), Table 25.8 presents top-ranked concerns raised by providers in terms of telehealth utilization.

TABLE 25.8 Top-Ranked Factors Influencing Provider Intention to Continue Practicing Telemedicine

Factor	N=305	%
Whether payers will continue to reimburse for telemedicine visits	120	39.3
Whether my division continues to offer telemedicine visits	62	20.3
Whether patients' families are interested in telemedicine visits	62	20.3
My own preference to use telemedicine in my practice	20	6.6
Other	14	4.6
No response	27	8.9

Adjusted from Schinasi et al. (2021).

25.6.2 PROVIDER RE-IMBURSEMENT ISSUES

The very issue that complicates the business process is the reimbursement for virtual visits, which consensus from both parties has not yet reached (NRC Health, 2021). Before the pandemic, insurance coverage for remotely provided healthcare services varied across insurers and geological regions. However, during the pandemic, CMS lifted the reimbursement limits. For example, in 2021, New Medicare-covered telehealth services are now available to patients. Specifically, Michael Elley, CIO of Baptist Health, postulated that to ensure continued utilization of telehealth, telehealth should be reimbursed in a consistent or nearly consistent manner as traditional in-person visits (Adams, 2021). Additionally, Mark Amey, CIO of Alameda Health, also proposed that apart from emergency virtual visits, phone visits are also supposed to be reimbursed at a basic rate. As such, the success of resolving the reimbursement issue can exert a positive influence on both parties and ultimately encourage healthcare businesses to make a long-term investment for a healthcare integration solution (Adams, 2021).

25.6.3 REGULATORY ISSUES POST PANDEMIC

While the pandemic lifted the regulatory limits regarding to reimbursement, privacy, and licensing; emerging legal, ethical, and social issues are likely to spillover to the post-COVID-19 era. In order to survive and grow in this ever-changing regulatory environment, telehealth companies must work in a flexible manner. Among the areas to focus on is the privacy and safety issues. COVID-19 enlarged the interoperability scaling and improved healthcare outcomes. However, due to the use of telehealth at a large scale, cybersecurity risks have arisen. This necessitates further regulations. Without legislation that encompasses the full range of data usage in telehealth, the risk of data misuse will increase. Such problems include "identity theft, employment, and insurance discrimination, invasive marketing, and cultural stigma" (Rokosh et al., 2021, p. 3). Therefore, new, and clarified rules and regulations are needed to protect patients' privacy after the current health crisis.

In addition to the privacy and safety regulations, reimbursement rules and practitioner licensure must also be considered and discussed for the post COVID-19 era. Table 25.9 provides the explanations and prospects on the three regulatory aspects.

TABLE 25.9 Regulatory Changes of Telehealth After the COVID-19

Regulatory Domains	Post COVID-19 Prospects
Payment	Low reimbursement for telehealth services was seen as a major deterrent. Regulatory policies that promote payment parity will need to be sustained after the pandemic. In addition, proper reimbursement for telehealth will be the key aspect in ensuring widespread telehealth usage.
Privacy	While during the health crisis, privacy appeared not to be the most crucial consideration, in the post-pandemic era, adequate regulations are necessary to ensure privacy. For example, security systems in the financial sectors (e.g., personal access to bank accounts) can be utilized to protect personal data security.
Licensing	In response to COVID-19, certain license requirements have been erased or deleted in various states. As a result of this trend, some clinicians from one state are able to provide medical service to patients in another state. While these rules make the environment more permissive, verification mechanisms are needed to ensure patients to approach to qualified clinicians.

Adjusted from Shachar, Engel, & Elwyn (2020).

25.7 ENGAGING PATIENTS IN TELEHEALTH

To ensure a continued demand on telehealth, ease of use and seamless technology are critical. Telehealth technology needs to be user-friendly and simple enough for patients to become more comfortable and seek service with virtual visits; for patients to connect to providers, a hospital-at-home multi-modal virtual care delivery system must be supplied seamlessly. Specifically, inclusive technologies need to be designed and embedded into the overall online delivery model to better assist elder patients and people with disabilities with clinical trial design, and customer experience reporting and testing (Noel & Ellison, 2020).

To provide parity, community-based healthcare organizations for under-served populations can be further developed to relieve health disparity issue after the pandemic. A phone call or a virtual conference can be used to perform a basic telehealth. On the flipside, Philips Health is pushing things to the next level with virtual care stations, which provide the individuals without internet in their homes, access to a community-based telehealth solution that connects patients and clinicians remotely through a secure clinical environment. In this way, telehealth serves as a particularly beneficial method for reaching out to disadvantaged communities. In 2019, thanks to technical breakthroughs in virtual care, the Department of Veterans Affairs' telehealth program serviced over 900,000 veterans. The technology could potentially be used to help close the gaps in health care that many communities still have, including tribal homelands and peripheral areas (Sweet, 2020).

Digital communication, particularly text messaging has also emerged as the new and preferred way that patients want to communicate with their doctors. A recent survey (SR Health, 2021) comprising of 250 qualified respondents, who have had at least one face-to-face visit, video, phone call, or text message with a healthcare provider within last 12 months, concluded that:

- 4-in-10 expect regular communication with their providers;
- 6-in-10 want to receive text messages from providers regarding reminders, notifications, pre-visit instructions;
- 1-in-2 prefer text messaging post-pandemic.

It is noteworthy that 6-in-10 of the subjects over 50-years old also liked to receive text messages from their healthcare providers, and 38% of the 50+ preferred to communicate with their healthcare providers post-pandemic. The most pervasive reason for text message use was convenience (87%), which was attributed to the simplicity of communication, as well as the speed with which inquiries could be asked and responses received. The most surprising response was that nearly 4-in-10 were agreeing to switch providers to receive modern communication.

25.8 DIGITAL INNOVATIONS AND THE FUTURE OF TELEHEALTH

In June 2021, The Healthcare Information and Management Systems Society (HIMSS) published the State of Healthcare Report. The study included 2,743 respondents from across the United States, including 225 health systems, 309 clinicians, 147 payers, and 2,062 patients (HIMSS, 2021). The study identified three major themes that will have a significant impact on the future of healthcare, including digital health, artificial intelligence and machine learning (AI/ML), and financial health. Some of the highlights of the survey are summarized in Table 25.10.

Other important findings:

- Telehealth offering heightened due to the pandemic and will remain as a competitive offering for healthcare systems. The uncertainty about reimbursement and regulatory issues remains high.
- Overall sentiment of patients is to go back to pre-COVID-19 times.
- The fact that Digital Health is here to stay is well acknowledged. 8-in-10 of providers now offer patient portal.
- Though most see various benefits of AI/ML, there are concerns about operationalizing AI/ML.

We expect big data collaborations, AI analysis, and machine learning to continue reinforcing innovation in clinical decision support (CDS) software, patient treatment, and pharmaceutical research in the coming years, ensuring that telehealth is delivered to the right patients for the right needs with increasing efficiency and efficacy (Baca, Combs, & Moundas, 2021).

Furthermore, companies such as Amazon, Google, and Apple are positioning linked smart home technologies to act as personal digital health assistants as well as the home hub for device connectivity and data flow. Danish Future Patient research project, which focuses on telerehabilitation (Healthtech Staff, 2018), patients with heart failure can utilize step counters, sleep sensors, and social media to assess illness progression and symptom management. As evident in these and other innovations, it is expected that virtual care will becomes more prevalent in daily life, and their physical and online visit interactions, or consumers' experiences with healthcare services will become increasingly merged. Therefore, to maintain uniform quality of healthcare services, establishing new and evidence-based practice guidelines and standardized operating methods will be indispensable in the near future.

TABLE 25.10 Healthcare Technology Trends

Topics	Digital Health	AI/ML	Financial Health
Health Systems	1/3 focus on tracking key performance indicators and outcomes, yet 52% still in pilot stage.	68% state they must grow investment, yet 70% have not done so.	68% of reimbursements coming through value-based care.
Clinicians	More than 80% trust their organization's ability to protect patients' information, yet such trust can be ruined by potential cybersecurity attacks.	77% are either utilizing or interested in AI/ML, yet only 20% received training so far.	70% posit that they are financially stable, and 12% saying they're some-what or highly financially unstable.
Payers	More than 60% agree that interoperability is a driver to enhance adoption, yet 40% worried about privacy and security issues.	62% believe that improve AI/ML adoption is of primary concern, and almost 50% already set up an innova-tion lab for company-wide adoption.	The aging population is the top reason (41%) for an increase in costs, followed by an increase in healthcare utilization (38%) and safety measures (38%).
Patients	71% of younger generations prefer telehealth for convenience; and 44% is likely to switch providers due to the discontinuation of telehealth.	Almost 50% express doubt about telehealth, and more than 75% are comfortable to share their personal health data for benefits such as accurate diagnosis.	79% believe the costs for healthcare in the U.S. are too expensive, yet 70% are satisfied with their health insurance, especially elder generation.

Adapted from HIMSS June 15, 2021, the Stated of Healthcare Report.

25.8.1 PITFALL AND CHALLENGES OF TELEHEALTH TECHNOLOGY

While telehealth carries a host of advantages and potentials, yet it is not without its pitfalls and challenges primarily relating to:

- ethical issues; and
- information security.

Let's consider ethical issues first. Telehealth utility produces paradoxical effects on the healthcare disparity issue. Despite the advances in technology, patient access is limited in underserved communities to reach out to certain health care providers (Sweet, 2020). Lack of such broadband access and smart tools also constrain the expansion of virtual care to the communities that have higher poverty rates (Lagasse, 2020). Additionally, with machine learning algorithms that is used to analyze patients' syndromes, the lack of information pertaining to underserved populations would possibly bias the source data or algorithm development, which in turn, worsens the existing health disparity problem (Rokosh et al., 2021).

The second double-edged sword challenge relates to the information security. While implementation of big data, AI, and other diagnostic telehealth tools facilitate a care delivery model with convenience and scientific benefits, such digital proliferation and data interoperability are likely to violate both patients and providers' privacy. As outlined by Davis (2021), as a result of the latest ransomware attack on Scripps Health, the health provider was forced to an EHR shutdown. Hackers hit early on a weekend morning when staffing was low, causing failure in communication between providers; several patient visits were pushed back; or rerouted to adjacent hospitals. In September 2020, Universal Health Services was the target of a similar ransomware attack. Both breaches spanned four weeks and resulted in outages at several care facilities, as well as disruptions to corporate operations and patient visits.

25.9 PATIENT EXPERIENCE AND TELEHEALTH ETIQUETTE

The Beryl institute refers to the patient experience as "the sum of all interactions, shaped by an organization's culture, that influence patient perceptions, across the continuum of care" (Wolf et al. 2014) and their comprehensive study of the literature revealed the following factors or attributes in the definition:

- emotional and physical lived experience;
- personal interactions;
- spanning across the continuum;
- shaped by the organization/culture; and
- importance of partnership/patient involvement.

As we will examine below, most all these inferences are clearly manifested in a telehealth experience. The goal is to make emotional connections at every touchpoint, and virtual connections should not create a barrier for the experience. With the help of technology today, such as iMotions's Activia, metrics are developed to record facial expressions, such as smile, joy, anger, disgust, surprise, fear, contempt, etc. (iMotions, 2021). Other metrics, such as eye-gaze tracking systems can analyze attention fluctuations, such as duration of eye-gaze relevant to the length of total conversation; frequency of eye-gaze; and patient response to provider behavior all advancing telehealth and providing a patient-centric experience and patient-perceived empathy.

Content-analysis (CA) tools are also important to identify words and behaviors leading to patient-perception of empathy. For example (Smith, K.A., Bishop, F.L., Dambha-Miller, H. et al. 2020) identified the following attribute using content analysis for empathy, with the top-5 most mentioned are highlighted:

- Explanations of treatment;
- Reassurance;
- Friendly manner;
- Non-verbal behaviors (gestures, looking, facial expression);
- Active listening;
- Describe evolution of disease;
- Instruct patient how to quantify symptoms;
- Elicit questions from patient;
- Patient-practitioner partnership;
- Consultation structure (appropriate to discussion);
- Lifestyle discussion (features outside immediate symptoms);
- Checking patient understands;
- Express positive expectation of treatment;
- More time;
- Non-specific conversation;
- Comply with patient wishes; and
- Emphasize comfort and well-being.

25.9.1 DISCONNECT IN PATIENT-PROVIDER TELEHEALTH EXPERIENCE

Although all these variables improve patient-perception of empathy, the issue of disconnection in virtual care may happen to both customers and providers. Patients are reported frequently multitasking, surfing on the Internet, and snaking (Wicklund, 2020). Patient preoccupations and distractions can have devastating effects, such as misunderstanding medication and missing important conversations, which can be somewhat unnoticed by providers during online visits versus in-person consultations. Therefore, it is advisable that providers set expectations and inform patients about some specific behaviors, such as making a list of questions in advance, muting their phones, finding a lit and quiet space for telehealth sessions, as well as focusing on patients and using teach-back methods assuring that patients understand what was discussed.

On the provider side, preparations need to be more comprehensive. Table 25.11 shows most important telehealth etiquette expected from providers:

TABLE 25.11 Telemedicine Etiquette Checklist

Etiquette Categories	Requirements	Check
Appearance	• Well-groomed • Lab coat when appropriate • Visible badge to identify self • Choose clothing – Not dark – Limited in patterns – No glitter • Limit jewelry – No bangles or dangling jewelry – Minimal jewelry	
Distractors	• Make sure equipment functioning (check 15 minutes prior) • Minimize outside noise • Remove clutter from room • No eating or drinking • Shut door and put sign up regarding meeting • Let others know telehealth visit is taking place • Assess lighting (both provider and patient sides) • Limit paper shuffling, pen tapping, etc. • Limit quick body and hand movements • Mute microphone when not in use • Turn off pagers and phones	

TABLE 25.11 *(Continued)*

Etiquette Categories	Requirements	Check
Privacy	• Assess security of environments • Introduce everyone in room (provider and patient side) • Determine if patient would like to proceed • Showroom with camera • Assure patient of equipment security • Do not record meeting	
Nonverbal communication	• Look at camera and not patient's face on screen • Check provider view screen – Eye contact – Position of body in picture – Distractors • Minimize gestures • Use facial expressions to express emotions • Minimize charting	
Verbal communication	• Use words to convey thoughts • Use motivational interviewing • Limit dead space	
Empathy	• Lean in • Use words to express empathy • Nod • Maintain eye contact	

Quoted from Gustin (2021).

For providers to communicate effectively Gustin (2021) has developed a comprehensive checklist that can providers review pre-telehealth event, at the start of the visit, and during the visit (see Table 25.12)

TABLE 25.12 Telemedicine Visit Checklist

Pre-Telehealth Event Checklist	Check
1. Check equipment	
2. Check environment for clutter and distracting items	
3. Check clothing, and get lab coat	
4. Mute phone and pager	
5. Assure the patient has signed the Telehealth Consent form and has an opportunity to ask questions or be provided with additional information	
6. The patient has an opportunity to request an interpreter or other support to enable maximum benefit of the encounter	
7. A backup communication plan (telephone, etc.) is developed and communicated to the provider	

TABLE 25.12 *(Continued)*

Pre-Telehealth Event Checklist	Check
8. Share provider and patient alternative contact information	
9. Place a Telehealth In-Session sign on the door	
10. Assure that both the patient and provider's sites are secure	
At the start of the visit	
1. Assure the patients know that the visit will be virtual	
2. Check that patient's setting is secure	
3. Seating should be checked to assure full visibility	
4. Assess for proper lighting and adjust as needed	
5. Place camera at the same elevation as the eyes (when possible)	
6. Assure that the face is clearly visible	
7. Begin visit with relationship building questions "small talk"	
During the visit	
1. Assure that patient is not recording visit	
2. Identify yourself when you speak (if there are multiple providers)	
3. Speak clearly with a strong voice	
4. Use meaningful empathetic word choices	
5. Avoid long periods of silence	
6. Allow for audio delay	
7. Allow time for questions	
8. Avoid interrupting others; give a visual cue for who should speak next	
9. Let the patient know if you are charting and looking down	

Quoted from Gustin (2021).

In conclusion, virtual healthcare model has further switched patient-provider relationship from a doctor-oriented to patient-oriented approach. In this model, with the help of technology, consumers' demographics are gathered, their past and present symptoms are analyzed, and their solicitations are further matched to a specific healthcare provider, who can then deliver them corresponding, thorough, on-demand, and continued service.

Finally, the rise of telehealth as a result of the pandemic could add a new dimension to medical travel, such as pre-trip consultations with physicians; connecting with the caregivers before traveling to an unknown destination; serving as digital triage before patients embarking on a trip; continuing access to providers pre-and-post-operative care, especially after transitioning a patient to an outpatient setting, such as a hotel; leveraging touchless travel for safety reasons and patient monitoring systems; and, ultimately shortening the travel time.

KEYWORDS

- **customer acceptance**
- **generational difference in telehealth usage**
- **regulatory issues**
- **reimbursement**
- **telehealth**
- **telehealth brand**
- **telehealth etiquette**
- **telehealth scope**
- **telehealth trends**

REFERENCES

Adams, K., (2021). How American attitudes on telehealth have changed since the start of the pandemic. *Becker's Hospital Review*. Retrieved from https://www.beckershospitalreview.com/telehealth/how-american-attitudes-on-telehealth-have-changed-since-the-start-of-the-pandemic.html (accessed on 25 August 2021).

Adams, K., (2021). Will telehealth have staying power after the pandemic? 9 CIOs weigh. *Becker's Hospital Review.* Retrieved from https://www.beckershospitalreview.com/telehealth/will-telehealth-have-staying-power-after-the-pandemic-9-cios-weigh-in.html (accessed on 25 August 2021).

Baca, M., Combs, K., & Moundas, C., (2021). *Digital Health 2021: Trends in Big Data, AI, Telehealth, and Beyond.* Bloomberg Law. Retrieved from Digital Heal https://news.bloomberglaw.com/health-law-and-business/digital-health-2021-trends-in-big-data-ai-telehealth-and-beyond (accessed on 25 August 2021).

Bestsennyy, O., Gilbert, G., Harris, A., & Rost, J., (2020). *Telehealth a Quarter-Trillion-Dollar Post-COVID-19 Reality.* Mckinsey & Company. Retrieved from https://www.mckinsey.com/~/media/McKinsey/Industries/Healthcare%20Systems%20and%20Services/Our%20Insights/Telehealth%20A%20quarter%20trillion%20dollar%20post%20COVID%2019%20reality/Telehealth-A-quarter-trilliondollar-post-COVID-19-reality.pdf (accessed on 25 August 2021).

Botta, M., (2021). *How Can Telemedicine Evolve in the Post-Covid Era.* MedCity News. Retrieved from https://medcitynews.com/2021/03/how-can-telemedicine-evolve-in-the-post-covid-era/ (accessed on 25 August 2021).

CDC, (2020). *Trends in the Use of Telehealth During the Emergence of the COVID-19 Pandemic—United States.* Retrieved from https://www.cdc.gov/mmwr/volumes/69/wr/mm6943a3.htm#suggestedcitation (accessed on 25 August 2021).

Davis, J., (2021). *What Happens After a Ransomware Attack in the Health IT Environment?* Health IT Security. Retrieved from https://healthitsecurity.com/news/what-happens-after-a-ransomware-attack-in-the-health-it-environment?eid=CXTEL000000391354&elqCampaignId=19584&utm_source=nl&utm_medium=email&utm_campaign=newsletter&elqTrackId=161ddfd34e0347a4947b34811af4280c&elq=6800c9be2da84319b7f44b95cc3e37f6&elqaid=20426&elqat=1&elqCampaignId=19584 (accessed on 25 August 2021).

Drees, J., (2021). Telehealth use falls across US for 3rd straight month: Three things to know. *Becker's Hospital Review*. Retrieved from: https://www.beckershospitalreview.com/telehealth/telehealth-use-falls-across-us-for-3rd-straight-month-3-things-to-know.html (accessed on 25 August 2021).

Dyrda, L., (2021). 260 + telehealth companies to know |2020. *Becker's Hospital Review.* Retrieved from https://www.beckershospitalreview.com/lists/260-telehealth-companies-to-know-2020.html (accessed on 25 August 2021).

Gandolf, S., (n.d.). *How to Market Your Telehealth Services.* Healthcare success. Retrieved from https://healthcaresuccess.com/blog/advertising/how-to-market-your-telehealth-services.html (accessed on 25 August 2021).

Grossmann, R., (2020). *Telemedicine Post COVID-19.* Health Management. Retrieved from https://healthmanagement.org/c/hospital/issuearticle/telemedicine-post-COVID-19 (accessed on 25 August 2021).

Gustin, T., (2021). Telemedicine etiquette. In: Atanda, Jr. A., & Lovejoy, III. J. F., (eds.), *Telemedicine in Orthopedic Surgery and Sports Medicine* (pp. 65–80). Springer, Cham. doi.org/10.1007/978–3-030–53879–8_6.

Hardy, (2021). *The Rise of Telehealth & its Impact on Strategy*. AMN Healthcare. Retrieved from https://nrchealth.com/fr12dmjkohbt/ (accessed on 25 August 2021).

Healthline, (2020). *Ten Best Telemedicine Companies.* Healthline. Retrieved from http://www.healthline.com/health/best-telemedicine-companies (accessed on 25 August 2021).

Healthtech Staff, (2018). *4 Ways to Advance Telehealth Through Innovation.* Healthtech. Retrieved from https://healthtechmagazine.net/article/2018/02/4-ways-advance-telehealth-through-innovation#:~:text=%204%20Ways%20to%20Advance%20Telehealth%20Through%20Innovation,Test-beds%20for%20Innovation.%20The%20United%20Kingdom%E2%80%99s...%20More%20 (accessed on 25 August 2021).

HIMSS, (2021). *State of Healthcare Report: Uncovering Healthcare Barriers and Opportunities*. Retrieved from https://www.himss.org/resources/state-healthcare-report-uncovering-healthcare-barriers-and-opportunities (accessed on 25 August 2021).

Kryuus, (2020). *2020 Patient Access Journal Report.* Retrieved from https://www.kyruus.com/hubfs/Kyruus%20Patient%20Access%20Journey%20Report%202020.pdf?hsLang=en (accessed on 25 August 2021).

Lagasse, J., (2021). *Most Consumers Want to Keep Telehealth After the COVID-19 Pandemic*. Healthcare Finance News. Retrieved from https://www.healthcarefinancenews.com/news/most-consumers-want-keep-telehealth-after-COVID-19-pandemic (accessed on 25 August 2021).

Lee, N., & Yaraghi, N., (2020). *How to Make Telehealth More Permanent After COVID-19.* Brookings. Retrieved from https://www.brookings.edu/blog/techtank/2020/09/04/how-to-make-telehealth-more-permanent-after-COVID-19/ (accessed on 25 August 2021).

Market Analysis Report, (2021). *Telemedicine Market Size, Share & Trends Analysis Report by Component, by Technology, by Application (Teleradiology, Telepsychiatry), by Delivery Mode, by Type, by End-Use, by Region, and Segment Forecasts, 2021–2028*. Grand View Research. Retrieved from https://www.grandviewresearch.com/industry-analysis/telemedicine-industry (accessed on 25 August 2021).

Michael, E., (2020). *CMS Makes Some Telehealth Services Permanent After COVID-19*. Healio. https://www.healio.com/news/primary-care/20201210/cms-makes-some-telehealth-services-permanent-after-covid19 (accessed on 25 August 2021).

Noel, K., & Ellison, B., (2020). Inclusive innovation in telehealth. *NPJ Digital Medicine, 3*(1), 1–3.

NRC Health, (2021). *2021 Healthcare Consumer Trends Report*. Retrieved from https://nrchealth.com/fr12dmjkohbt (accessed on 25 August 2021).

Rokosh, R., Lewis, II. W., Chaikof, E., & Kavraki, L., (2021). *How Should We Prepare for the Post-Pandemic World of Telehealth and Digital Medicine?* National Academy of Medicine. Retrieved from https://nam.edu/how-should-we-prepare-for-the-post-pandemic-world-of-telehealth-and-digital-medicine/ (accessed on 25 August 2021).

Sauceman, R., (n.d.). *10 Marketing Ideas if You are Launching Telemedicine at Your Practice*. Full Media. Retrieved from https://www.fullmedia.com/10-marketing-ideas-if-you-are-launching-telemedicine-at-your-practice (accessed on 25 August 2021).

Schinasi, D. A., Foster, C. C., Bohling, M. K., Barrera, L., & Macy, M. L., (2021). Attitudes and perceptions of telemedicine in response to the COVID-19 pandemic: A survey of naïve healthcare providers. *Front. Pediatr., 9*, 647937. doi: 10.3389/fped.2021.647937.

Shachar, C., Engel, J., & Elwyn, G., (2020). Implications for telehealth in a post-pandemic future: Regulatory and privacy issues. *JAMA, 323*(23), 2375, 2376. doi:10.1001/jama.2020.7943

Smith, K. A., Bishop, F. L., Dambha-Miller, H., et al., (2020). Improving empathy in healthcare consultations—a secondary analysis of interventions. *J. Gen. Intern. Med., 35*, 3007–3014. doi.org/10.1007/s11606–020–05994-w.

SR Health, (2021). *Patient Communication Preferences in 2021*. Retrieved from https://f.hubspotusercontent30.net/hubfs/2332687/SRH-Patient-Communication-Preferences-2021-survey.pdf?__hstc=&__hssc=&hsCtaTracking=cf91d403-e16a-429f-a28e-a5d8273182d5%7Cf0468e08–3f58–4de6-a792-a7887b823ed8 (accessed on 25 August 2021).

Sweet, J., (2020). *How One Company is Preparing the Health Care System for the Post-Pandemic World*. Philips. Retrieved from https://www.usatoday.com/story/sponsor-story/philips/2020/08/28/how-one-company-preparing-health-care-system-post-pandemic-world/3437691001/ (accessed on 25 August 2021).

The Business Research Company, (2020). *COVID-19 is Behind the Remarkable Telemedicine Market Growth Rate of 40% Through 2023*. KSU| The Sentinel Newspaper. Retrieved from https://ksusentinel.com/2021/06/07/telehealth-telemedicine-market-by-component-software-services-rpm-real-time-application-hardware-end-user-delivery-forecast-to-2025/ (accessed on 25 August 2021).

Verma, S., (2020). *Early Impact of CMS Expansion of Medicare Telehealth During COVID-19*. Health Affairs. Retrieved from https://www.healthaffairs.org/do/10.1377/hblog20200715.454789/full/ (accessed on 25 August 2021).

Wicklund, E., (2020). *Guidelines for Marketing a Practice's New Telehealth Platform*. mHealth Intelligence. Retrieved from https://mhealthintelligence.com/news/guidelines-for-marketing-a-practices-new-telehealth-platform (accessed on 25 August 2021).

Wicklund, E., (2020). *Telehealth Faces New Challenges: Surfing, Snacking and Social Hour*. mHealth Intelligence. Retrieved from https://mhealthintelligence.com/news/telehealth-faces-new-challenges-surfing-snacking-and-social-hour?eid=CXTEL000000391354&elq=149ca15533d540ed9e1b7c1a92fbdf64&elqCampaignId=16633&elqCampaignId=16633&elqTrackId=fb6e2270e04c49a7980867b56ca6d11b&elqaid=17333&elqat=1&utm_source=pocket_mylist (accessed on 25 August 2021).

Wolf, J. A., Niederhauser, V., Marshburn, D., & LaVela, S. L., (2014). Defining patient experience. *Patient Experience Journal, 1*(1), Article 3. doi: 10.35680/2372–0247.1004 Available at: https://pxjournal.org/journal/vol1/iss1/3 (accessed on 25 August 2021).

INDEX

E

F

I

M

S

T

U

X

Y

Z